Biomass Burning in South and Southeast Asia

Biomass Burning in South and Southeast Asia

Impacts on the Biosphere, Volume Two

Edited by
Krishna Prasad Vadrevu,
Toshimasa Ohara, and Christopher Justice

CRC Press
Taylor & Francis Group
Boca Raton London New York

CRC Press is an imprint of the
Taylor & Francis Group, an **Informa** business

First edition published 2021
by CRC Press
6000 Broken Sound Parkway NW, Suite 300, Boca Raton, FL 33487-2742

and by CRC Press
2 Park Square, Milton Park, Abingdon, Oxon, OX14 4RN

© 2021 Taylor & Francis Group, LLC

CRC Press is an imprint of Taylor & Francis Group, LLC

ISBN: 978-0-367-07604-7 (hbk)
ISBN: 978-0-429-02203-6 (ebk)
ISBN: 978-1-032-01353-4 (pbk)

Typeset in Times
by codeMantra

Contents

SECTION I   *Biomass Burning and Regional Air Quality*

SECTION II   *Biomass Burning Emissions*

SECTION III   *Aerosol Pollution and Biomass Burning*

Foreword

It is my pleasure to write a foreword to the two-volume book on the topic of vegetation fires and biomass burning edited by Dr. Krishna Prasad Vadrevu, Dr. Toshimasa Ohara, and Prof. Christopher Justice. I had the privilege to write a foreword to their earlier 2018 book on Land Atmospheric Interactions in South and Southeast Asia, which attracted broad attention. While society is facing more environmental problems with disasters, such as fires, floods, droughts, and landslides, scientists work hard on their detection, spatiotemporal variability, and impacts and quantify the uncertainties. Remote sensing and geospatial technologies offer unique opportunities in mapping and monitoring natural disasters and their impacts and can provide useful information for land management.

The current two-volume compilation of articles focuses on biomass burning processes, the tools for its mapping and monitoring, and quantifying impacts on atmospheric composition including air quality. The contributions in this book are a result of several international workshops organized in Asia under the NASA South/Southeast Asia Research Initiative (SARI; http://sari.umd.edu), funded under the Land-Cover/Land-Use Change Program (http://lcluc.umd.edu), which I manage at NASA Headquarters. Prior to SARI, the NASA LCLUC program contributed to other regional initiatives in Africa, northern Eurasia, and South America by soliciting, selecting, and funding research projects and supporting NASA scientists to attend regional science workshops. Recognizing the pervasive land-use changes driven by rapid population growth and economic development in Asia, the NASA LCLUC program launched the SARI, with Dr. Krishna Prasad Vadrevu as the lead, a few years ago. Since then, SARI workshops have been organized annually, with more than 150 scientists in attendance, and where important regional science issues have been addressed and research needs and priorities identified. One of the topics requiring immediate attention is biomass burning and the associated pollution widespread in the region.

As Land-Cover/Land-Use Change Program Manager, during my 21-year tenure at NASA Headquarters, I have funded multiple projects that involved land–atmospheric interactions and included analysis of drivers and impacts. The NASA LCLUC program is a truly interdisciplinary science program within the Science Mission Directorate. The program is distinct from other NASA discipline programs as it integrates both physical and socioeconomic sciences with the use of remote sensing observations in an interdisciplinary framework to address environmental and societal issues. The LCLUC program aims to develop and use NASA space- and airborne remote sensing technologies, relying on US and non-US satellite data sources, to improve our understanding of human interactions with the environment. One of the critical questions addressed in the program is: "What are the causes and consequences of LCLUC?" Several decades of research in the SARI region have revealed the drivers of extensive modifications of the natural environment and environmental impacts of LCLUC, ranging from changes in atmospheric composition and air quality, biomass burning being a significant contributor.

My scientific interests in biomass burning extend over 30 years back when I worked on my research with remote sensing applications over land and in the atmosphere. During the early 1990s, while working at NOAA, I collaborated with Israeli scientists on remote sensing of atmospheric processes with Dr. U. Dayan on Dust Intrusion Events into the Mediterranean Basin (*J. Appl. Met.*, 1991, 30, 1185–1199) and with Prof. D. Rosenfeld on Retrieving Microphysical Properties Near the Tops of Potential Rain Clouds by Multispectral Analysis of AVHRR data (*Atmos. Research J.*, 1994, 34, 259–283). However, most of my research was on detecting/screening clouds to study land surface. I have used remote sensing to address atmosphere–vegetation feedbacks across diverse geographical regions and at multiple scales, including land–climate interactions using long-term observations from satellites. I have always promoted the use of satellite data in climatology even though 30 years ago the length of time series might not have been sufficient but hoped that continuous observations would one day provide a decent observational base for analyzing climate-related land-cover anomalies. For example, I specifically published my paper of 2000 on 1997–1998 fires in Indonesia (*Bulletin Amer. Met. Soc.* 2000, 81, 1189–1205) in a meteorological journal in an attempt to attract the attention of meteorologists and climatologists to the richness of the Advanced Very High-Resolution Radiometer (AVHRR) observations for land–climate applications. In that paper, I demonstrated how many land and atmospheric products can be derived from AVHRR alone and highlighted the compound effect of fires from slash-and-burn agriculture and one of the strongest El Niño events on the atmosphere (smoke and cloud microphysics) and land (surface reflectance, temperature, and greenness).

This two-volume book is a collection of papers from several SARI workshops organized in South/Southeast Asian countries. More than 85 authors contributed to the book chapters. Scientists from different disciplines gathered systematic information and knowledge guided by robust theories and models useful for fire management and analyzing atmospheric impacts. Each chapter is unique, highlighting the drivers of fire regime to study its characteristics and biomass burning impacts in the region, integrating remote sensing technologies with ground-based measurements. The material in this book will provide information pertinent to fire mitigation, preparedness, response, and monitoring vegetation recovery in various ecosystems. Several chapters also highlight the impacts of biomass burning pollution on the environment in South/Southeast Asian countries. Overall, the scientific results will be of interest to the research community and useful to policymakers, emergency managers, and other practicing professionals in the field. With contributions from several international scholars and professionals, and edited by three eminent scientists, this two-volume book is a major addition to the research and applications literature on fires and biomass burning in South/Southeast Asia (S/SEA). I am impressed with the quality of cutting-edge science and contributions from regional scientists. The book is a timely contribution to the NASA LCLUC SARI science and reflects the current status in this research field.

The Editors of this book are highly active and reputed researchers. I commend all the team members, particularly the Chief Editor, Dr. Krishna Prasad Vadrevu, on accomplishing the exhaustive task of creating this comprehensive book, rich with material and latest references. I am confident that it will motivate researchers on the pertinent topic and trigger innovative ideas. I welcome more such contributions from the US and regional scientists and wish you all an exciting and informative reading.

Dr. Garik Gutman
Land-Cover/Land-Use Change Program Manager
NASA Headquarters, Washington DC, USA

Preface

The impacts of biomass burning pollutants on the atmosphere, climate, air quality, health, and overall environment are poorly understood. Biomass burning emissions include a complex mixture of both gases and particulate matter (PM). The gaseous species include carbon dioxide (CO_2), carbon monoxide (CO), hydrocarbons, and various trace gases that can impact atmospheric chemistry. The black carbon (BC) PM can influence Earth's radiation budget directly by absorbing radiation and indirectly by changing the cloud formation and the surface albedo in snow- and ice-covered areas. In particular, the smoke released during burning can cause severe health problems. In cities and other industrialized landscapes, the pollution sources are mostly from transport, industrial, and domestic sectors. In rural areas, pollution mainly results from land management activities such as open biomass burning through slash-and-burn agriculture, residential wood burning, prescribed burning of forests, agricultural residue burning, and garbage burning. The pollutants from both urban and rural sources often get intermixed, causing severe air pollution problems. Such phenomena are commonly observed in several countries of South/Southeast Asia (S/SEA). Some countries such as Indonesia, Myanmar, Laos, and Cambodia still have significant areas of slash-and-burn agriculture resulting in biomass burning smoke emissions that get transported long distances, impacting air quality, visibility, and human health. The spatial and temporal variations of biomass burning emissions and their sources, transportation, and impacts on the biosphere are not well understood. Therefore, it is crucial to quantify and understand biomass burning sources and impacts. The term biosphere is broadly used in our book, "where all life occurs on Earth, comprising the lithosphere, atmosphere, and hydrosphere."

This two-volume book (Volume 1: "Biomass Burning in South and Southeast Asia – Mapping and Monitoring"; and Volume 2: "Biomass Burning Impacts in South and Southeast Asia – Impacts on the Biosphere") is a collection of papers from several South/Southeast Asia Research Initiative (SARI; sari.umd.edu) workshops organized in Asia since 2015. SARI is a NASA Land-Cover/Land-Use Change Program (lcluc.umd.edu)-funded research activity. SARI's goal is to develop an innovative regional research, education, and capacity-building program involving state-of-the-art remote sensing, natural sciences, engineering, and social sciences to enrich LCLUC science in S/SEA. To address LCLUC science, SARI has been utilizing a systems approach to problem-solving that examines both biophysical and socio-economic aspects of land systems, including the interactions between land use and climate and the interrelationships among policy, governance, and land use. During the last few decades, fire events in S/SEA have attracted international attention due to the significant transboundary pollution they have been causing, impacting air quality and human health. Several SARI meetings identified fires and biomass burning as a vital issue in the region, requiring immediate attention. This two-volume book series on biomass burning was planned to address community research and application needs. All three editors of the current volumes have published significant papers on fires, biomass burning, and emissions.

Volume 2 is divided into three sections: (1) Biomass Burning and Regional Air Quality, (2) Biomass Burning Emissions, and (3) Aerosol Pollution and Biomass Burning.

Section 1, "Biomass Burning and Regional Air Quality," includes four different contributions. In Chapter 1, the authors investigate the emissions from forest fires, garbage burning, and agricultural residue burning and their air quality impacts in several South/Southeast Asian (S/SEA) countries. The authors present several policy recommendations, including creating educational programs and affordable garbage and crop residue disposal systems and investing in renewable energies to improve S/SEA's air quality. In Chapter 2, the author details biomass burning emissions inventory from different sources and the impacts of the emissions on air quality in Thailand. Emissions from forest fires and the crop residue burning were studied, including rice, sugarcane, corn, soybean, cassava, and potato. Results suggest that more than 56% of the emissions are due to crop residue burning. The author also stresses the need for quantifying physicochemical characteristics of PM to address health concerns. In Chapter 3, the authors review the biomass burning emissions in South China. They show higher springtime air quality deterioration due to massive burning in the Indochina Peninsula and long-range transport. In Chapter 4, the authors report MOPITT-derived CO variations during 2001–2017 over Lahore and Karachi, Pakistan, including other South Asian megacities. They attribute enhanced CO to crop residue burning.

Section 2, "Biomass Burning Emissions," has seven chapters. In Chapter 5, the authors present a high-resolution ($0.1° \times 0.1°$), multi-year, monthly biomass burning emissions inventory, including all land types in S/SEA during 2001–2017. The authors report high CO_2 emissions in northwest India, western Myanmar, northern Thailand, etc., with Myanmar as the largest emitter. In Chapter 6, the authors review the current state of knowledge on BC emitted from crop residue burning and forest and peat fires in Southeast Asia (SEA) and discuss the impacts of BC on climate and human health. The review can serve as a useful reference for environmental policymakers to prioritize and implement BC mitigation measures. In Chapter 7, the authors estimate the global CO_2 emissions from biomass burning using satellite data. They also integrate the aboveground biomass and soil organic matter values using a process-based model. The authors report carbon emissions of 1.25 ± 0.12 PgC/year from fires. In Chapter 8, the authors report that the maritime continent contributes to ~13% of the biomass burning mercury emissions' annual burden. They also report that during the El Niño and La Niña years, emissions change by a factor of ~1.5 and 0.27, respectively. Chapter 9 briefly reviews the biomass burning emissions resulting from oil palm expansion and peatlands' destruction in Indonesia and highlights some policy measures for effective fire management. The authors infer that smallholder farmers should have more incentives and alternate financial sources to implement a zero-burning policy. In Chapter 10, the authors quantify the estimates of total fine particulate matter ($PM_{2.5}$) emissions in S/SEA from biomass burning of shrublands, tropical forests, agriculture, peatlands, and temperate forests. The authors report India, Indonesia, Bangladesh, Myanmar, and Vietnam as important sources of $PM_{2.5}$. In Chapter 11, the authors quantify $PM_{2.5}$ emissions from rice residue burning in

Punjab and Haryana, India. They report 15.75 Mt/year of greenhouse gas emissions with a global warming potential (GWP) of 17.71 Mt/year of CO_2 equivalent and 0.194 Mt/year of total particulate matter (TPM).

Section 3, "Aerosol Pollution and Biomass Burning," focuses on biomass burning aerosols and comprises seven chapters. Chapter 12 addresses the influence of biomass burning aerosols on radiative forcing in South Asia (Kanpur and Gandhi College in Indo-Ganges region) and Southeast Asian cities (Singapore). Agro-residue burning emissions influence Indo-Ganges region, and emissions from forest and peat fires influence Singapore. The authors conclude that the aerosol radiative forcing at the surface and in the atmosphere is significantly higher (~25 W/m^2) when aerosol characteristics are influenced by biomass burning over Indo-Ganges. In Chapter 13, the authors review several methods to estimate PM from ground and satellite data. They also review several approaches grouped into scaling factor models, physical analysis models, geostatistical interpolation techniques, and empirical statistical models useful for PM estimation. The authors stress the need for exploring advanced models such as support vector machines and hierarchical models for robust estimation of PM in SEA. In Chapter 14, the authors report an overview of biomass burning emissions and their impacts on surface chemistry, aerosol optical properties, radiation budget, and public health over northern peninsular SEA and downwind regions from the 7-SEAS campaign. The results highlight biomass burning sources from Myanmar, Laos, Thailand, Vietnam, and Cambodia. In Chapter 15, the authors attempt a source apportionment study of $PM_{2.5}$ samples in Phuket, Thailand, to infer various pollution sources. Specifically, the authors quantify carbonaceous species, polycyclic aromatic hydrocarbons (PAHs), and water-soluble ionic species and conclude that these levels are much lower in Phuket compared to cities of Beijing, Shanghai, Qingdao, Hong Kong, etc.

In Section 3, the last three chapters use atmospheric models to address the biomass burning impacts. In Chapter 16, the authors use surface visibility data for estimating haze frequency in the major cities of SEA. Further, the authors also conduct numerical model simulations to identify the contribution of fire-aerosol-caused low visibility (visibility < 10 km). The authors found that biomass burning aerosols account for ~90% of severe haze events (visibility < 7 km) in SEA's major cities and that convective systems in the maritime continent are also significantly influenced by biomass burning aerosols during the monsoons. In Chapter 17, the authors address the impacts of biomass burning on the regional and long-range transport of $PM_{2.5}$ resulting from intense fires in Northeast China during the autumn of 2014. The authors used the Community Multiscale Air Quality (CMAQ) model for addressing the transport and infer that agricultural residue burning impacts the air quality of the region. In the final chapter (Chapter 18), the authors use a coupled WRF–HYSPLIT modeling system to estimate PM_{10} concentrations in SEA during the dry season. The model results show that biomass burning and anthropogenic emissions account for 56% and 44% of PM_{10} concentrations in the region, respectively.

The above chapters highlight the impact of biomass burning on the biosphere and reflect extensive research by interdisciplinary teams of experts. The contents of the book will appeal to students and professionals using remote sensing and geospatial technologies, including geographers, ecologists, atmospheric and environmental scientists, and all interested in biomass burning pollution. The Editors are grateful to all the contributing authors and anonymous reviewers for their time and expertise. Acknowledgments are due to Britton Irma and Rebecca Pringle, CRC Press, and codemantra team for their encouragement, patience, and support. We wish everyone a stimulating read.

Editors
Krishna Prasad Vadrevu, Huntsville, Alabama, USA
Toshimasa Ohara, Tsukuba, Japan
Christopher Justice, College Park, Maryland, USA

Editors

Dr. Krishna Prasad Vadrevu is a remote sensing scientist at NASA Marshall Space Flight Center, Huntsville, Alabama, USA. His research focuses on land-cover and land-use change (LCLUC) studies, fires, and biomass burning emissions. He has 20 years of research experience in satellite remote sensing. He is currently serving as the Deputy Program Manager for the NASA LCLUC Program (lcluc.umd. edu) and leading the South/Southeast Asia Research Initiative (www.sari.umd.edu).

Dr. Toshimasa Ohara is a scientist at the National Institute of Environmental Studies, Tsukuba, Japan. He has 32 years of research experience in air quality modeling, emission inventories, and pollution research. He is a lead developer for Regional Emission Inventory in Asia (REAS) and one of the highly cited researchers on the emissions. He is currently working on linking top-down and bottom-up approaches for emission quantification from different sectors in Asia.

Dr. Christopher Justice is Chair of the Department of Geographical Sciences, University of Maryland, College Park, USA. He has 40 years of research experience in remote sensing. His current research is on land-cover and land-use change and global agricultural monitoring using remote sensing. He is an authority on satellite remote sensing of fires. He serves as Project Scientist for the NASA LCLUC Program and the Land Discipline Lead for the NASA MODIS and the Suomi-NPP VIIRS Science Team. He is the Co-Chair of the GEO Global Agricultural Monitoring (GEOGLAM) Initiative, Chief Scientist for NASA HARVEST, and Chair of the international Global Observations of Forest and Land Use Dynamics (GOFC-GOLD) program.

Dedication

We dedicate the book to Dr. Badarinath Venkata Srinivasa Kandalam, an internationally renowned physicist for his significant research contributions in remote sensing, vegetation fires, aerosols, and climate change studies. He was popularly known as KVSB among the research community.

Born on the 18th May 1959, Badarinath received his bachelor's degree in Physics, Chemistry and Mathematics (1977), Master's degree in Physics with Electronics specialization from Andhra University, Visakhapatnam (1979), and a Ph.D. degree in Experimental Solid State Physics from Indian Institute of Technology, Madras (1984).

From 1986 till May 2011, he worked as a scientist at different levels at the National Remote Sensing Agency (Department of Space-Government of India), Hyderabad, India. He focused on remote sensing research applications to environmental problems. As a part of the same, he led the International Geosphere-Biosphere Program Section and the Atmospheric Science section at the NRSA. During his career, he received numerous awards. He received the International START visiting scientist award to work at National Resources Ecology Laboratory, Colorado State University, Colorado, U.S.A., during 1996–1997. He also worked as a visiting Professor at the Centre for Climate System Research headed by Prof. Nakajima, University of Tokyo, Japan, from July 2010 to October 2010. For several years, he also served as an expert on satellite remote sensing of vegetation fires from India and contributed to the international GOFC-GOLD Fire Implementation team.

Dr. Badarinath supervised 10 Ph.D. students and more than 50 M.Tech and B.Tech project studies on Digital Image Processing, Atmospheric sciences, Land surface temperature estimation studies, Environmental Monitoring and modeling, Mesoscale Model studies for dust prediction, cyclone track prediction, and heatwave conditions. He had an excellent publication record, International Journals: 134; National Journals: 73; Total Publications: 207. He served as an editorial member for five various journals.

Dr. K.V.S. Badarinath (1959–2012)

Several students received various awards under his supervision, including IGBP-START Young Scientist Award, Indian Science Congress Young Scientist Award, ISRS-OPTOMECH Award, Kalpana Chawla Memorial Award, and others. Dr. Krishna Vadrevu, the main editor of this book, is Dr. Badarinath's first Ph.D. student.

Dr. Badarinath's contributions to Indian science were exemplary. Several of his publications are replete with examples of outstanding contributions to remote sensing and atmospheric science. His research focused on integrating bottom-up ground-based measurements with top-down remote sensing methodologies to address scientific questions in atmospheric science. His work produced huge amounts of aerosol and greenhouse gas emissions data over the Indian region that was not previously available; for example, Multiwavelength Radiometer for Aerosol Optical Depth, MFRSR, UV-MFRSR, MICROTOPS-II, QCM/GRIMM Aerosol particle analyzer, PREDE Skyradiometer, Nephelometer, LIDAR, and various gas analyzers-CO/CO2/NOx/O3/SO2. He was the pioneer in designing and implementing a satellite-based fire alert system over the Indian region. He collaborated with several international researchers from countries such as U.K., U.S.A., Italy, Canada, Germany, Japan, Greece, and others.

He was extremely creative, unusually prolific, and highly elegant in his choice of research topics. He was fully committed to science. As a mentor, he made things simple with great enthusiasm and caring empathy. He was sincere in his approach to solving student issues. He was gifted with the ability to recognize and clarify the key concepts that are ambiguous. He was also in constant demand as an invited lecturer and journal article reviewer. Most of his students recognized for undergoing rigorous training in fundamentals under his able guidance have been sought after by reputed labs. They are pursuing successful careers in academia and industry.

He was a devoted teacher, husband, father, and a good scientist. His students, colleagues, and well-wishers will remember him as a humble, down to the earth, helping, and kind person. He will continue to inspire us. We miss him sorely.

Contributors

Max Gerrit Adam
Department of Civil and Environmental
    Engineering
National University of Singapore
Singapore, Singapore

Teerachai Amnuaylojaroen
Department of Environmental
    Science, School of Energy and
    Environment
University of Phayao
Phayao, Thailand

Nungruethai Anuma
Atmospheric Pollution and Climate
    Change Research Unit, School of
    Energy and Environment
University of Phayao
Phayao, Thailand

Priyadarshini Babu
Bennett University
Greater Noida, India

Rajasekhar Balasubramanian
Department of Civil and Environmental
    Engineering
National University of Singapore
Singapore, Singapore

Varaprasad Bandaru
Department of Geographical Sciences
University of Maryland
College Park, Maryland, USA

Hung Quang Bui
Center of Multidisciplinary Integrated
    Technologies for Field Monitoring
Vietnam National University
Hanoi, Vietnam

Junji Cao
State Key Laboratory of Loess and
    Quaternary Geology, Key Lab of
    Aerosol Chemistry & Physics
Institute of Earth Environment, Chinese
    Academy of Sciences (IEECAS)
Xi'an, China

Chomsri Choochuay
Faculty of Environmental
    Management
Prince of Songkla University
Songkhla, Thailand

Katie Cush
Environmental Sciences Department
Emory University
Atlanta, Georgia

Woranuch Deelaman
Faculty of Environmental Management
Prince of Songkla University
Songkhla, Thailand

Rajkumar Dhakar
Indian Council of Agricultural Research
Indian Agricultural Research Institute
New Delhi, India

Xingfa Gu
Aerospace Information Research
    Institute, Chinese Academy of
    Sciences, China

Yongming Han
State Key Laboratory of Loess and
    Quaternary Geology, Key Lab of
    Aerosol Chemistry & Physics
Institute of Earth Environment, Chinese
    Academy of Sciences (IEECAS)
Xi'an, China

Muhammad Zaffar Hashmi
Meteorology
COMSATS University
Islamabad, Pakistan

Zia ul Haq
Remote Sensing and GIS
    Group, Department
    of Space Science
University of the Punjab
Lahore, Pakistan

Niveta Jain
Indian Council of Agricultural
    Research
Indian Agricultural Research Institute
New Delhi, India

Kasturi Devi Kanniah
TropicalMap Research Group, Faculty
    of Built Environment and Surveying
Universiti Teknologi Malaysia
Johor Bahru, Malaysia

Aaron Kaulfus
Department of Atmospheric and
    Earth Science
University of Alabama in Huntsville
Huntsville, Alabama, USA

Kristina Koh
Environmental Sciences Department
Emory University
Atlanta, Georgia, USA

Akira Kondo
Graduate School of Engineering
Osaka University
Osaka, Japan

Om Kumar
Indian Council of Agricultural Research
Indian Agricultural Research Institute
New Delhi, India

Yun Fat LAM
Department of Geography
University of Hong Kong
Pokfulam, Hong Kong

Kristofer Lasko
Geospatial Research Laboratory
Engineer Research and Development
    Center
Alexandria, Virginia, USA

Hsiang-He Lee
Atmospheric, Earth, and Energy
    Division
Lawrence Livermore National
    Laboratory
Livermore, California

Guohui Li
State Key Laboratory of Loess and
    Quaternary Geology (SKLLQG),
    Key Lab of Aerosol Chemistry &
    Physics
Institute of Earth Environment, Chinese
    Academy of Sciences (IEECAS)
Xi'an, China

Zhengqiang Li
Aerospace Information Research
    Institute, Chinese Academy of
    Sciences, China

Neng-Huei Lin
Cloud and Aerosol Laboratory,
    Department of Atmospheric
    Sciences
National Central University
Taoyuan, Taiwan

Tsuneo Matsunaga
National Institute for Environmental
    Studies
Tsukuba, Japan

Sandhi Imam Maulana
Research, Development and
    Innovation Agency-The Ministry of
    Environment and Forestry
Jakarta, Indonesia

Udaysankar Nair
Department of Atmospheric and Earth
    Science
University of Alabama in
    Huntsville
Huntsville, Alabama, USA

Thanh Nhat Thi Nguyen
Center of Multidisciplinary
    Integrated Technologies
    for Field Monitoring
Vietnam National University
Hanoi, Vietnam

Shantanu Kumar Pani
Cloud and Aerosol Laboratory,
    Department of Atmospheric Sciences
National Central University
Taoyuan, Taiwan

Haemi Park
Institute of Industrial Science
The University of Tokyo
Meguro, Japan

Himanshu Pathak
Indian Council of Agricultural Research
Indian Agricultural Research Institute
New Delhi, India

Worradorn Phairuang
Faculty of Environmental Management
Prince of Songkla University
Songkhla, Thailand

Siwatt Pongpiachan
NIDA Center for Research &
    Development of Disaster
    Prevention & Management, School
    of Social and Environmental
    Development
National Institute of Development
    Administration (NIDA)
Bangkok, Thailand

Shimul Roy
Department of Geography, University of
    Hong Kong and Mawlana Bhashani
    Science and Technology
University, Bangladesh

S. Ramachandran
Physical Research Laboratory
Ahmedabad, India

Eri Saikawa
Environmental Sciences Department
Emory University
Atlanta, Georgia, USA

Vinay Kumar Sehgal
Indian Council of Agricultural Research
Indian Agricultural Research Institute
New Delhi, India

Yusheng Shi
Aerospace Information Research
    Institute, Chinese Academy of
    Sciences, China

Hikari Shimadera
Graduate School of Engineering
Osaka University
Osaka, Japan

Lailan Syaufina
Faculty of Forestry
Bogor Agricultural University (IPB
    University)
Bogor, Indonesia

Wataru Takeuchi
Institute of Industrial Science
The University of Tokyo
Tokyo, Japan

Salman Tariq
Remote Sensing and GIS Group,
    Department of Space Science
University of the Punjab
Lahore, Pakistan

Yu Tian
Graduate School of Arts and
    Sciences
The University of Tokyo
Tokyo, Japan

Danai Tipmanee
Faculty of Technology
    and Environment
Prince of Songkla University
Phuket, Thailand

Katsushige Uranishi
Graduate School of Engineering
Osaka University
Osaka, Japan

Krishna Prasad Vadrevu
Earth Science Office
NASA Marshall Space Flight Center
Huntsville, Alabama, USA

Chien Wang
Laboratoire d'Aerologie/CNRS
University of Toulouse
Toulouse, France

Qiyuan Wang
State Key Laboratory of Loess and
    Quaternary Geology, Key Lab of
    Aerosol Chemistry & Physics
Institute of Earth Environment, Chinese
    Academy of Sciences (IEECAS)
Xi'an, China

Li Xing
State Key Laboratory of Loess and
    Quaternary Geology, Key Lab of
    Aerosol Chemistry & Physics
Institute of Earth Environment, Chinese
    Academy of Sciences (IEECAS)
Xi'an, China

Yasushi Yamaguchi
Graduate School of Environmental Studies
Nagoya University
Nagoya, Japan

**Nurul Amalin Fatihah Kamarul
Zaman**
TropicalMap Research Group, Faculty
    of Built Environment and Surveying
Universiti Teknologi Malaysia
Johor Bahru, Malaysia

Shuying Zang
School of Geographical Sciences
Harbin Normal University
Harbin, China

Section I

Biomass Burning and
Regional Air Quality

1 Impacts of Biomass and Garbage Burning on Air Quality in South/ Southeast Asia

Katie Cush, Kristina Koh, and Eri Saikawa
Emory University, USA

CONTENTS

INTRODUCTION

Biomass burning results in emissions that decrease air quality, impact climate, and harm human health on regional, local, and global scales (Arola et al., 2007; Nastos et al., 2010; Ramachandran, 2018). Biomass burning is the largest source of fine particulate matter with an aerodynamic diameter of 2.5 µm or less ($PM_{2.5}$) and the second-largest source of trace gases (Akagi et al., 2011). However, studying and understanding the impact of biomass burning can be difficult due to the seasonal and spatial variability of fire intensities, including the type of biomass burnt (Yadav et al., 2017; Lasko et al., 2018; Vadrevu et al., 2019). Consequently, to fully understand the impacts of biomass burning, the aerosols and trace gases emitted from different burning biomass types in specific locations must be identified (Kumar et al., 2015; Xu et al., 2018). Despite the widespread practice of open burning, there are unknown environmental and health implications. Researchers in Hyderabad, India, estimated that in 2006, 3000 premature deaths and other health problems including chronic bronchitis, asthma, and more respiratory symptom days resulted from air pollution (Guttikunda and Kopakka, 2014). Biomass burning emissions are especially high in South and Southeast Asia compared to the rest of the Northern Hemisphere, where fossil fuel burning dominates (Lawrence and Lelieveld, 2010). This chapter focuses on the three main types of biomass burning: forest burning in Southeast Asia, especially in Indonesia, and garbage and crop residue burning in South Asia, specifically in India, Nepal, and Pakistan.

One of the ways to categorize the ambient air quality is the Air Quality Index (AQI) or Pollution Standards Index (PSI), which is an index based on concentrations of particulate matter (PM), ozone (O_3), nitrogen dioxide (NO_2), sulfur dioxide (SO_2), and carbon monoxide (CO). PM is categorized into different particle size distributions, such as $PM_{2.5}$. Categories of air quality levels are "good," "moderate," "unhealthy for sensitive groups," "unhealthy," "very unhealthy," and "hazardous." Different countries have varying numeral intervals for air quality. In addition to $PM_{2.5}$, another common PM category is coarse PM, with an aerodynamic diameter of 10 µm or less (PM_{10}). After a forest burning event, there are often elevated AQI levels, indicating unhealthy levels of ambient air. Transboundary pollution is also very prominent in Southeast Asia, where smoke haze reaches other countries and increases their AQI levels.

Researchers have used satellite imagery to capture and study the impacts of biomass burning worldwide (Sharma et al., 2010; Vadrevu and Justice, 2011; Mishra and Shibata, 2012, Tariq et al., 2015). Specifically, aerosol optical depth (AOD) images can be used to examine aerosol load over burning areas. NASA's Moderate Resolution Imaging Spectroradiometer (MODIS) on the Terra satellite measures the way aerosol particles absorb and reflect infrared radiation to calculate AOD (Przyborski, 2019). In South Asia, MODIS captures high AOD during the crop residue burning months of October to November (Figure 1.1). The MODIS image, coupled with a fire satellite image, can better understand the origin and patterns of fires and aerosols in South Asia (Badarinath et al., 2008a, b).

Understanding the extent and impact of biomass burning in South/Southeast Asian countries is important to combat the degradation of air quality, health, and the environment (Itahashi et al., 2018; Vadrevu et al., 2017, 2018; Vadrevu and Lasko 2018). Because biomass burning is widespread in South and Southeast Asia, in this study, we do a literature review to assess the impact of forest fires in Indonesia, open garbage burning, and crop residue burning on air quality in India, Nepal, and Pakistan. By understanding the sources and the magnitude of biomass burning in South and Southeast Asia, it will be possible to present practical recommendations for improving the region's air quality.

SOUTHEAST ASIAN FOREST FIRES

As Southeast Asia grew its population and improved its economy, forest burning became a popular method for clearing the land. The felling of trees and burning them is a low-cost method with high efficiency (Islam et al., 2016). This slash-and-burn

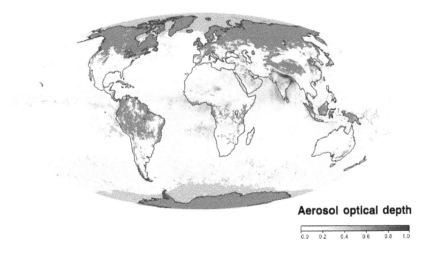

Aerosol optical depth

FIGURE 1.1   Aerosol optical depth based on observation from MODIS during November 2018. Obtained from NASA's Earth Observatory website.

technique is often used after logging timber forests or to clear land for palm oil plantations, which are valuable cash crops. Although forest burning is efficient and cheap, environmental impacts can be devastating. Forest fires result in greenhouse gas emissions, such as carbon dioxide (CO_2) and methane (CH_4) contributing to climate change, and air pollutants such as PM. Other environmental consequences include soil erosion, water contamination, and biodiversity (Miller and Spoolman, 2015; Pavuluri and Kawamura, 2018). The poor air quality resulting from forest burning indicates unhealthy environments and has detrimental effects on human health.

In Southeast Asia, biomass burning is prevalent in Indonesia (Israr et al., 2018; Hayasaka et al., 2014; Hayasaka and Sepriando, 2018). Therefore, its surrounding countries, Singapore and Malaysia, often experience the aftermath of burning events, impacting their air quality. Impacts of forest fires and their air quality effects on human health are reviewed to depict the environmental consequences of burning events. Forest fire events from 1994 to 2015 are discussed along with their air pollutant emission estimates.

INDONESIA

Indonesia has a history of destroying peatlands and forests for clearing land. In 1994 from August to October, over 50,000 km^2 of the forest, plantations, bush, grassland, and peat swamp was burned (Nichol, 1997). The majority of the fires occurred in Sumatra and Kalimantan. In 1997, the slash-and-burn was used again in Sumatra and Kalimantan, with 30% of fires occurring in forests and bushes coinciding with the severe drought due to El Niño, prolonging the smoke haze inversions in urban areas. Air quality was often unhealthy in Southeast Asia during these months without rainfall. It was not until the wet inter-monsoon season after October when the fires extinguished, and air quality improved. By then, around 70 million people were exposed to low air quality in Southeast Asia. Several researchers have estimated the total air pollutant emissions from these fires in Sumatra and Kalimantan. For example, Page et al. (2002) in Central Kalimantan estimated 190–230 Tg of carbon, including CO_2 and CH_4, merely from peat combustion, contributing to around 13%–40% of the mean annual global carbon emissions from fossil fuels in 1997. From August 1997 to March 1998, estimated emission values from forest burning in Indonesia were 11.1 Tg of CO_2, 942 Gg of CO, 35 Gg of CH_4, 5.85 Tg of NO_x, 12 Gg of ammonia (NH_3), and 547 Gg of total suspended particles (Levine, 1999).

Smoke/haze, mostly present before the inter-monsoon seasons, contained poly-cyclic aromatic hydrocarbons (PAHs) and PM, both attributable to respiratory and cardiovascular disease, including acute morbidity effects (Sastry, 2002). Long-term health consequences from the resulting air pollution were investigated and shown to depend on gender and age, possibly. The air pollution impacts affected men four times more often than women regardless of age. Women also experience consistently poor indoor air pollution. Women were more often are at home where there was a lack of proper ventilation for burning stoves or heating, so the impacts of poor ambient air quality from the fires were less severe than what they were regularly exposed to. Older men and women over 42 were more negatively impacted by air pollution than people aged 9–21 in 1997. This may have been due to the longer recovery times needed for older individuals (Kim et al., 2015, 2017).

During 1st week in June of 2013, another forest and peat burning event in Central Sumatra, Indonesia, burned 1.6% of Indonesia's land and removed 12 ha of plantations, bush, forest, grassland, and peat swamp. This occurred during a non-drought year, so unlike the previous burning event in 1997, there was more immediate precipitation to remove air pollutants. Estimated atmospheric emissions solely from forest fires from June 18 to 24 were 1.49 ± 0.07 Tg of CO_2, 100 ± 1 Gg of CO, 10 Gg of CH_4, and 1.5 ± 0.5 Gg of NO_x (Gaveau et al., 2014).

Fires during 2015 set record-high carbon emissions and exceeded the severity of burning compared to the fires in 1997. From September to October of 2015, forest and peatland fires spread throughout Southeast Asia but were mostly concentrated in southern Kalimantan, Sumatra, and Papua, Indonesia. An emission rate of 11.3 Tg of CO_2 per day was estimated from the burning in the 2 months, totaling 692 Tg CO_2. Total estimated carbon emissions were 227 ± 67 Tg, where 83% was in the form of CO_2, 16% CO, and 1% CH_4 (Huijnen et al., 2016). In Palembang, Indonesia, PM_{10} concentrations peaked on September 28, 2015, with 606 µg/m^3 (Agustine et al., 2018).

High concentrations of PM_{10} often represent decreased visibility and increased sensitivity in people with asthma or respiratory illnesses. During an El Niño year, there were drought-like conditions throughout Southeast Asia, further prolonging the unhealthy conditions (Agustine et al., 2018).

SINGAPORE

Singapore is located south of Malaysia and north of Indonesia. While the Sumatra swamp forests in Indonesia burned in 1994, the wind carried the smoke to Singapore, which caused air quality to deteriorate for months quickly. The majority of the pollution was black carbon (BC), which lasts approximately a week in the atmosphere. Singapore's PSI is usually between 0 and 50, representing healthy air quality.

However, after the Indonesian fire, Singapore's PSI, which averages 24-hour concentrations of PM_{10}, SO_2, CO, O_3, and NO_2, reached over 150 for a few days (Nichol, 1997). The transboundary pollution contained mostly PM_{10}. Wind shifts provided temporary relief before monsoons' precipitation removed the smoke haze, and Singapore's air quality returned to normal healthy levels. This relief, however, may have affected surrounding countries.

There are potentially lagged effects of PM_{10}, taking weeks for PM_{10} to affect human health. Reduction in pulmonary function and increased respiratory symptoms are commonly studied as PM_{10} impacts human health (Ransom and Pope, 1992). In late September 1997, concentrations of PM_{10} rose from 50 to 150 $\mu g/m^3$ from Indonesia's forest fires. PSI levels reached a level of 226 (Islam et al., 2016). Furthermore, during the haze period, there was a 12% increase in upper-respiratory-tract illness, a 19% increase in asthma cases, and a 26% increase in rhinitis. However, there was no increase in hospital admission or mortality (Emmanuel, 2000).

Singapore's air quality also suffered from Indonesia's forest clearing in 2013. Its 24-hour PSI was ranked "very unhealthy" at a maximum PSI level of 246 on June 22, 2013. There were three consecutive days where levels were more significant than 236 as well. Even though it was not a drought year in Singapore, they were experiencing dry weather for 2 months during this time (Gaveau et al., 2014).

MALAYSIA

Geographically, Malaysia lies near the equator, so there is less variation in temperature or humidity than other Southeast Asian countries. During the transboundary haze from Indonesia's forest burning from July to December 1997, the number of outpatient visits increased two to three times in Kuching and Sarawak. In Kuala Lumpur's general hospitals, respiratory diseases' diagnosis increased from 250 to 800 per day (Sastry, 2002).

Another notable forest burning incident in Indonesia occurred in August of 2005 and affected the Malay Peninsula's air quality. Malaysian Ambient Air Quality Guidelines (MAAQG) are 150 $\mu g/m^3$ of PM_{10} for 24-hour averages. At the air quality stations in Klang and Kuala Lumpur, PM_{10} concentrations exceeded these guidelines for seven consecutive days from August 6 to 12, while the Penang and Langkawi stations exceeded them on August 13. There was no strong correlation between mortality cases and PM_{10} concentrations in this particular study because from Klang, Kuala Lumpur, Penang, and Langkawi, there were 106, 217, 209, and 4 mortality cases, respectively (Jie, 2017). However, there were more respiratory and cardiovascular mortality cases from people aged 56 and above than any other age range. The youngest age-group, 15 years old and under, had the least number of mortality cases (Jie, 2017). These findings are similar to those of Kim et al. (2015, 2017), highlighting the health consequences from the 1997 Indonesian forest fire. Those aged between 9 and 21 had no long-run health effects.

THAILAND

Thailand lies north of Malaysia with recurrent agriculture, biomass, and forest burning events. A study from Phairuang et al. (2017) reports emissions from five different stations during the biomass burning peak period, which includes suburban Chiang Mai, urban Chiang Mai, Nakhon Sawan, Khon Kaen, and Nakhon Ratchasima. Their annual emission estimates from forest fires for Thailand in 2014 were 69.64 Gg of PM_{10}, 69.64 Gg of $PM_{2.5}$, 7.312 Gg of NO_x, and 1.516 Gg of SO_2. The forest fire emissions were closely correlated with ambient PM levels, compared to the agricultural activities. NO_x and SO_2 displayed poorer correlation between emissions from forest fires (Phairuang et al., 2017), mainly due to large amounts of these gaseous pollutants emitted from motor vehicles and industries.

SOUTH ASIAN GARBAGE BURNING

The common practice of garbage burning takes place both in the Global North and South. In the USA, residential open garbage burning is more economical in many rural areas, as garbage disposal costs continue to rise (Lighthall and Kopecky, 2000). In the Global South, garbage burning occurs in the streets, homes, and local markets of countries that lack the infrastructure for garbage collection and management (Kumar et al., 2015). Open garbage burning also occurs in landfills that have no other infrastructure for garbage disposal. Out of the 2400 Gt of garbage produced annually globally, 41% of this is disposed of through burning (Wiedinmyer et al., 2014). Waste burning releases pollutants, especially BC, from the low-temperature combustion (US EPA, 2006). Open garbage burning emits reactive trace gases, PM, greenhouse gases, and toxic compounds (Li et al., 2012). In 2010, a study showed that 5% of all global anthropogenic emissions were attributed to open garbage burning (Wiedinmyer et al., 2014). While open burning is a significant source of emissions, there is still a great deal of uncertainty about the exact characteristics of the pollutants emitted from different sources (Gupta et al., 2001). The composition of emissions varies based on the burning method and the makeup of waste itself (Wiedinmyer et al., 2014). Due to the uncertainty about the combustion of garbage and the burning methods, it is difficult to obtain accurate emission factors (Wiedinmyer et al., 2014).

Wiedinmyer et al. (2014) present the first global estimates of emissions from garbage burning, although many assumptions led to high uncertainty in the estimates. Past studies typically only focused on a few pollutants, and there has consistently been a lack of global studies (Wiedinmyer et al., 2014). Consequently, very few emission inventories include estimates for garbage burning, and therefore, they are missing from climate and chemistry models (Wiedinmyer et al., 2014). Without these emissions, models greatly underestimate emissions (Li et al., 2012); therefore, more research is increasingly done to understand global garbage burning. A better understanding of open burning will also allow for more productive and effective mitigation strategies.

INDIA

As one of the world's largest growing countries, India has a severe air pollution problem. The Environmental Performance Index (EPI) created by Yale University and Columbia University ranks 180 countries based on environmental health and ecosystem vitality (Wendling et al., 2018). In 2018, India was ranked 178th for overall air quality (Wendling et al., 2018). This low ranking is mainly attributed to the burning of fossil fuels; however, the burning of municipal solid waste (MSW) too contributes significantly to the poor air quality (Li et al., 2014). Coinciding with India's growing population, waste production has increased by 61% from 1996 to 2002 (Kawamura and Pavuluri, 2010). Wiedinmyer et al. (2014) estimated that globally, at the residential level and dumpsites, 620 Tg and 350 Tg of garbage are burned each year.

Rising rates of plastic consumption have led to an increase in the amount of plastics in MSW. In 2000, plastics' consumption reached 150 million tons per year, with an annual growth rate of 5.5% (Simoneit et al., 2005). In the Global North, plastics are often recycled, but in the Global South, they are typically mixed with MSW and consequently burned at incineration and residential sites to prevent odor and disease-carrying mosquitoes (Kawamura and Pavuluri, 2010). Greater than 90% of MSW is disposed of into open landfills that lack regulations (Kawamura and Pavuluri, 2010). A 2010 study conducted in Chennai, India, found that terephthalic acid (t-Ph), industrial material for making plastics known to cause bladder cancer, was in high concentrations at sampling areas far from MSW incineration sites (Kawamura and Pavuluri, 2010). Due to the high rate of generation of MSW and the growing consumption of plastics, the authors professed an urgent call of action to better understand the health and environmental impacts of solid waste burning in South Asia and for regulations to account for the associated risks (Kawamura and Pavuluri, 2010). Additionally, a total suspended particle winter sampling campaign conducted in New Delhi, India, and Xi'an, China, in 2014 indicated the extent of garbage burning in one of India's megacities. Consequently, the emissions from the open burning of plastic and solid waste, phthalates, and bisphenol A are more abundant in New Delhi than in Xi'an (Li et al., 2014).

In Agra, India, MSW burning is also highly prevalent. A 2016 study estimated that chronic exposure to $PM_{2.5}$ released from waste burning increased premature deaths by 713 per year (Lal et al., 2016). This was because the yearly impact of $PM_{2.5}$ from waste burning was up to 33 $\mu g/m^3$ (Lal et al., 2016). A 2016 pollution episode in Delhi, in which the pollution index was deemed "severe," caused limited outdoor activity and forced schools to close, resulting in a dire need to understand the sources to take regulatory action (Nagar et al., 2017). A 2017 study conducted in Delhi due to this pollution episode found that waste burning accounts for 7.7% of the total winter $PM_{2.5}$ and 7.2% of the total summer $PM_{2.5}$ (Nagar et al., 2017).

Nepal

The Kathmandu Valley in Nepal is historically burdened with episodic air pollution due to the growing population and the topographic and meteorological characteristics of the area. Wan et al. (2017) used MODIS to map out the spatial distribution of biomass burning sites throughout the IGP, showing that there is a high concentration of fires during the pre-monsoon period in April 2013. As a bowl-shaped basin, the Kathmandu Valley is surrounded by four mountains with altitudes of 2000–2800 m above sea level (Sarkar et al., 2016). Urbanization and land-use change are major sources of air pollution, and their effects are exacerbated during the winter months due to inversion that causes the suppression of mixing along with the dry weather (Putero et al., 2015; Sarkar et al., 2016).

Recently, two separate air pollution campaigns have been launched in Nepal in order to better understand the pollution sources and to identify effective mitigation strategies. The Sustainable Atmosphere for the Kathmandu Valley-Atmospheric Brown Clouds (SusKat-ABC) was conducted from December 2012 to June 2013 in order to characterize the chemical composition of the Kathmandu ambient air (Mahata et al., 2017). In 2015, a study utilized the SusKat-ABC data and found that during winter, there are four primary sources of particulate carbon in the Kathmandu Valley: brick kilns, motor vehicles, soil dust, and biomass/garbage burning (Kim et al., 2015). Biomass/garbage burning was estimated to explain 32% of the organic carbon (OC) concentration during the sampling period (Kim et al., 2015). Another 2015 study investigating O_3 and BC in Kathmandu used back-trajectory modeling to understand atmospheric circulation scenarios (Putero et al., 2015). However, garbage burning and other short-lasting open fires were not included because they could not be detected by satellites (Putero et al., 2015).

In April 2015, the Nepal Ambient Monitoring and Source Testing Experiment (NAMaSTE) was conducted with the intention of characterizing prevalent but under-sampled emission sources in Nepal such as cookstoves, brick kilns, open waste burning, generators, irrigation pumps, crop burning, and motorcycles through monitoring and *in situ* sampling in Bode (Jayarathne et al., 2018). There have been a few studies utilizing these data (Stockwell et al., 2016; Goetz et al., 2018; Jayarathne et al., 2018; Zhong et al., 2019). One study highlighted the impacts of various garbage compositions on the characterization of emissions (Stockwell et al., 2016). After conducting controlled burns, researchers found high emission factors for formic and acetic acids and formaldehyde, possibly indicating that garbage burning is a major source for these oxygenated volatile organic compounds (Stockwell et al., 2016). When comparing the optical properties of these undersampled emission sources, garbage burning had the second lowest absorption Ångström exponent and single-scattering albedo (Stockwell et al., 2016). Another study utilizing the NAMaSTE data was published in 2018 (Jayarathne et al., 2018). Previous studies have indicated that the moisture content of garbage affects the volatile matter combustion (Saito et al., 2001). Jayarathne et al. (2018) characterized emissions from different compositions of garbage and observed that damp garbage had higher emission factors for $PM_{2.5}$ compared to dry garbage; thus, avoiding waste burning after rainfall or other wet conditions could help improve air quality (Jayarathne et al., 2018). Also, the relative standard deviation across five garbage burning samples for the $PM_{2.5}$ emission factor was 63%, suggesting that there was large variability in the emissions of various garbage compositions (Jayarathne et al., 2018). This is an additional obstacle when quantifying global garbage burning emissions.

SOUTH ASIAN CROP RESIDUE BURNING

While crop residue burning is banned in parts of the Global North, this practice still occurs in many agrarian countries due to poor agricultural waste management (Ravindra et al., 2019). As the populations grow in India, there will be a parallel increase in agricultural production and crop waste (Jain et al., 2014). In Asia, crop residue burning accounts for 34% of the burning activities (Streets et al., 2003). The main reasons for burning include clearing for the next season, controlling diseases and pests, nutrient regeneration, and weed removal (Gadde et al., 2009). Widespread agricultural burning occurs in the post-monsoon season from October to November in South Asia (Sarkar et al., 2013). The open burning of crop residue degrades air quality by releasing PM, CO, SO_2, NO_x, NH_3, CH_4, BC, OC, volatile organic compounds, and oxy-PAHs (Awasthi et al., 2011; Jain et al., 2014; Zhang et al., 2018). However, the specific composition of pollutants is determined by the crop type and burning conditions (Mittal et al., 2009; Zhang et al., 2011). Crop residue burning produces $PM_{2.5}$ that can stay in the air for a long duration and travel hundreds of miles in the wind (Hao and Liu, 1994). Due to the complex array of pollutants, there are many health and environmental risks associated with open crop burning (Awasthi et al., 2010).

INDIA

Streets et al. (2003) estimated that India is responsible for 18% of the total biomass burned in Asia. As an agrarian country, India produced an estimated 97 Tg of rice residue from 1999 to 2005, of which about 23% was burned in fields releasing harmful pollutants into the atmosphere (Gadde et al., 2009; Vadrevu et al., 2011). Every monsoon season, extensive rice burning occurs in the northwestern Indian states of Punjab, Haryana, and western Uttar Pradesh (Sarkar et al., 2013; Kaskaoutis et al., 2014). After 1960, northwest India played a major role in the food security of India as there was a significant increase in the production of the paddy–wheat crop in northwestern India, especially Haryana and Punjab, contributing ~40% of wheat and ~30% of rice to the stocks of India (Hira 2009; Lohan et al., 2015, 2018).

The aerosols produced from crop residue burning contribute to the atmospheric heating over the area through surface radiative forcing (Carmichael et al., 2009; Kaskaoutis et al., 2014). Using CALIPSO and Cloud-Aerosol Lidar, researchers found that the smoke plumes created by crop burning over Punjab are mostly between 800 and 900 m in altitude and are transported below 2.5 km over the Indo-Gangetic Plain (IGP) (Kaskaoutis et al., 2014). In 2017, out of the 488 Mt of crop residue produced in India, 24% was burned in the fields (Ravindra et al., 2019). Consequently, 824 Gg of $PM_{2.5}$, 58 Gg of BC, 239 Gg of OC, and 211 Tg of greenhouse gases were released into the atmosphere during 2017 (Ravindra et al., 2019). Jain et al. (2014) found that the crop burning emissions differed from state to state based on the material and quantity being burned. The maximum amount of crop residue burnt on farms was in Punjab, Uttar Pradesh, Haryana, and Maharashtra, with rice, wheat, and sugarcane being the major crops burnt (Jain et al., 2014).

Delhi, a megacity in India, experiences smog events associated with high $PM_{2.5}$ pollution (Sawlani et al., 2018). A study investigated a severe smog episode in Delhi that lasted from October 30 to November 7, 2016, with average $PM_{2.5}$ concentrations of 703 $\mu g/m^3$ after the Diwali festival (Sawlani et al., 2018). By using variations in chemical and isotopic signatures of $PM_{2.5}$ along with meteorological data, the study found that the smog episode could be the result of transported carbonaceous material from crop residue burning in northwestern India, weaker northerly winds, shallower boundary layer, cooler temperatures, increased humidity, and emissions from fire-crackers from the festival (Sawlani et al., 2018). Based on long-term fire data in the Punjab area, this study suggests that delayed burning of crop waste after 2010 could cause smog conditions in northern India (Sawlani et al., 2018). Out of the total carbon present in paddy straw, 70% is released as CO_2, 7% is released as CO, and 0.66% is released as CH_4 (Hays et al., 2005; Lohan et al., 2018). Additionally, 2.09% of the nitrogen in the paddy straw is released as N_2O (Galanter et al., 2000; Lohan et al., 2018).

A study conducted in the city of Patiala in northern India found background levels of total suspended particulates (TSP) to be at about National Ambient Air Quality Standards; however, during two separate rice burning episodes from September 2006 to January 2008, the levels of suspended PM, NO_2, and SO_2 greatly increased, dete-riorating the air quality of the region (Singh et al., 2010). Sarkar et al. (2013) found average enhancements in benzenoid levels of more than 300% during post-harvest paddy residue burning with in situ, satellite, and back-trajectory data. If these lev-els of benzenoids persist for 1–2 months each year, it could enhance cancer rates in northwestern India (Sarkar et al., 2013).

A study conducted in northwest India also found that the increase in PM caused by crop residue burning has a negative effect on pulmonary function tests (PFTs) of healthy subjects, especially children (Awasthi et al., 2010). In some cases, exposure to high pol-lution levels caused permanent decreases in children's PFTs, and women are at higher risk for decreased pulmonary functions (Awasthi et al., 2010; Agarwal et al., 2012; Gupta et al., 2016). Ravindra et al. (2019) estimated that if India does not change its crop residue burning practices, emissions from crop burning will increase by 45% by 2050.

Nepal

During the dry season, crop residue burning occurs widely in the Kathmandu Valley as a means to enable faster crop rotation (Stockwell et al., 2016). Additionally, the brick kilns in the Kathmandu Valley burn large amounts of crop residues, potentially adding acetonitrile and benzene to the atmosphere (Sarkar et al., 2016). A 2015 study conducted in Nepal's Kathmandu Valley used MODIS data to investigate the impacts of open vegetation fires on the air quality of the region (Putero et al., 2015). Putero et al. (2015) indicate that the open vegetation fires can be a "non-negligible source" of O_3 precursors during the Kathmandu Valley pre-monsoon season.

Stockwell et al. (2016) created the first measurements of the trace gas chemistry and aerosol properties of burning common Nepali crop residues. The study looked at rice, wheat, mustard, grass, and a mixture of all residues (Stockwell et al., 2016). The results indicated significant SO_2 emissions of 2.54 g/kg from crop residue burn-ing, and the authors suggest that this finding is critical to include in updated emission inventories because previous inventories have used lower local and global values for SO_2 emissions (Stockwell et al., 2016). The NAMaSTE project collected data from the residues of rice, wheat, mustard, lentils, and grasses during the pre-monsoon in the Tarai to better estimate the emission factors (Jayarathne et al., 2018). It was found that moisture content can impact the emissions from biomass (Jayarathne et al.,

2018). Such data will help to provide more accurate records of pollution sources in South Asia (Jayarathne et al., 2018).

In 2018, researchers studied the concentrations, composition, and sources of nitro-PAHs and oxy-PAHs (OPAH's) in indoor air and dust of four cities in Nepal (Yadav et al., 2018). Crop residue burning was a significant source of OPAHs in indoor air (Yadav et al., 2018). Because PAH derivatives are potentially cancer-causing agents, the study assessed that the primary exposure route for OPAHs is through dermal contact via dust in children and adults in Nepal (Yadav et al., 2018). Using data collected from the SusKat-ABC site at Bode, Nepal, and two Indian sites, Bhardwaj et al. (2018) showed that regional pollution caused by crop residue burning in the northwestern IGP led to increases in O_3 and CO levels at Bode and Nainital in May 2013. HYSPLIT backward air trajectories indicated that the polluted air masses in late April and early May arrive from northern India as this is the harvest period and crop residue burning is a common practice (Bhardwaj et al., 2016, 2018).

PAKISTAN

Pakistan's economy is reliant upon its agricultural sector. Specifically, in Punjab, Pakistan, rice–wheat cropping is a dominant practice (Amir and Aslam, 1992). A study done by Ahmed et al. (2015) found that farmers in Pakistan burn crop residue because of farm machinery's inconveniences and the short time frame between harvesting rice and sowing wheat. Additionally, farmers believe that burning crop residues increases future yields and improves soil quality (Ahmed et al., 2015). A 2015 study found that there was a large spatiotemporal variability of CO over Pakistan and its neighboring regions as a result of crop residue burning, vehicles, energy generation, and local meteorology from 2003 to 2012 (ul-Haq et al., 2015). In Pakistan, 1.7 Mt of agricultural waste is burned in the rice–wheat cropping system per year, releasing significant amounts of NO_2 into the atmosphere (Ahmed et al., 2015). Intense burning episodes occur in the post-monsoon months of October and November. During this time, the meteorological conditions can cause the smoke from these crop residue fires to cover the entire IGP (Mishra and Shibata, 2012; Singh and Kaskaoutis, 2014). The region's air quality has drastically decreased due to these agricultural fires (Goss et al., 2014). The smog formed can cause adverse health impacts such as asthmas, irritated respiratory tracts, or lung cancer (Goss et al., 2014).

A study done by Tariq et al. (2015) used ground data from the Aerosol Robotic Network (AERONET), satellite data from MODIS, and CALIPSO instruments to characterize the aerosols of a residue burning event in October 2010 in Lahore, Pakistan. Researchers found a maximum daily mean AOD of 2.75 observed on October 20, 2010, suggesting that the crop burning caused heavy aerosol loading in the area (Tariq et al., 2015). Based on the MODIS images, Tariq et al. (2015) inferred that the southeast region of Lahore is dominated by biomass burning activities (Tariq et al., 2015).

In 2018, Ullah et al. (2018) investigated the microphysical and optical properties of aerosols during a haze event in Lahore, Pakistan, also using data from MODIS, CALIPSO, and AERONET. The source of the haze event was found to be the large-scale crop burning and urban–industrial emissions (Ullah et al., 2018). The AERONET measurements from this study indicated a daily mean AOD of 2.36, a value eight times higher than normal, and the fine-mode volume concentration was more than 1.5 times greater than the coarse-mode volume concentration during the haze event (Ullah et al., 2018).

DISCUSSION AND CONCLUSION

This study's focus was to review the literature on biomass burning, specifically on forest fires in Southeast Asia and garbage and crop residue burning in South Asia, to understand the impact on air quality. Forest burning is a large part of Southeast Asia's economy as it is a rapid method to clear land for farming and housing, but the environmental consequences offer a negative trade-off. Transboundary pollution also causes surrounding-country issues, experiencing low air quality from biomass burning in neighboring countries. Furthermore, forest fires during drought seasons further decrease air quality as it prolongs the smoke haze. Health impacts from forest burning include respiratory illnesses, asthma symptoms, and cardiovascular mortality. The emissions and area burnt were mostly found through modeling, so all emissions and areas burnt were estimates. Therefore, the uncertainties may be large. However, the impact on air quality is clear, and the relationship between poor air quality and illnesses can be portrayed from these forest fire events.

While there were many biomass burning studies in India, Nepal, and Pakistan, South Asia lacks research on the impacts of biomass burning. Due to the health and environmental risks associated with garbage and crop residue burning emissions, more research and monitoring are needed in South Asia (Wester et al., 2019). Garbage burning emissions play a role as a global pollutant source, but little research has addressed its impacts. Additionally, new policies and societal changes must be implemented to combat the poor air quality associated with biomass burning. Biomass burning is abundant because of the lack of infrastructure to accommodate garbage and crop residue generation (Kumar et al., 2015). Consequently, to deter people from burning biomass, where possible, funds must be allotted to create alternate garbage and crop residue using affordable and realistic mechanisms for people. Ravindra et al. (2019) recommend instituting easy credit schemes to buy advanced agricultural machinery and innovation research to help modernize agrarian practices. Education programs should be established for farmers that teach alternative uses of crop residues (Ravindra et al., 2019). Bringing more media attention to the negative impacts of biomass burning on health and air quality would educate people on the dangers of biomass burning while also creating pressure for policy interventions (Saikawa et al., 2019). Countries such as India and Nepal should continue investing in solar power, wind power, and hydropower to improve air quality and decrease biomass burning reliance. Overall, we infer that there is a global knowledge gap on the prevalence and impact of biomass burning; however, it is possible to improve air quality in the future through policies and education.

ACKNOWLEDGEMENTS

Authors thank the Codemantra and CRC team for manuscript edits and the final publication.

REFERENCES

Agarwal, R., Awasthi, A., Singh, N., Gupta, P.K., & Mittal, S.K. 2012. Effects of exposure to rice-crop residue burning smoke on pulmonary functions and Oxygen Saturation level of human beings in Patiala (India). *Science of the Total Environment*, 429, 161–166. doi:10.1016/j.scitotenv.2012.03.074.

Agustine, I., Yulinawati, H., Gunawan, D., & Suswantoro, E. 2018. Potential impact of particulate matter less than 10 micron (PM_{10}) to ambient air quality of Jakarta and Palembang. *IOP Conference Series: Earth and Environmental Science*, 106, 012057. doi:10.1088/1755-1315/106/1/012057.

Ahmed, T., Ahmad, B., & Ahmad, W. 2015. Why do farmers burn rice residue? Examining farmers' choices in Punjab, Pakistan. *Land Use Policy*, 47, 448–458. https://doi.org/10.1016/j.landusepol.2015.05.004- good for backgeiybd.

Akagi, S.K., Yokelson, R.J., Wiedinmyer, C., Alvarado, M.J., Reid, J.S., Karl, T., Crounse, J.D., & Wennberg, P.O., 2011. Emission factors for open and domestic biomass burning for use in atmospheric models. *Atmospheric Chemistry and Physics*, 11(9), 4039–4072.

Amir, P., & Aslam, M. 1992. The rice–wheat system of the Punjab: A conflict in crop management. In: Byerlee, D., Husain, T. (Eds.), *Farming Systems of Pakistan*. Vanguard Books, Lahore.

Arola, A., Lindfors, A., Natunen, A., & Lehtinen, K.E.J. 2007. A case study on biomass burning aerosols: effects on aerosol optical properties and surface radiation levels. *Atmospheric Chemistry and Physics*, 7(16), 4257–4266.

Awasthi, A., Agarwal, R., Mittal, S.K., Singh, N., Singh, K., & Gupta, P.K. 2011. Study of size and mass distribution of particulate matter due to crop residue burning with seasonal variation in rural area of Punjab, India. *Journal of Environmental Monitoring*, 13(4), 1073. doi:10.1039/c1em10019j.

Awasthi, A., Singh, N., Mittal, S., Gupta, P.K., & Agarwal, R. 2010. Effects of agriculture crop residue burning on children and young on PFTs in North West India. *Science of the Total Environment*, 408(20), 4440–4445. doi:10.1016/j.scitotenv.2010.06.040.

Badarinath, K.V.S., Kharol, S.K., Krishna Prasad, V., Kaskaoutis, D.G., & Kambezidis, H.D. 2008a. Variation in aerosol properties over Hyderabad, India during intense cyclonic conditions. *International Journal of Remote Sensing*, 29(15), 4575–4597.

Badarinath, K.V.S., Kharol, S.K., Prasad, V.K., Sharma, A.R., Reddi, E.U.B., Kambezidis, H.D., & Kaskaoutis, D.G. 2008b. Influence of natural and anthropogenic activities on UV Index variations–A study over tropical urban region using ground based observations and satellite data. *Journal of Atmospheric Chemistry*, 59(3), 219–236.

Bhardwaj, P., Naja, M., Kumar, R., & Chandola, H.C. 2016. Seasonal, interannual, and long-term variabilities in biomass burning activity over South Asia. *Environmental Science and Pollution Research*, 23(5), 4397–4410.

Bhardwaj, P., Naja, M., Rupakheti, M., Lupascu, A., Mues, A., Panday, A.K., Kumar, R., Mahata, K.S., Lal, S., Chandola, H.C., & Lawrence, M.G., 2018. Variations in surface ozone and carbon monoxide in the Kathmandu Valley and surrounding broader regions during SusKat-ABC field campaign: Role of local and regional sources. *Atmospheric Chemistry & Physics*, 18(16), 11949–11971.

Carmichael, G.R., Adhikary, B., Kulkarni, S., D'Allura, A., Tang, Y., Streets, D., Zhang, Q., Bond, T.C., Ramanathan, V., Jamroensan, A., & Marrapu, P. 2009. Asian aerosols: current and year 2030 distributions and implications to human health and regional climate change. *Environmental Science & Technology*, 43(15), 5811–5817.

Emmanuel, S.C. 2000. Impact to lung health of haze from forest fires: The Singapore experience. *Respirology*. 5(2), 175–182. doi:10.1046/j.1440-1843.2000.00247.x.

Gadde, B., Bonnet, S., Menke, C., & Garivait, S. 2009. Air pollutant emissions from rice straw open field burning in India, Thailand and the Philippines. *Environmental Pollution*, 157(5), 1554–1558. doi:10.1016/j.envpol.2009.01.004.

Galanter, M., Levy, H., & Carmichael, G.R. 2000. Impacts of biomass burning on tropospheric CO, NO_x, and O_3. *Journal of Geophysical Research: Atmospheres*. 105(D5), 6633–6653.

Gaveau, D.L., Salim, M.A., Hergoualc'h, K., Locatelli, B., Sloan, S., Wooster, M., Marlier, M.E., Molidena, E., Yaen, H., DeFries, R., & Verchot, L. 2014. Major atmospheric emissions from peat fires in Southeast Asia during non-drought years: Evidence from the 2013 Sumatran fires. *Scientific Reports*, 4, 6112.

Goetz, J.D., Giordano, M.R., Stockwell, C.E., Christian, T.J., Maharjan, R., Adhikari, S., Bhave, P.V., Praveen, P.S., Panday, A.K., Jayarathne, T., & Stone, E.A. 2018. Speciated online PM1 from South Asian combustion sources-Part 1: Fuel-based emission factors and size distributions. *Atmospheric Chemistry and Physics*, 18, 14653–14679.

Goss, P.E., Strasser-Weippl, K., Lee-Bychkovsky, B.L., Fan, L., Li, J., Chavarri-Guerra, Y., Liedke, P.E., Pramesh, C.S., Badovinac-Crnjevic, T., Sheikine, Y., & Chen, Z. 2014. Challenges to effective cancer control in China, India, and Russia. *The Lancet Oncology*, 15(5), 489–538.

Gupta, S., Agarwal, R., & Mittal, S.K. 2016. Respiratory health concerns in children at some strategic locations from high PM levels during crop residue burning episodes. *Atmospheric Environment*, 137, 127–134.

Gupta, P.K., Prasad, V.K., Sharma, C., Sarkar, A.K., Kant, Y., Badarinath, K.V.S., & Mitra, A.P. 2001. CH_4 emissions from biomass burning of shifting cultivation areas of tropical deciduous forests–Experimental results from ground-based measurements. *Chemosphere-Global Change Science*, 3(2), 133–143.

Guttikunda, S.K., & Kopakka, R.V. 2014. Source emissions and health impacts of urban air pollution in Hyderabad, India. *Air Quality, Atmosphere & Health*, 7(2), 195–207.

Hao, W.M., & Liu, M.H. 1994. Spatial and temporal distribution of tropical biomass burning. *Global Biogeochemical Cycles*, 8(4), 495–503.

Hayasaka, H., & Sepriando, A. 2018. Severe Air Pollution Due to Peat Fires During 2015 Super El Niño in Central Kalimantan, Indonesia. In: *Land-Atmospheric Research Applications in South/Southeast Asia*. Vadrevu, K.P., Ohara, T., and Justice, C. (Eds.). Springer, Cham, pp. 129–142.

Hayasaka, H., Noguchi, I., Putra, E.I., Yulianti, N., & Vadrevu, K. 2014. Peat-fire-related air pollution in Central Kalimantan, Indonesia. *Environmental Pollution*. 195, 257–266. doi:10.1016/j.envpol.2014.06.031.

Hays, M.D., Fine, P.M., Geron, C.D., Kleeman, M.J., & Gullett, B.K. 2005. Open burning of agricultural biomass: physical and chemical properties of particle-phase emissions. *Atmospheric Environment*, 39(36), 6747–6764.

Hira, G.S. 2009. Water management in northern states and the food security of India. *Journal of Crop Improvement*, 23(2), 136–157.

Huijnen, V., Wooster, M.J., Kaiser, J.W., Gaveau, D.L., Flemming, J., Parrington, M., Inness, A., Murdiyarso, D., Main, B., & Van Weele, M., 2016. Fire carbon emissions over maritime southeast Asia in 2015 largest since 1997. *Scientific Reports*, 6, 26886.

Islam, M., Pei, Y.H., & Mangharam, S. 2016. Trans-boundary haze pollution in Southeast Asia: Sustainability through plural environmental governance. *Sustainability*, 8(5), 499. doi:10.3390/su8050499.

Israr, I., Jaya, S.N.I, Saharjo, H.S., Kuncahyo, B., & Vadrevu, K.P. 2018. Spatio-temporal analysis of land and forest fires in Indonesia using MODIS active fire dataset. In: *Land-Atmospheric Research Applications in South/Southeast Asia*. Vadrevu, K.P., Ohara, T., and Justice, C. (Eds.). Springer, Cham, pp. 105–128.

Itahashi, S., Uno, I., Irer, H., Kurokawa, J.-I., & Ohara, T. 2018. Impacts of biomass burning emissions on tropospheric NO2 vertical column density over continental Southeast Asia. In: *Land-Atmospheric Research Applications in South/Southeast Asia*. Vadrevu, KP, Ohara, T., and Justice, C. (Eds.). Springer, Cham, pp. 67–82.

Jain, N., Bhatia, A., & Pathak, H. 2014. Emission of air pollutants from crop residue burning in India. *Aerosol and Air Quality Research*, 14(1), 422–430. doi:10.4209/aaqr.2013.01.0031.

Jayarathne, T., Stockwell, C.E., Bhave, P.V., Praveen, P.S., Rathnayake, C.M., Islam, M.R., Panday, A.K., Adhikari, S., Maharjan, R., Goetz, J.D., & DeCarlo, P.F. 2018. Nepal Ambient Monitoring and Source Testing Experiment (NAMaSTE): Emissions of particulate matter from wood-and dung-fueled cooking fires, garbage and crop residue burning, brick kilns, and other sources. *Atmospheric Chemistry and Physics*, 18(3), 2259.

Jie, Y. 2017. Air pollution associated with Sumatran forest fires and mortality on the Malay Peninsula. *Polish Journal of Environmental Studies*, 26(1), 163–171. doi:10.15244/pjoes/64642.

Kaskaoutis, D.G., Kumar, S., Sharma, D., Singh, R.P., Kharol, S.K., Sharma, M., Singh, A.K., Singh, S., Singh, A., & Singh, D., 2014. Effects of crop residue burning on aerosol properties, plume characteristics, and long-range transport over northern India. *Journal of Geophysical Research: Atmospheres*, 119(9), 5424–5444.

Kawamura, K., & Pavuluri, C.M. 2010. New Directions: Need for better understanding of plastic waste burning as inferred from high abundance of terephthalic acid in South Asian aerosols. *Atmospheric Environment*, 44(39), 5320–5321. http://hdl.handle.net/2115/48561

Kim, M.B., Park, J., Kim, S., Kim, H., Jeon, H., Cho, C., Lim, J., Hong, S., Rupakheti, M., Panday, A.K., Park, R.J., Hong, J., & Yoon, S. 2015. Source apportionment of PM10 mass and particulate carbon in the Kathmandu Valley, Nepal. *Atmospheric Environment*, 123, 190–199. doi:10.1016/j.atmosenv.2015.10.082.

Kim, Y., Knowles, S., Manley, J., & Radoias, V., 2017. Long-run health consequences of air pollution: Evidence from Indonesia's forest fires of 1997. *Economics & Human Biology*, 26, 186–198.

Kumar, S., Aggarwal, S.G., Gupta, P.K., & Kawamura, K. 2015. Investigation of the tracers for plastic-enriched waste burning aerosols. *Atmospheric Environment*, 108, 49–58.

Lal, R.M., Nagpure, A.S., Luo, L., Tripathi, S.N., Ramaswami, A., Bergin, M.H., & Russell, A.G. 2016. Municipal solid waste and dung cake burning: discoloring the Taj Mahal and human health impacts in Agra. *Environmental Research Letters*, 11(10), 104009.

Lasko, K., Vadrevu, K.P., & Nguyen, T.T. 2018. Analysis of air pollution over Hanoi, Vietnam using multi-satellite and MERRA reanalysis datasets. *Plos One*, 13(5). doi:10.1371/journal.pone.0196629.

Lawrence, M.G., & Lelieveld, J. 2010. Atmospheric pollutant outflow from southern Asia: A review. *Atmospheric Chemistry and Physics*, 10, 11017–11096.

Levine, J.S. 1999. The 1997 fires in Kalimantan and Sumatra, Indonesia: Gaseous and particulate emissions. *Geophysical Research Letters*, 26(7), 815–818. doi:10.1029/1999gl900067.

Li, G., Lei, W., Bei, N., & Molina, L.T. 2012. Contribution of garbage burning to chloride and PM2.5 in Mexico City. *Atmospheric Chemistry and Physics*, 12, 8751–8761. doi:10.5194/acp-12-8751-2012.

Li, J., Wang, G., Aggarwal, S.G., Huang, Y., Ren, Y., Zhou, B., Singh, K., Gupta, P.K., Cao, J., & Zhang, R., 2014. Comparison of abundances, compositions and sources of elements, inorganic ions and organic compounds in atmospheric aerosols from Xi'an and New Delhi, two megacities in China and India. *Science of the Total Environment*, 476, 485–495.

Lighthall, D.R., & Kopecky, S. 2000. Confronting the problem of backyard burning: The Case for a national ban. *Society & Natural Resources*, 13(2), 157–167.

Lohan, S.K., Jat, H.S., Yadav, A.K., Sidhu, H.S., Jat, M.L., Choudhary, M., Peter, J.K., & Sharma, P.C. 2018. Burning issues of paddy residue management in north-west states of India. *Renewable and Sustainable Energy Reviews*, 81, 693–706.

Lohan, S.K., Narang, M.K., Manes, G.S., & Grover, N. 2015. Farm power availability for sustainable agriculture development in Punjab state of India. *Agricultural Engineering International: CIGR Journal*, 17(3), 196–207.

Mahata, K.S., Panday, A.K., Rupakheti, M., Singh, A., Naja, M., & Lawrence, M.G. 2017. Seasonal and diurnal variations in methane and carbon dioxide in the Kathmandu Valley in the foothills of the central Himalayas. *Atmospheric Chemistry & Physics*. 17(20), 12573–12596.

Miller, G.T., & Spoolman, S. 2015. *Living in the Environment: Principles, Connections, and Solutions*. Stamford, CT: Cengage Learning.

Mishra, A.K., & Shibata, T. 2012. Synergistic analyses of optical and microphysical properties of agricultural crop residue burning aerosols over the Indo-Gangetic Basin (IGB). *Atmospheric Environment*, 57, 205–218. doi:10.1016/j.atmosenv.2012.04.025.

Mittal, S.K., Susheel, K., Singh, N., Agarwal, R., Awasthi, A., & Gupta, P.K. 2009. Ambient Air Quality during Wheat and Rice Crop Stubble Burning Episodes in Patiala. *Atmospheric Environment*, 43, 238–244.

Nagar, P.K., Singh, D., Sharma, M., Kumar, A., Aneja, V.P., George, M.P., Agarwal, N., & Shukla, S.P. 2017. Characterization of PM2.5 in Delhi: Role and impact of secondary aerosols, burning of biomass, and municipal solid waste and crustal matter. *Environmental Science and Pollution Research*, 24, 25179–25189.

Nastos, P.T., Paliatsos, A.G., Anthracopoulos, M.B., Roma, E.S., & Priftis, K.N. 2010. Outdoor particulate matter and childhood asthma admissions in Athens, Greece: A time-series study. *Environmental Health*, 9(1), 45.

Nichol, J. 1997. Bioclimatic impacts of the 1994 smoke haze event in Southeast Asia. *Atmospheric Environment*, 31(8), 1209–1219. doi:10.1016/s1352-2310(96)00260-9.

Page, S.E., Siegert, F., Rieley, J.O., Boehm, H.V., Jaya, A., & Limin, S. 2002. The amount of carbon released from peat and forest fires in Indonesia during 1997. *Nature*, 420(6911), 61–65. doi:10.1038/nature01131.

Pavuluri, C.M., & Kawamura, K. 2018. Organic aerosols in South and East Asia: Comparison and sources. In: *Land-Atmospheric Research Applications in South/Southeast Asia*. Vadrevu, KP, Ohara, T., and Justice, C. (Eds). Springer, Cham, pp.105–128, 379–408

Phairuang, W., Hata, M., & Furuuchi, M. 2017. Influence of agricultural activities, forest fires and agro-industries on air quality in Thailand. *Journal of Environmental Sciences*, 52, 85–97. doi:10.1016/j.jes.2016.02.007.

Przyborski, P. 2019. Aerosol Optical Depth. https://earthobservatory.nasa.gov/global-maps/ MODAL2_M_AER_OD.

Putero, D., Cristofanelli, P., Marinoni, A., Adhikary, B., Duchi, R., Shrestha, S.D., Verza, G.P., Landi, T.C., Calzolari, F., Busetto, M., Agrilo, G., Biancofiore, F, Di Carlo, P., Panday, A.K., Rupakheti, M., & Bonsanoni, P. 2015. Seasonal variation of ozone and black carbon observed at Paknajol, an urban site in the Kathmandu Valley, Nepal. *Atmospheric Chemistry and Physics*, 15, 13957–13971.

Ramachandran, S. 2018. Aerosols and climate change: Present understanding, challenges and future outlook. In: *Land-Atmospheric Research Applications in South/Southeast Asia*. Vadrevu, K.P., Ohara, T., and Justice, C. (Eds.). Springer, Cham, pp. 341–378.

Ransom, M.R., & Pope, C.A. 1992. Elementary school absences and PM10 pollution in Utah Valley. *Environmental Research*. 58(1–2), 204–219. doi:10.1016/s0013-9351(05)80216-6.

Ravindra, K., Singh, T., & Mor, S. 2019. Emissions of air pollutants from primary crop residue burning in India and their mitigation strategies for cleaner emissions. *Journal of Cleaner Production*, 208, 261–273. doi:10.1016/j.jclepro.2018.10.031.

Saikawa, E., Panday, A., Kang, S., Gautam, R., Zusman, E., Cong, Z., ... & Adhikary, B. (2019). Air pollution in the Hindu Kush Himalaya. In: *The Hindu Kush Himalaya Assessment*, Wester, P., Mishra, A., Mukherji, A., Shrestha, A. (Eds.) Springer, Cham, pp. 339–387.

Saito, M., Amagai, K., Ogiwara, G., & Arai, M. 2001. Combustion characteristics of waste material containing high moisture. *Fuel*, 80(9), 1201–1209.

Sarkar, C., Sinha, V., Kumar, V., Rupakheti, M., Panday, A., Mahata, K.S., Rupakheti, D., Kathayat, B., & Lawrence, M.G. 2016. Overview of VOC emissions and chemistry from PTR-TOF-MS measurements during the SusKat-ABC campaign: High acetaldehyde, isoprene and isocyanic acid in wintertime air of the Kathmandu Valley. *Atmospheric Chemistry and Physics*, 16, 3979–4003. doi:10.5194/acp-16-3979-2016.

Sarkar, C., Kumar, V., & Sinha, V. 2013. Massive emissions of carcinogenic benzenoids from paddy residue burning in north India. *Current Science*, 104, 1703–1709.

Sastry, N. 2002. Forest fires, air pollution, and mortality in Southeast Asia. *Demography*, 39(1) 1. doi:10.2307/3088361.

Sawlani, R., Agnihotri, R., Sharma, C., Patra, P.K., Dimri, A.P., Ram, K., & Verma, R.L. 2018. The severe Delhi SMOG of 2016: A case of delayed crop residue burning, coincident firecracker emissions, and atypical meteorology. *Atmospheric Pollution Research*. doi:10.1016/j.apr.2018.12.015.

Sharma, A.R., Kharol, S.K., Badarinath, K.V.S., & Singh, D. 2010. Impact of agriculture crop residue burning on atmospheric aerosol loading—A study over Punjab State, India. *Annales Geophysicae*. 28(2), 09927689.

Simoneit, B.R.T, Medeiro, P.M., & Didyk, B.M. 2005. Combustion products of plastics as indicators for refuse burning in the atmosphere. *Environmental Science & Technology*, 39(18) 6961–6970.

Singh, N., Agarwal, R., Awasthi, A., Gupta, P.K., & Mittal, S.K. 2010. Characterization of atmospheric aerosols for organic tarry matter and combustible matter during crop residue burning and non-crop residue burning months in Northwestern region of India. *Atmospheric Environment*, 44(10), 1292–1300. doi:10.1016/j.atmosenv.2009.12.021.

Singh, R.P., & Kaskaoutis, D.G. 2014. Crop residue burning: A threat to South Asian air quality. *Eos, Transactions American Geophysical Union*, 95(37), 333–334. doi:10.1002/2014EO370001.

Stockwell, C.E., Christian, T.J., Goetz, J.D., Jayarathne, T., Bhave, P.V., Praveen, P.S., Adhikari, S., Maharjan, R., DeCarlo, P.F., Stone, E.A., & Saikawa, E. 2016. Nepal ambient monitoring and source testing experiment (NAMaSTE): Emissions of trace gases and light-absorbing carbon from wood and dung cooking fires, garbage and crop residue burning, brick kilns, and other sources. *Atmospheric Chemistry and Physics*. 16(17), 11043–11081.

Streets, D.G., Yarber, K.F., Woo, J.H., & Carmichael, G.R. 2003. Biomass burning in Asia: Annual and seasonal estimates and atmospheric emissions. *Global Biogeochemical Cycles*. 17(4). doi:10.1029/2003gb002040.

Tariq, S., Ul-Haq, Z., & Ali, M. 2015. Analysis of optical and physical properties of aerosols during crop residue burning event of October 2010 over Lahore, Pakistan. *Atmospheric Pollution Research*, 6(6), 969–978. doi:10.1016/j.apr.2015.05.002.

ul-Haq, Z., Rana, A.D., Ali, M., Mahmood, K., Tariq, S., & Qayyum, Z. 2015. Carbon monoxide (CO) emissions and its tropospheric variability over Pakistan using satellite-sensed data. *Advances in Space Research*, 56(4), 583–595. doi:10.1016/j.asr.2015.04.026.

Ullah, A., Khan, D., Khan, I., & Zheng, S. 2018. Does agricultural ecosystem cause environmental pollution in Pakistan? Promise and menace. *Environmental Science and Pollution Research*, 25(14), 13938–13955. doi:10.1007/s11356-018-1530-4.

US EPA (Environmental Protection Agency). 2006. An inventory of sources and environmental releases of dioxin-like compounds in the USA for the years 1987, 1995, and 2000.

Vadrevu, K.P., Ellicott, E., Badarinath, K.V.S., & Vermote, E. 2011. MODIS derived fire char-acteristics and aerosol optical depth variations during the agricultural residue burning season, north India. *Environmental Pollution*, 159(6), 1560–1569.

Vadrevu, K.P. and Justice, C.O. 2011. Vegetation fires in the Asian region: Satellite observational needs and priorities. *Global Environmental Research*, 15(1), 65–76.

Vadrevu, K.P., Lasko, K., Giglio, L., Schroeder, W., Biswas, S., & Justice, C. 2019. Trends in vegetation fires in south and southeast Asian countries. *Scientific Reports*, 9(1), 7422. doi:10.1038/s41598-019-43940-x.

Vadrevu, K.P., Ohara, T., & Justice, C. 2017. Land cover, land use changes and air pollution in Asia: A synthesis. *Environmental Research Letters*, 12(12), 120201.

Vadrevu, K.P., & Lasko, K. 2018. Intercomparison of MODIS AQUA and VIIRS I-Band fires and emissions in an agricultural landscape—Implications for air pollution research. *Remote Sensing*, 10(7), 978. doi:10.3390/rs10070978.

Vadrevu, K.P., Ohara, T., & Justice, C. (Eds.) 2018. *Land-Atmospheric Research Applications in South and Southeast Asia*. Springer, Cham.

Wan, X., Kang, S., Li, Q., Rupakheti, D., Zhang, Q., Guo, J., Chen, P., Tripathee, L., Rupakheti, M., Panday, A.K., & Wang, W. 2017. Organic molecular tracers in the atmospheric aero-sols from Lumbini, Nepal, in the northern Indo-Gangetic Plain: Influence of biomass burning. *Atmospheric Chemistry & Physics,* 17(14), 8867–8885.

Wendling, Z.A., Emerson, J.W., Esty, D.C., Levy, M.A., de Sherbinin, A., et al. 2018. *2018 Environmental Performance Index*. Yale Center for Environmental Law & Policy, New Haven, CT. https://epi.yale.edu/.

Wester, P., Mishra, A., Mukherji, A., & Shrestha, A.B. 2019. *The Hindu Kush Himalaya Assessment: Mountains, Climate Change, Sustainability and People*. Springer International Publishing, New York.

Wiedinmyer, C., Yokelson, R.J., & Gullett, B.K. 2014. Global emissions of trace gases, par-ticulate matter, and hazardous air pollutants from open burning of domestic waste. *Environmental Science & Technology*. doi:10.1021/es502250z.

Xu, R., Tie, X., Li, G., Zhao, S., Cao, J., Feng, T., & Long, X. 2018. Effect of biomass burning on black carbon (BC) in South Asia and Tibetan Plateau: The analysis of WRF-Chem modeling. *Science of the Total Environment*, 645, 901–912.

Yadav, I.C., Devi, N.L., Li, J., Syed, J.H., Zhang, G., & Watanabe, H. 2017. Biomass burn-ing in Indo-China peninsula and its impacts on regional air quality and global climate change-a review. *Environmental Pollution*, 227, 414–427.

Yadav, I.C., Devi, N.L., Singh, V.K., Li, J., & Zhang, G. 2018. Concentrations, sources and health risk of nitrated- and oxygenated-polycyclic aromatic hydrocarbon in urban indoor air and dust from four cities of Nepal. *Science of the Total Environment*, 643, 1013–1023. doi:10.1016/j.scitotenv.2018.06.265.

Zhang, H., Hu, D., Chen, J., Ye, X., Wang, S.X., Hao, J.M., Wang, L., Zhang, R., & An, Z., 2011. Particle size distribution and polycyclic aromatic hydrocarbons emissions from agricultural crop residue burning. *Environmental Science & Technology*, 45(13), 5477–5482.

Zhang, X., Lu, Y., Wang, Q., & Qian, X. 2018. A high-resolution inventory of air pollutant emissions from crop residue burning in China. *Atmospheric Chemistry and Physics Discussions*, 1–19. doi:10.5194/acp-2017-1113.

Zhong, M., Saikawa, E., Avramov, A., Chen, C., Sun, B., Ye, W., Keene, W.C., Yokelson, R.J., Jayarathne, T., Stone, E.A., & Rupakheti, M. 2019. Nepal Ambient Monitoring and Source Testing Experiment (NAMaSTE): Emissions of particulate matter and sulfur dioxide from vehicles and brick kilns and their impacts on air quality in the Kathmandu Valley, Nepal. *Atmospheric Chemistry & Physics*, 19(12), 8209–8228.

2 Biomass Burning and Their Impacts on Air Quality in Thailand

Worradorn Phairuang
Prince of Songkla University, Thailand

CONTENTS

INTRODUCTION

Biomass burning is an important source of air pollution in several countries (Andreae and Merlet, 2001). The emissions from the biomass burning release a large amount of aerosol species and toxic gases including nitrogen oxides (NO_x), carbon monoxides (CO), sulfur dioxide (SO_2), and polycyclic aromatic hydrocarbons (PAHs) into the atmosphere (Badarinath et al., 2008a, b; Gupta et al., 2001; Prasad et al., 2000, 2002; Oanh et al., 2002; Prasad et al., 2003; 2005). Biomass combustion can affect not only the environment but also human health. In developing countries, especially in South/Southeast Asia, open biomass burning is commonly practiced to clear the forests for agriculture, raising commercial plantations, disposing crop residues before or after harvesting. All these activities release a significant amount of pollutions to the atmosphere(Kant et al., 2000; Prasad et al., 2001; 2008; Lasko et al., 2017; Lasko and Vadrevu, 2018; Lasko et al., 2018a, b; Garivait et al., 2004; Vadrevu et al., 2006;

Vadrevu, 2008; Vadrevu et al., 2008; Zhang et al (2011); Vadrevu and Lasko, 2015; Phairuang et al., 2017). The countries that are most affected due to biomass burning pollution are India, Myanmar, Thailand, Indonesia, Laos, Cambodia, etc. (Vadrevu et al., 2012, 2013; Hayasaka et al., 2014; Vadrevu et al., 2014a, b, 2015, 2017, 2018; Biswas et al., 2015a, b; Justice et al., 2015; Hayasaka and Sepriando, 2018; Israr et al., 2018; Saharjo and Yungan, 2018; Ramachandran, 2018; Inoue, 2018; Vadrevu and Lasko, 2018; Vadrevu et al., 2019). Specifically, Thailand is an agriculture-based country and generates a large amount of agricultural residues. In the future, crop residues are bound to increase with rising population as well as growing agriculture sector (Kasem and Thapa, 2012; Oanh et al., 2018). Thus, air pollution from crop residue burning might increase in the coming years.

Developing an Emission Inventory (EI) is a standard approach to report the total air pollutants and greenhouse gases in a geographical area useful to support air quality management (U.S. EPA, 2010; Phairuang et al., 2017). EIs are developed for primary pollutants such as PMand greenhouse gases. For a large-scale EI, mostly, top-down methods are used when the local data are not available. For instance, Global Emission Inventory Activity (GEIA) (Graedel et al., 1993), Transport and Chemical Evolution over the Pacific (TRACE-P) (Streets et al., 2003), and Regional Emission inventory in ASia (REAS) (Ohara et al., 2007) all use top-down approaches. Instead, the bottom-up approach is used when the domestic data are available. Specifically, for Thailand, the bottom-up EI data are very limited. For example, EIs reported by governmental sectors and some academic researchers such as the Pollution Control Department in Thailand (PCD, 1994; 2005) and researchers addressing the open biomass burning studies are mostly based on top-down approaches (Kanabkaew and Oanh, 2011; Vadrevu and Justice (2011); Vadrevu and Choi (2011); Cheewaphongphan and Garivait, 2013; Junpen et al., 2013). Most of the previous studies on EIs in Thailand focused on the open biomass burning in the field; however, biomass burning emissions from agro-industries are also very important in Thailand.

Air pollution is one of the critical environmental concerns in Thailand, specifically, PM_{10} (particulate matter with an aerodynamic diameter smaller than $10\,\mu m$). Important sources of PM_{10} includes biomass combustion (40.2%), industrial activities (31.5%), traffic emissions (17.1%), and power plants (10.9%) (Vongmahadlek et al., 2009). The sources of pollution vary in different regions of Thailand; hence, there is a strong need to quantify them.

In northern Thailand, biomass burning is an important source of pollution (Pengchai et al., 2009; Phairuang et al., 2017; Moran et al., 2019). In addition to biomass burning, the other main contributors include motor vehicles which contribute to significant pollution, such as in the Bangkok Metropolitan Region (BMR) in the central part of Thailand (Oanh, 2017; Phairuang et al., 2019a). The recent study of air quality in the BMR showed that for $PM_{2.5}$ (particulate matter with an aerodynamic diameter smaller than $2.5\,\mu m$), road traffic and biomass combustion are the two essential sources (Narita et al., 2019). In southern Thailand, Pongpiachan et al. (2014) indicated that the maritime aerosols and biomass combustion from crop residue burning are the main sources of PM_{10} related to carbon components. Moreover, fuelwood burning at the rubber factory is the main source of PAHs and other pollutants (Phairuang et al., 2019b). The burning period and the moisture content in the wood is the main contributing factor to smoke particles and associated PAHs in the workplace and surrounding areas (Chomanee et al., 2018).

Overall, forest fires, crop residue burning, and agro-industries are the main contributors to the ambient PMs (coarse- and fine-mode particles) in Thailand, although emissions from traffic in an urban area, i.e., the BMR, are an important source of particles (Oanh, 2017). There has been a lack of studies on the spatiotemporal characteristics of atmospheric aerosols in Thailand. Consequently, PMs' physical and chemical characteristics in Thailand must be investigated for the main emission sources and air quality.

In the north and northeastern regions of Thailand, an important economic crop is sugarcane that is increasing rapidly because of the promotion of renewable energy, e.g., bioethanol and gasohol by the Thai government (OAE, 2006; 2009). For example, the total output of sugarcane in Thailand increased from 49.58 million tons (Mt) in 2005 to 100.09 Mt in 2013. Sugarcane production increased by more than 100% in 9 years. Most of the production is transported to factories for generating sugar and molasses (OCSB, 2014). An enormous amount of biomass fuel in the sugarcane plant, which is a sugarcane residue product after sugar processing, has been used as an energy source. However, the energy consumption and related pollutant emissions generated during the sugarcane production process may be significant, releasing huge air pollutants such as PMs since they are not controlled in many cases.

METHODOLOGY

CROP RESIDUE BURNING

The amount of agricultural residues burned and related emissions is reported by many researchers (Table 2.1). Emissions from the crop residue burning are calculated by Equations 2.1 and 2.2:

$$E = \Sigma_{crop} T \times EF \tag{2.1}$$

and

$$T = A \times R \times D \times \beta \times F \tag{2.2}$$

where E is the emission of each pollutant (tons), T is the total amount of biomass burning, and EF is the emission factor of different pollutants from crop residue burning type (tons/unit activity) available from the domestic values of the literature. In the above equation, A is an annual crop production (tons); the data were obtained from the Office of Agricultural Economics in Thailand (OAE) based year 2013. This information is corrected for each of the provinces in Thailand. R represents residue-to-crop ratio; D is the dry matter fraction, β is the fraction burned in the field, and F is a fraction of residue oxidized.

FOREST FIRES

PM from forest fires was estimated following the Global Atmospheric Pollution Forum Air Pollutant Emission Inventory Manual (GAPF) version 5.0. Emissions

TABLE 2.1
Summary of the parameters used for Emission Inventory in 2013

Crop	Annual Production (Mt/year)	Residue-to-Crop Ratio	Dry Matter Fraction	Fraction Burnt	Fraction Oxidized
Total rice	38.78	1.19[a]	0.85[b]	0.48[c]	0.89[b]
Corn	5.06	0.19[a]	0.40[b]	0.61[d]	0.92[b]
Sugarcane	100.09	0.37[e]	0.71[b]	0.55[d]	0.68[b]
Cassava	30.22	0.12[a]	0.71[b]	0.41[d]	0.68[b]
Soybeans	0.07	1.50[f]	0.71[b]	0.76[g]	0.68[b]
Potatoes	0.10	050[f]	0.45[h]	1.00	0.90[h]
Total	174.32				

[a] DEDE (2007). Biomass potential for energy production in Thailand.
[b] Streets et al. (2003). Default value from several area sources.
[c] DEDE (2003). Non-exploited rice straw in Thailand.
[d] EFE (2009). Non-exploited crop residues in Thailand in 2007.
[e] Sornpoon et al. (2014). Estimation of emissions from sugarcane field burning in Thailand.
[f] Yang et al. (2008). Default value from crop residue burning in China.
[g] Sajjakulnukit et al. (2005). Surplus availability of crop residues in Thailand.
[h] IPCC (2006). Default value from the IPCC EI manual.

from forest fires in Thailand were calculated from the following equation (Giglio et al., 2006):

$$E = \Sigma M \times EF \tag{2.3}$$

where E = the emission of each pollutant (g) from a forest fire; M = total amount of biomass consumption; and EF = emission factor of different pollutants (g/kg$_{dry\ mass}$); EF was taken from national data in Thailand (Chaiyo and Garivait, 2014).

$$M = A \times B \times C \tag{2.4}$$

where A = burned area (km^2); B = biomass density (kg dry mass/km^2) from forest area in Thailand; and C = combustion efficiency. The burned areas directly come from the Forest Fire Control Division in Thailand (http://www.dnp.go.th/forest fire/Eng/indexing.htm). EF is an emission factor of different pollutants emitted from vegetation fires (tons/unit activity) (GAPF version 5.0). In general, forest types in Thailand are classified as the tropical forest that consists of tropical evergreen and deciduous species (Thawatchai, 2012).

Active fires or open burning of forest fires was detected by Moderate Resolution Imaging Spectroradiometer (MODIS) satellite remote sensing imagery, accessible from http://earthdata.nasa.gov/data/near-real-time-data/firms. The resolution of active fires is represented at a 1 km × 1 km resolution. Tanpipat et al. (2009) validated the MODIS hotspot data over Thailand for the fire seasons from 2007 to 2009 with a 95% accuracy. The fire hotspots from MODIS are very useful for providing for-est fires, including spatial and temporal distribution (Junpen et al., 2011; Oanh and Leelasakultum, 2011).

AGRO-INDUSTRIES

Agro-industries are an essential source of emissions in Thailand. In our study, we used the sugarcane production rates from the Office of the Cane and Sugar Board (OCSB). There are 47 sugar mills distributed mainly in the lower northern and north-eastern regions of Thailand in 2012. The consumption of solid biomass fuel in sug-arcane factories for generating power to generate sugar production process has been calculated as,

$$(W_{fuel}) / (year) \times J / (W_{fuel}) \times (W) / J = W / year$$

where W_{fuel}/year is the total biomass residue consumption, namely bagasse in the sugar factory (tons/year), retrieved from the OCSB in Thailand; J/W_{fuel} is an average lower heating value in bagasse equal to 7600 kJ/kg (Jenjariyakosoln et al., 2014); and W/J is an EF of different pollutants emitted from the boiler in the agro-industry from GAPF version 5.0.

VALIDATION OF EMISSION INVENTORY (EI)

We compared the EI from biomass burning emissions during 2013 with the ambient air pollutant data from various Air Quality Monitoring (AQM) stations in Thailand, established by the Pollution Control Department (PCD). There were 66 AQM stations in Thailand during 2013; however, they didn't cover all of the provinces. Nearly half of AQM stations (28 stations) were in the BMR of central Thailand. We compared the EI data and AQM station data in northern and northeastern Thailand. The AQM stations were carefully chosen to represent each area for this study, including two stations in Chiang Mai, upper northern Thailand, and one station each in Nakhon Sawan, lower northern Thailand, and in Khon Kaen and Nakhon Ratchasima, north-eastern Thailand. In this study, we excluded pollutant emissions from the BMR. Thailand Environment Monitor 2002 indicates that air pollutant sources in Bangkok mainly come from the transportation, industry, and power sectors. As we focused mostly on biomass combustion, exclusion of BMR data is justified

RESULTS AND DISCUSSION

ANNUAL EMISSIONS FROM OPEN BIOMASS BURNING

Annual crop production in Thailand is ~174.32 Mt/year. The highest crop production is from sugarcane (100.09 Mt), followed by rice (38.78 Mt), cassava (30.22 Mt), and minor production of corn, soybean, and potato (5.23 Mt) (Table 2.1). The rice resi-due burning-generated air pollutants are highest with ~71%, followed by sugarcane (21%), cassava (1%), and others (1% with corn, soybean, and potato); the contribution from forest fires is ~6% (Figure 2.1). The annual emissions for the year 2013, including crop residue burning and forest fires, were as follows: PM_{10}, 210,908 tons; $PM_{2.5}$, 192,367 tons; NO_x, 87,865 tons; and SO_2, 13,896 tons (Table 2.2). In total, air pollutants from the burning of rice residue contribute to the largest emissions, followed by sugarcane and cassava. Others such as corn, soybean, and potato residue burning contribute very less to the emissions. Also, emissions from forest fires are relatively small (6%) compared to residue burning (94%) in Thailand.

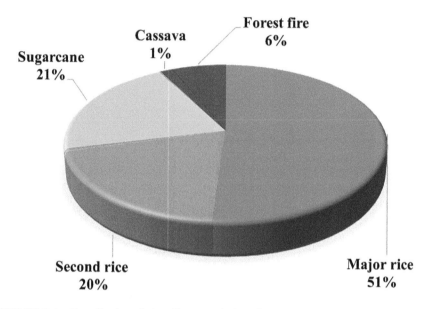

FIGURE 2.1   Contribution of air pollutant emissions from open biomass burning in 2013.

TABLE 2.2

Emissions of Air Pollutants from Forest Fires and Crop Residue Burning in Thailand for the Year 2013

Type	PM_{10}	$PM_{2.5}$	NO_x	SO_2
		(tons/year)		
Major rice	110,188	100,501	41,532	5,812
Second rice	42,335	38,641	15,957	2,233
Corn	842	842	659	104
Sugarcane	38,354	36,387	25,569	4,720
Soybean	316	316	138	39
Potato	83	83	36	10
Cassava	2,800	2,800	1,221	345
Forest fire	15,990	12,797	2,753	633
All types	210,908	192,367	87,865	13,896

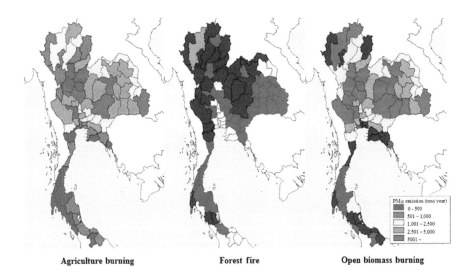

Agriculture burning   Forest fire   Open biomass burning

FIGURE 2.2 Spatial distribution of PM_{10} emissions from open biomass burning in Thailand in 2013.

Mostly, the emissions from forest fires mainly came from the northern part of Thailand. Although forest fire emissions are relatively less, the forest fire risk is very high, i.e., ~80%, in Thailand (Forest Fire Control Division, 2014). The spatial variation in emissions of PM_{10}, $PM_{2.5}$, NO_x, and SO_2 for open biomass burning in each province are shown in Figure 2.2. Emissions in Thailand are mostly released from the northeast and north parts, followed by central and southern regions, respectively. The rice burning in paddy is high in the central and northeast regions.

Figure 2.3 depicts the land-utilization map of Thailand for the year 2013. In the northeastern region, rain-fed lowland rice is mainly cultivated (Bridhikitti and Overcamp, 2012). The rice production was excessive, with a significant amount of residues generated and burning. For this reason, the northeastern area had a considerable amount of PM and greenhouse gas emissions. Furthermore, sugarcane is an important economic crop in the lower northern, central (excepting the BMR), and northeastern regions of Thailand. In contrast, in the southern part of Thailand, Para rubber and palm oil are the dominant crops, mostly used as biofuel in agro-industries for primary production. In our study, detailed information on biomass fuel utilization in Songkhla Province, southern Thailand, was obtained by the author through interviews with industry personnel and EI analysis. In southern Thailand, emissions are not from crop residue burning in the field; however, they are from direct combustion of biomass fuel used in the boiler for agro-industries.

EMISSIONS FROM AGRO-INDUSTRIES

Table 2.3 shows the emissions from sugarcane burning, from both the open burning and sugar mills. Sugarcane production in Thailand is the highest. The harvest statistics for 2013 came from the Office of Agricultural Economics, which is ~100,095,580

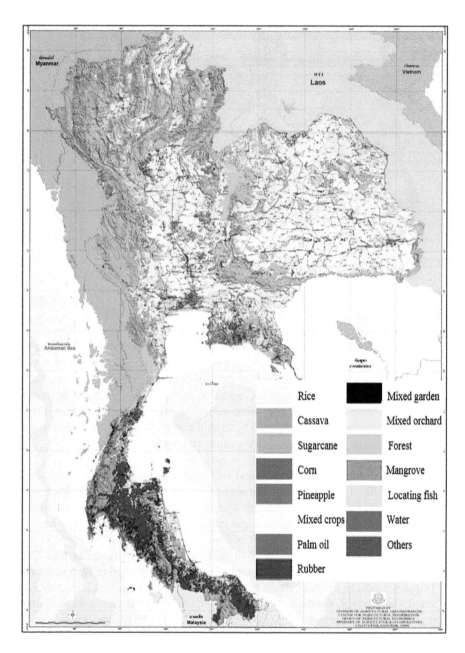

FIGURE 2.3   The land-use map in Thailand in 2013.

TABLE 2.3

PM$_{10}$ Emissions from Open Biomass Burning and Agro-Industries from Sugarcane Mills (2013)

Province	Harvest Production	Sugar Factories Consuming	Open Biomass Burning	Agro-Industry (Number of Factories)	Total Biomass Burning
	(tons/year)	(tons/5 months)	(tons/ year)	(tons/5 months)	(tons/year)
Nakhon Sawan (LN)	8,283,369	7,816,034	3,174	2,495 (2)	5,669
Kanchanaburi (C)	8,129,166	10,182,056	3,115	3,250 (8)	6,365
Khon Kaen (NE)	7,248,278	6,737,499	2,777	2,151 (2)	4,928
Suphanburi (C)	7,215,403	6,684,082	2,730	2,134 (3)	4,864
Nakhon Ratchasima (NE)	7,040,709	9,050,113	2,698	2,889 (3)	5,587
Kamphaeng Phet (LN)	6,092,926	8,146,786	2,337	2,600 (3)	4,937
Udon Thani (NE)	5,159,341	4,850,297	1,997	1,548 (4)	3,545
Seven provinces	49,169,192	53,466,867	18,828	17,067 (25)	35,895
Total Thailand	100,095,580	100,002,515	38,354	31,920 (47)	70,274

tons. Furthermore, the sugarcane used in sugar factories is ~100,002,515 tons (OCSB, Thailand). 99.91% of sugarcane is processed by the 47 sugar mills in Thailand to produce sugar and molasses. Pre-harvest burning of sugarcane accounts for 82%, and post-harvest burning contributes to the remaining 18% (Sornpoon, 2013) of emissions. The pre-harvesting is a standard method to remove sharp foliage and eliminate insects and snakes (Junginger et al., 2001).

The peak sugarcane cultivation province is Nakhon Sawan in lower northern Thailand, followed by Kanchanaburi, Khon Kaen, Suphanburi, Nakhon Ratchasima, Kamphaeng Phet, and Udon Thani. However, the air pollutants from sugar mills are slightly different based on the number of sugar mills in each province. The total PM$_{10}$ emitted by sugarcane residue burning is 70,274 tons/year, with open biomass burning contributing to ~38,354 tons (54.58%) and agro-industries to ~31,920 tons (45.42%).

UNCERTAINTY IN THE EMISSION INVENTORY (EI)

In our study, uncertainty in each pollutant emission came from burned areas, dry biomass burnt fraction, and emission factors (EFs). Most satellite burned area products, such as MODIS, underestimate burnt agricultural areas due to coarser resolution (Lasko et al., 2017). Besides, the field size of paddy areas in Thailand is small, for example, ~100 m, and the residues are burning within a short time of 2–3 hours (Kanabkaew and Oanh, 2011). Thus, there is a high possibility that crop residue burning events are not captured by satellite. Thus, we recommend the use of ground-based statistics on crop burning to reduce the uncertainties. Specific to the EFs, each crop species can have a different EF, resulting in uncertainty. Both the IPCC (2006) and GAPF (2012) suggested EFs are more general in nature and default values for open biomass burning; however, country-specific EFs are needed to reduce the uncertainty. In this study, we used the EFs from Thailand to estimate EIs. When EFs or default parameters are unavailable in Thailand, we preferred EFs from the other Asian countries, followed by others depending on the similar climate and crop cultivation (Cheewaphongphan and Garivait, 2013).

VALIDATION OF EMISSION INVENTORY (EI) WITH GROUND-BASED MONITORING DATA

Air quality is affected by emission source intensity and meteorological conditions, including wind direction, wind speed, temperature, and precipitation. The author validated the EI from the total biomass burning in Thailand with air quality data from ground-based monitoring stations in Thailand. Figure 2.4a and b shows the correlations between EI and average PM_{10} concentrations in each province of Thailand from January to December 2013. Chiang Mai 1 and Chiang Mai 2, in upper northern Thailand, showed high correlations between average monthly PM_{10} concentration and monthly EI. The correlations were $R^2 = 0.9431$ in Chiang Mai station 1 and 0.9198 in Chiang Mai station 2. In this region, forest fires are the primary source of air pollutants. Higher emissions from the biomass burning were noted during the dry season (January–May), whereas lower emissions were noted during the wet season (June–November). Moreover, December had emissions mostly from paddy residue burning.

In contrast, Nakhon Sawan in lower north Thailand and Khon Kaen and Nakhon Ratchasima in northeast Thailand were dominated by emissions from both crop residue burning and agro-industries (R^2 is 0.6725, 0.6150, and 0.6099, respectively). The forest fires' emissions contribute around 1%–4% in Nakhon Sawan, Khon Kaen, and Nakhon Ratchasima. In Nakhon Sawan, higher emissions were attributed to the open biomass burning from crop residues and agro-industries in the dry season (from January to April and December). In contrast, lower emissions were observed during the wet season (June–November). December had the highest emissions from rice and sugarcane residue burning, including emissions from sugar mills. Nakhon Sawan is an important province cultivating sugarcane and uses a large quantity of sugarcane in sugar mills for almost 5 months (from January to April and December). For that reason, air quality in Nakhon Sawan is deteriorated by the sugarcane residue burning in both the field and factories.

Khon Kaen and Nakhon Ratchasima have the same pattern as Nakhon Sawan. Higher EI was found in the dry season (from January to May and December), whereas lower EI was found during the wet season (from June to November). Khon Kaen has high EI from paddy field burning followed by open sugarcane burning and sugar

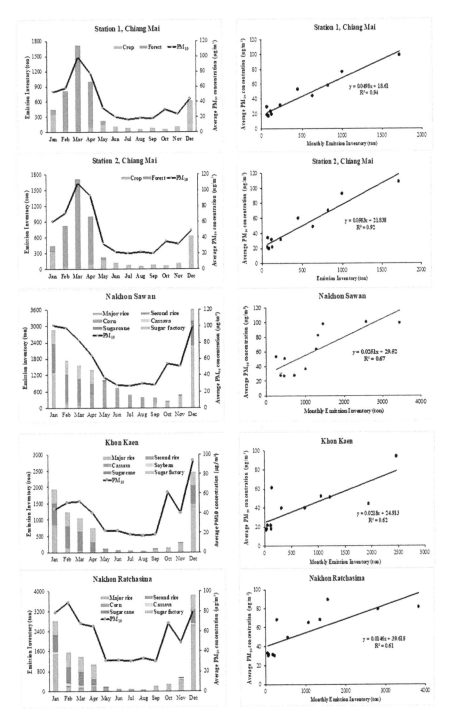

FIGURE 2.4   (a and b) Correlation between EI and PM_{10} concentration for different air monitoring stations in Thailand.

mills, and Nakhon Ratchasima is affected by rice, sugarcane, and cassava residue burning, including emissions from sugar mills. The good correlation between total EI and monthly average PM_{10} concentrations in Thailand supports the fact that biomass burning from crop residue burning and sugar mills during the dry season is a significant emission source. However, differences exist; upper northern Thailand is mostly affected by forest fires, whereas the lower north and northeastern regions of Thailand are affected by agricultural residue burning, including emissions from agro-industries.

EFFECT OF METEOROLOGICAL CONDITIONS ON AIR QUALITY

Figure 2.5 demonstrates the monthly average PM_{10} concentrations in Chiang Mai and Nakhon Sawan provinces. The monthly mean PM_{10} concentrations in Chiang Mai are relatively large during the dry season, especially in March. The highest

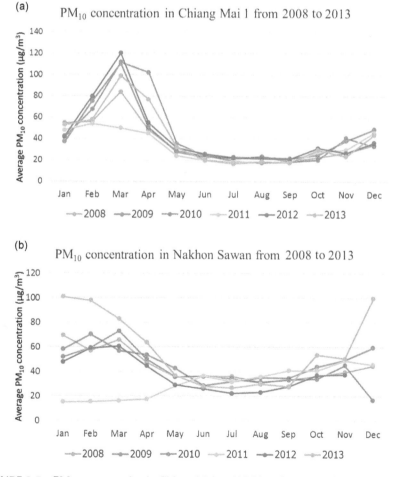

FIGURE 2.5 PM_{10} concentration in Chiang Mai and Nakhon Sawan ambient air from 2008 to 2013.

PM_{10} concentration in Chiang Mai came from the forest fire in this area from 2008 to 2013, except in 2011, where the PM_{10} is slightly less compared with others. Late 2010 to early 2011 are La Niña years. The La Niña period caused a drop in sea surface temperatures in Southeast Asia. It also caused heavy rains, which decreased the forest fires at this time. Further, the monthly average PM_{10} concentrations in Nakhon Sawan are high during the dry season from December to April. After the same, the PM_{10} concentration decreases from May till September. The PM_{10} concentrations from December 2010 till April 2011 are very low compared to other regions, mainly due to robust precipitation and floods in Thailand. Nakhon Sawan in lower northern Thailand is substantially affected by flood during this time.

CONCLUSIONS AND RECOMMENDATIONS

Air quality in Thailand is influenced by open biomass burning and indoor burning. In this study, emissions from total biomass combustion in Thailand were quantified from January to December 2013. The results were compared with the local monitoring stations. Results suggested air quality in Thailand was significantly affected by biomass burning activities during the dry season (January–April and December). However, the emission sources are different in each region. In upper northern Thailand (Chiang Mai), most of the PM_{10} came from the forest fires. In contrast, in lower northern (Nakhon Sawan) and northeastern (Khon Kaen and Nakhon Ratchasima) Thailand, PM_{10} emissions came from crop residue burning agro-industries, specifically sugarcane. Continuous monitoring of PM in terms of mass and number concentrations is needed to address air quality and health concerns in Thailand. Future research should focus on the identification and quantification of atmospheric PM, including sources. Better knowledge about PMs' physicochemical characteristics from biomass burning in Thailand and other Asian countries will help improve air quality through effective management and mitigation options.

REFERENCES

Andreae, M.O. and Merlet, P. 2001. Emission of trace gases and aerosols from biomass burning. *Global Biogeochem Cycles*. 15: 955–966.

Badarinath, K.V.S., Kharol, S.K., Krishna Prasad, V., Kaskaoutis, D.G. and Kambezidis, H.D. 2008a. Variation in aerosol properties over Hyderabad, India during intense cyclonic conditions. *International Journal of Remote Sensing*. 29(15): 4575–4597.

Badarinath, K.V.S., Kharol, S.K., Prasad, V.K., Sharma, A.R., Reddi, E.U.B., Kambezidis, H.D. and Kaskaoutis, D.G. 2008b. Influence of natural and anthropogenic activities on UV Index variations–a study over tropical urban region using ground based observations and satellite data. *Journal of Atmospheric Chemistry*. 59(3): 219–236.

Biswas, S., Lasko, K.D. and Vadrevu, K.P. 2015a. Fire disturbance in tropical forests of Myanmar—Analysis using MODIS satellite datasets. *IEEE Journal of Selected Topics in Applied Earth Observations and Remote Sensing*. 8(5): 2273–2281.

Biswas, S., Vadrevu, K.P., Lwin, Z.M., Lasko, K. and Justice, C.O. 2015b. Factors controlling vegetation fires in protected and non-protected areas of Myanmar. *PLoS One*. 10(4): e0124346.

Bridhikitti, A. and Overcamp, T.J. 2012. Estimation of Southeast Asian rice paddy areas with different ecosystems from moderate-resolution satellite imagery. *Agriculture, Ecosystems & Environment*. 146(1): 113–120.

Chaiyo, U. and Garivait, S. 2014. Estimation of black carbon emissions from dry dipterocarp forest fires in Thailand. *Atmosphere*. 5(4): 1002–1019

Cheewapnongphan, P. and Garivait, S. 2013. Bottom-up approach to estimate air pollution of rice residue open burning in Thailand. *Asia-Pacific Journal of Atmospheric Sciences*. 49(2): 139–149.

Chomanee, J., Tekasakul, S., Tekasakul, P. and Furuuchi, M. 2018. Effect of irradiation energy and residence time on decomposition efficiency of polycyclic aromatic hydrocarbons (PAHs) from rubber wood combustion emission using soft X-rays. *Chemosphere*. 210: 417–423.

DEDE. 2003. Rice in Thailand. Bangkok: Department of Alternative Energy Development and Efficiency (DEDE).

DEDE. 2007. Biomass energy. Bangkok: Department of Alternative Energy Development and Efficiency (DEDE).

EFE. 2009. Final Report: Study on Biomass Resources Management for Alternative Energy in Macro level. Bangkok: Energy for Environment Foundation (EFE).

Forest Fire Control Division. 2013. Department of National Park in Thailand. http://www.dnp.go.th /forest fire/Eng/ indexing.htm.

Forest Fire Control Division. 2014. Department of National Park in Thailand. http://www.dnp.go.th /forest fire/Eng/ indexing.htm.

GAPF. 2012. The Global Atmospheric Pollution Forum Air Pollutant Emission Inventory Manual Version 5.0. http://sei-international.org/rapidc/gapforum/html/emissions-manual.php.

Garivait, S., Bonnet, S., Sorapipith, V. and Chaiyo, U. 2004. Estimation of air pollutant emission from open biomass burning in Thailand. *Proc. The Joint Inter Conf on Sustain Energy and Environ*, Hua Hin, Phachuapkhirikhan, Thailand, 723–726.

Giglio, L., Van der Werf, G.R., Randerson, J.T., Collatz, G.J. and Kasibhatla, P. 2006. Global estimation of burned area using MODIS active fire observations. *Atmospheric Chemistry and Physics*. 6(4): 957–974.

Gupta, P.K., Prasad, V.K., Sharma, C., Sarkar, A.K., Kant, Y., Badarinath, K.V.S. and Mitra, A.P. 2001. CH_4 emissions from biomass burning of shifting cultivation areas of tropical deciduous forests–experimental results from ground-based measurements. *Chemosphere-Global Change Science*. 3(2): 133–143.

Graedel, T.E., Bates T.S., Bouwman, A.F., Cunnold, D., Dignon, J., Fung, I., Jacob, D.J., Lamb B.K., Logan, J.A., Marland, G., Middleton, P., Pacyna, J.M., Placet, M. and Veldt, C. 1993. Compilation of inventories of emissions to the atmosphere. *Global Biogeo Cycles*. 7: 1–26.

Hayasaka, H., Noguchi, I., Putra, E.I., Yulianti, N. and Vadrevu, K. 2014. Peat-fire-related air Pollution in Central Kalimantan, Indonesia. *Environmental Pollution*. 195: 257–266. doi:10.1016/j.envpol.2014.06.031.

Hayasaka, H. and Sepriando, A. 2018. Severe air pollution due to peat fires during 2015 super El Niño in Central Kalimantan, Indonesia. In: *Land-Atmospheric Research Applications in South/Southeast Asia*. Vadrevu, K.P., Ohara, T. and Justice, C. (Eds.). Springer, Cham, pp. 129–142.

IPCC-Intergovernmental Panel on Climate Change. 2006. IPCC Guidelines for National Greenhouse Gas Inventories, Vol. 4, Eggleston, H.S., L. Buendia, K. Miwa, T. Ngara and K. Tanabe (Eds.), IGES, Japan.

Inoue, Y. 2018. Ecosystem carbon stock, atmosphere and food security in slash-and-burn land use: A geospatial study in mountainous region of Laos. In: *Land- Atmospheric Research applications in South/Southeast Asia*. Vadrevu, K.P., Ohara, T. and Justice, C. (Eds.). Springer, Cham, pp. 641–666.

Israr, I., Jaya, S.N.I, Saharjo, H.S., Kuncahyo, B. and Vadrevu, K.P. 2018. Spatio-temporal analysis of land and forest fires in Indonesia using MODIS active fire dataset. In: *Land Atmospheric Research Applications in South/Southeast Asia*. Vadrevu, K.P., Ohara, T. and Justice, C. (Eds.). Springer, Cham, pp.105–128.

Jenjariyakosoln, S., Gheewala, H., Sajjakulnukit, B. and Savitri, G. 2014. Energy and GHG emission reduction potential of power generation from sugarcane residues in Thailand. *Energy for Sustainable Development*. 23:32–45.

Junginger, M., Faaij, A., van den Broek, R., Koopmans, A. and Hulscher, W. 2001. Fuel supply strategies for large-scale bio-energy projects in developing countries. Electricity generation from agricultural and forest residues in Northeastern Thailand. *Biomass and Bioenergy*. 21: 259–275.

Junpen, A., Garivait, S., Bonnet, S. and Pongpullponsak, A. 2011. Spatial and temporal distribution of forest fire PM10 emission estimation by using remote sensing information. *International Journal of Environmental Science and Development* 2: 156–61.

Junpen, A., Garivait, S. and Bonnet, S. 2013. Estimating emissions from forest fires in Thailand using MODIS active fire product and country specific data. *Asia-Pacific Journal of Atmospheric Sciences*. 49: 389–400.

Justice, C., Gutman, G. and Vadrevu, K.P. 2015. NASA land cover and land use change (LCLUC): An interdisciplinary research program. *Journal of Environmental Management*. 148(15): 4–9.

Kanabkaew, T. and Oanh, K.N.T. 2011. Development of spatial and temporal emission inventory for crop residue field burning. *Environmental Modeling & Assessment*. 16: 453–464.

Kant, Y., Ghosh, A.B., Sharma, M.C., Gupta, P.K., Prasad, V.K., Badarinath, K.V.S. and Mitra, A.P. 2000. Studies on aerosol optical depth in biomass burning areas using satellite and ground-based observations. *Infrared Physics & Technology*. 41(1): 21–28.

Kasem, S. and Thapa, G.B. 2012. Sustainable development policies and achievements in the context of the agriculture sector in Thailand. *Sustainable Development*. 20(2): 98–114.

Lasko, K., Vadrevu, K.P., Tran, V.T., Ellicott, E., Nguyen, T.T., Bui, H.Q. and Justice, C. 2017. Satellites may underestimate rice residue and associated burning emissions in Vietnam. *Environmental Research Letters*. 12(8): 085006.

Lasko, K. and Vadrevu, K.P. 2018. Improved rice residue burning emissions estimates: Accounting for practice-specific emission factors in air pollution assessments of Vietnam. *Environmental Pollution*. 236(5): 795–806.

Lasko, K., Vadrevu, K.P. and Nguyen, T.T.N. 2018a. Analysis of air pollution over Hanoi, Vietnam using multi-satellite and MERRA reanalysis datasets. *PloS One*. 13(5): e0196629.

Lasko, K., Vadrevu, K.P., Tran, V.T. and Justice, C. 2018b. Mapping double and single crop Paddy rice with Sentinel-1A at varying spatial scales and polarizations in Hanoi, Vietnam. *IEEE Journal of Selected Topics in Applied Earth Observations and Remote Sensing*. 11(2): 498–512.

Moran, J., NaSuwan, C., and Poocharoen, O.O. 2019. The haze problem in northern Thailand and policies to combat it: A review. *Environmental Science and Policy*. 97: 1–15.

Narita, D., Oanh, N.T.K., Sato, K., Huo, M., Permadi, D.A., Chi, N.N.H., Ratanajaratroj, T. and Pawarmart, I. 2019. Pollution characteristics and policy actions on fine particulate matter in a growing Asian economy: The case of Bangkok metropolitan region. *Atmosphere*. 10(5): 227.

Oanh, N.T.K., Bghiem, L.H. and Phyu, Y.L. 2002. Emission of polycyclic aromatic hydrocarbon, toxicity and mutagenicity from domestic cooking using sawdust briquettes, wood, and kerosene. *Environmental Science & Technology*. 36: 833–839.

Oanh, N.T.K., Bich, T.L., Tippayarom, D., Manadhar, R., Pongkiatkul, P., Simpson, C.D. and Liu, L-J.S. 2011. Characterization of particulate matter emission from open burning of rice straw. *Atmospheric Environment*. 45: 493–502.

Oanh, N.T.K. 2017. A Study in Urban Air Pollution Improvement in Asia. JICA-Research Institute. Available at: https://www.jica.go.jp/jicri/publication/booksandreports/ 175nbg00000kjwkk-att/Final_report.pdf. (Accessed 25 August 2019).

Oanh, N.T.K., Permadi, D.A., Dong, N.P., Nguyet, D.A. 2018. Emission of toxic air pollutants and greenhouse gases from crop residue open burning in Southeast Asia. In: *Land-Atmospheric Research applications in South/Southeast Asia*. Vadrevu, K.P., Ohara, T. and Justice, C. (Eds.). Springer, Cham, pp.47–68.

OAE-Office of Agricultural Economics. 2006. Agricultural statistic 2005 Thailand, Office of Agricultural Economics.

OAE-Office of Agricultural Economics. 2009. Agricultural statistic 2008 Thailand, Office of Agricultural Economics.

OCSB. 2014. Office of the Cane and Sugar Board. Statistics of sugarcane production in 2002–2014 (in Thai). [Online]. Available: http://www.ocsb.go.th/th/=production.

Ohara, T., Akimoto, H., Kurokawa, J., Horii, N., Yamaji, K., Yan, X. and Hayasaka, T. 2007. An Asian emission inventory of anthropogenic emission sources for the period 1980–2020. *Atmospheric Chemistry and Physics*, 7: 4419–4444.

PCD, Pollution Control Department. 1994. Air emission database of vehicles and industry in Bangkok metropolitan region 1992. Prepared by Chulalongkorn University, Bangkok, Thailand.

PCD-Pollution Control Department. 2005. National Master Plan for Open Burning Control, Pollution Control Department, Ministry of Natural Resource and Environment, Thailand.

Pengchai, P., Chantara, S., Sopajaree, K., Wangkarn, S., Tengcharoenkul, U. and Rayanakorn, M. 2009. Seasonal variation, risk assessment and source estimation of PM 10 and PM10-bound PAHs in the ambient air of Chiang Mai and Lamphun, Thailand. *Environmental Monitoring and Assessment*. 154(1–4): 197.

Phairuang, W., Hata, M. and Furuuchi, M. 2017. Influence of agricultural activities, forest fires and agro-industries on air quality in Thailand. *Journal of Environmental Sciences*. 52: 85–97.

Phairuang, W., Suwattiga, P., Chetiyanukornkul, T., Hongtieab, S., Limpaseni, W., Ikemori, F., Hata, M. and Furuuchi, M. 2019a. The influence of the open burning of agricultural biomass and forest fires in Thailand on the carbonaceous components in size-fractionated particles. *Environmental Pollution*. 247, 238–247.

Phairuang, W., Tekasakul, P., Hata, M., Tekasakul, S., Chomanee, J., Otani, Y. and Furuuchi, M. 2019b. Estimation of air pollution from ribbed smoked sheet rubber in Thailand exports to Japan as a pre-product of tires. *Atmospheric Pollution Research*. 10(2): 642–650.

Pongpiachan, S., Ho, K. and Cao, J. 2014. Effects of biomass and agricultural waste burnings on diurnal variation and vertical distribution of OC/EC in Hat-Yai City, Thailand. *Asian Journal of Applied Sciences*. 7(5): 360–374.

Prasad, V.K., Gupta, P.K., Sharma, C., Sarkar, A.K., Kant, Y., Badarinath, K.V.S., Rajagopal, T. and Mitra, A.P. 2000. NO_x emissions from biomass burning of shifting cultivation areas from tropical deciduous forests of India–estimates from ground- based measurements. *Atmospheric Environment*. 34(20): 3271–3280.

Prasad, V.K., Kant, Y. and Badarinath, K.V.S., 2001. Century ecosystem model application for quantifying vegetation dynamics in shifting cultivation areas: A case study from Rampa Forests, Eastern Ghats (India). *Ecological Research*. 16(3): 497–507.

Prasad, V.K., Kant, Y., Gupta, P.K., Elvidge, C. and Badarinath, K.V.S. 2002. Biomass burning and related trace gas emissions from tropical dry deciduous forests of India: A study using DMSP-OLS data and ground-based measurements. *International Journal of Remote Sensing*. 23(14)2837–2851.

Prasad, V.K., Lata, M. and Badarinath, K.V.S. 2003. Trace gas emissions from biomass burning from northeast region in India—estimates from satellite remote sensing data and GIS. *Environmentalist*. 23(3): 229–236.

Prasad, V.K., Anuradha, E. and Badarinath, K.V.S. 2005. Climatic controls of vegetation vigor in four contrasting forest types of India—evaluation from National Oceanic and atmospheric administration's advanced very high resolution radiometer datasets (1990–2000). *International Journal of Biometeorology.* 50(1): 6–16.

Prasad, V.K., Badarinath, K.V.S. and Eaturu, A. 2008. Biophysical and anthropogenic controls of forest fires in the Deccan Plateau, India. *Journal of Environmental Management.* 86(1): 1–13.

Ramachandran, S. 2018. Aerosols and climate change: Present understanding, challenges and future outlook. In: *Land-Atmospheric Research Applications in South/Southeast Asia.* Vadrevu, K.P., Ohara, T. and Justice, C. (Eds.). Springer, Cham, pp. 341–378.

Saharjo, B.H. and Yungan, A. 2018. Forest and land fires in Riau province; A Case study in fire prevention, policy implementation with local concession holders. In: *Land-Atmospheric Research Applications in South/Southeast Asia.* Vadrevu, K.P., Ohara, T. and Justice, C. (Eds.). Springer, Cham, pp. 143–170.

Sajjakulnukit, B., Yingyuad, R., Maneekhao, V., Pongnarintasut, V., Bhattacharya, S.C. and Abdul Salam, P. 2005 Assessment of sustainable energy potential of non-plantation biomass resources in Thailand. *Biomass and Bioenergy.* 3: 214–224.

Sornpoon, W. 2013. Greenhouse Gas Balance under Burned and Unburned Sugarcane Plantation in Thailand. Ph.D. Thesis, King Mongkut's University of Technology Thonburi, Bangkok, Thailand.

Sornpoon, W., Bonnet, S., Kasemsap, P., Prasertsak, P. and Garivait, S. 2014. Estimation of emissions from sugarcane field burning in Thailand using bottom-up country- specific activity data. *Atmosphere,* 5, 669–685.

Streets, D.G., Yarber, K.F., Woo, J.H. and Carmichael, G.R. 2003. An Inventory of Gaseous and Primary Aerosol Emissions in Asia in the Year 2000. *Journal of Geophysical Research,* 108: 8809–8823.

Tanpipat, V., Honda, K. and Nuchaiya, P. 2009. MODIS Hotspot Validation over Thailand. *Remote Sensing.* 1(4): 1043–1054.

Thawatchai, S. 2012. *Forests in Thailand.* National Office of Buddhism Press, Bangkok, Thailand (in Thai).

U.S. EPA. 2010. Our nation's air: Status and trends through 2008. Report Number EPA 524 454/R-09-002; prepared by U.S. Environmental Protection Agency, Research 525 Triangle Park, NC, http://www.epa.gov/airtrends/2010/report/fullreport.pdf.

Vadrevu, K.P., Eaturu, A. and Badarinath, K.V.S. 2006. Spatial distribution of forest fires and controlling factors in Andhra Pradesh, India using spot satellite datasets. *Environmental Monitoring and Assessment.* 123(1–3): 75–96.

Vadrevu, K.P., 2008. Analysis of fire events and controlling factors in eastern India using spatial scan and multivariate statistics. *Geografiska Annaler: Series A, Physical Geography.* 90(4): 315–328.

Vadrevu, K.P., Badarinath, K.V.S. and Anuradha, E. 2008. Spatial patterns in vegetation fires in the Indian region. *Environmental Monitoring and Assessment.* 147(1–3): 1. doi:10.1007/s10661-007-0092-6.

Vadrevu, K.P. and Justice, C.O. 2011. Vegetation fires in the Asian region: satellite observational needs and priorities. *Global Environmental Research.* 15(1): 65–76.

Vadrevu, K.P. and Choi, Y. 2011. Wavelet analysis of airborne CO_2 measurements and related meteorological parameters over heterogeneous landscapes. *Atmospheric Research.* 102(1–2): 77–90.

Vadrevu, K.P., Csiszar, I., Ellicott, E., Giglio, L., Badarinath, K.V.S., Vermote, E. and Justice, C. 2012. Hotspot analysis of vegetation fires and intensity in the Indian region. *IEEE Journal of Selected Topics in Applied Earth Observations and Remote Sensing.* 6(1): 224–238.

Vadrevu, K.P., Giglio, L. and Justice, C. 2013. Satellite based analysis of fire–carbon monoxide relationships from forest and agricultural residue burning (2003–2011). *Atmospheric Environment*. 64: 179–191.

Vadrevu K.P., Ohara T, Justice C. 2014a. Air pollution in Asia. *Environmental Pollution*. 12: 233–235.

Vadrevu, K.P., Lasko, K., Giglio, L. and Justice, C. 2014b. Analysis of Southeast Asian pollution episode during June 2013 using satellite remote sensing datasets. *Environmental Pollution*. 12: 45–256.

Vadrevu, K.P., Lasko, K., Giglio, L. and Justice, C. 2015. Vegetation fires, absorbing aerosols and smoke plume characteristics in diverse biomass burning regions of Asia. *Environmental Research Letters*. 10(10): 105003.

Vadrevu, K.P. and Lasko, K.P., 2015. Fire regimes and potential bioenergy loss from agricultural lands in the Indo-Gangetic Plains. *Journal of Environmental management*. 148:10–20.

Vadrevu, K.P., Ohara, T. and Justice, C. 2017. Land cover, land use changes and air pollution in Asia: a synthesis. *Environmental Research Letters*. 12(12): 120201.

Vadrevu, K.P., Ohara, T. and Justice, C. eds., 2018. *Land-Atmospheric Research Applications in South and Southeast Asia*. Springer, Cham.

Vadrevu, K.P. and Lasko, K. 2018. Intercomparison of MODIS AQUA and VIIRS I-Band fires and emissions in an agricultural landscape—Implications for air pollution research. *Remote Sensing*. 10(7): 978. doi:10.3390/rs10070978.

Vadrevu, K.P., Lasko, K., Giglio, L., Schroeder, W., Biswas, S. and Justice, C. 2019. Trends in vegetation fires in South and Southeast Asian countries. *Scientific Reports*. 9(1): 7422. doi:10.1038/s41598-019-43940-x.

Vongmahadlek, C., Pham, T.B.T., Satayopas, B. and Thongboonchu, N. 2009. A compilation and development of spatial and temporal profiles of high-resolution emissions inventory over Thailand. *Journal of the Air & Waste Management Association*. 5(7): 845–856.

Yang, S., He, H., Lu, S., Chen, D. and Zhu, J. 2008. Quantification of crop residue burning in the field and its influence on ambient air quality in Suqian, China. *Atmospheric Environment*. 42(9): 1961–1969.

Zhang, H., Hu, D., Chen, J., Ye, X., Wang, S.X., Hao, J., Wang, L., Zhang, R. and Zhisheng, A., 2011. Particle size distribution and polycyclic aromatic hydrocarbons emissions from agricultural crop residue burning. *Environmental Science & Technology*. 45: 5477–5482.

3 Impact of Biomass Burning on Local Air Quality in South China

Yun Fat LAM
University of Hong Kong, Hong Kong

Shimul Roy
University of Hong Kong
and
Mawlana Bhashani Science and Technology
University, Bangladesh

CONTENTS

INTRODUCTION

The term "biomass burning" (BB) generally refers to organic materials' combustion, both living and dead plants and vegetation. The burning mainly results from two reasons: (1) the forest fires due to lightning or auto-ignition of plants under high temperatures and (2) the anthropogenic burning, which involves agricultural land clearing, crop residue burning, and biofuel uses for heating and cooking. Yadav and Devi (2018) describe BB as a complex process that involves both physical and chemical reactions and transfer of mass and heat. Air pollutants and greenhouse gases are emitted as by-products of the burning. These include carbon dioxide (CO_2), carbon monoxide (CO), nitrogen oxide ($NO_x = NO + NO_2$), methane (CH_4), non-methane volatile organic compounds (NMVOCs), ammonia (NH_3), black carbon (BC), organic carbon (OC), particulate matters (PMs; e.g., PM_{10}, PM_{25}), and trace pollutants (e.g., heavy metals) (Prasad et al., 2000, 2002, 2003; Vadrevu et al., 2014a, b, 2015, 2017, 2018, 2019; Chen et al., 2016). Depending on the type of biomass and soils being burned, the composition of pollutants may vary widely. For example, household BB (e.g., charcoal, wood pellets, sawdust) emits large quantities of BC, CO, and NO_x, while infield straw burning predominantly releases OC, $PM_{2.5}$, and NMVOCs (Zhang et al., 2013).

BB has attracted much attention in South China (SA) due to increased burning activities and contributes to global warming in recent years. A more intense burning from both local and Southeast Asian countries (e.g., Vietnam, Myanmar, Thailand, Cambodia, Laos, and Indonesia) was observed, and BB is recognized as one of the key contributors to frequent haze and ozone episodes in the region (Biswas et al., 2015a, b; Lasko et al., 2017; 2018a, b; Hayasaka et al., 2014; Hayasaka and Sepriando, 2018; Inoue, 2018; Israr et al., 2018; Lasko and Vadrevu, 2018; Oanh et al., 2018; Saharjo and Yungan, 2018). The increase in primary pollutants (e.g., VOCs, NO_x, and CO) from local and long-range transport acted as an important precursor, contributing to the formation of secondary PM and ozone in spring and autumn, and caused elevated pollution levels (Chan et al., 2000; Chan et al., 2003; Wu et al., 2005, 2006; Liu et al., 2008; Zheng et al., 2011; Wang et al., 2009; Zheng et al., 2010; Bi et al., 2011; Ramachandran, 2018). Several researchers reported that numerous health problems, including asthma, chronic obstructive pulmonary diseases, and respiratory allergies, are associated with BB-related air pollutants. Bi et al. (2011) found that inhaling large quantities of biomass-related carbonaceous aerosols and organic constituents such as polycyclic aromatic hydrocarbons (PAHs) potentially increases the risk of developing cancer (Kim et al., 2013).

In this study, a thorough investigation of sources of BB affecting South China has been addressed. At first, the Global Fire Emissions Database (GFED) is summarized to provide an overview of BB's change in recent years in the Asian region, followed by discussing local and regional causative factors of pollution in South China and their seasonal variations. We also synthesized the local and long-range transport of BB emissions better to understand the influence of air quality in South China.

BIOMASS BURNING IN ASIA

Like other continents (Africa, South America, and North America), BB in Asia occurs with varying magnitude and is impacted by regional climate and meteorology (Yan et al., 2017; Yadav and Devi, 2018). The weather system substantially influences human activities and agricultural practices, resulting in distinct types of burning with seasonal variations. As Asia covers more than seven different climatic zones, agrarian practices, types of burning, and the seasonal and spatial distribution may vary hugely across the territory (Van der Werf et al., 2006). In Asia, the common burning types are forest fires, crop residue burning, and savanna/grassland burning, which account for ~45%–65%, 13%–34%, and 5%–30%, respectively (Chang and Song, 2010). In general, forest fires and savanna/grassland burning occur from

October to May during the dry season of Southeast Asia (SEA), while residue burning occurs in autumn and spring after crop harvesting in East Asia. Also, biofuel burning from residential cooking and heating occurs in all regions throughout the year. Besides the weather system, regional land use also affects the location and types of burning in Asia. For example, in East Asia, specifically in northern China, crop residue and savanna/grassland burning is the major type of burning, which accounts for more than 70% of overall burning in the region, while in SEA, an abundance of undisturbed forests with traditional agricultural practices (i.e., burning for clearing agricultural fields) and deforestation cause frequent forest fires. Table 3.1 shows the percent contribution of Asian BB emissions to the global emissions and distribution of total burned areas and the amount of biomass being burned within Asia from 2006 to 2015 (GFED, 2018). On an average, Asia contributes to about 10%–27% (depending on the year) of BB to the global fires, and its magnitude of burning varies between 445 and 855 million tons per year, in which large amounts of CO and carbonic aerosols were released to the environment (i.e., ~12 and ~4 Tg/year of CO and $PM_{2.5}$ emissions, respectively in 2006).

SEA always had the highest burning among the Asian regions from the satellite-derived burned areas, followed by South Asia and East Asia. Moreover, it has the largest interannual variability of burning throughout the years. It is clear that the variation of burning has mainly resulted from the variation of seasonal temperature induced from the global climate oscillations (i.e., El Niño), which causes a higher temperature and low-humidity condition; hence, it results in more frequent and intense fires in certain years (highlighted in gray). Figure 3.1 shows the annual BB emissions with the values of the Multivariate El Niño–Southern Oscillation (ENSO) Index from 2006 to 2015. The data were extracted from MEI.v2 during July–December in each year, and the positive ENSO index indicates a warmer year in the region (MEI, 2019). It should be noted that the years (i.e., 2006, 2009, 2014, and 2015) with a positive ENSO index (i.e., El Niño condition is shown in red color) are indeed having much higher emissions than those with a negative index. It echoes the same finding in Table 3.1, where a substantial amount of biomass was burned in those selected years. With the noticeably higher emissions of CO, CH_4, OC, and $PM_{2.5}$, shown in Figure 3.1, it indicates those emissions were coming from forest fires where high organic fraction was burned under the condition of incomplete combustion. During the low-burning years (e.g., 2008), the total burning in SEA can be as low as 17.6 Tg/year, while in the high-burning years (e.g., 2006 and 2015), the total burning can reach above 6.0 Tg/year. The value is three times higher than the burning in 2008 for SEA and six times higher than the average burning in East Asia and South Asia. It constitutes around ~70% of overall burning in entire Asia. BB's contribution from Southeast Asian forest fires during El Niño can be more than 50% of overall emissions in Asia, resulting in vast amounts of BB-related emissions injected into the atmosphere.

To better understand the geospatial pattern of the burning, Figure 3.2 shows examples of the geographic distribution of selected pollutants (i.e., OC, BC, CO, and CO_2) from 2006 burning. The data were extracted from the sector-specified gridded emissions from EDGAR v4.3.2 with 0.1°×0.1° resolution (EDGAR, 2017). It includes all types of burning, including boreal forest fires; savanna, grassland, and shrubland fires; temperate forest fires; tropical deforestation and degradation; peat fires; and agricultural waste

TABLE 3.1

Amounts of Total Burned Area and BB Being Burned in Asian Regions

Year	Contribution[a] (%)	Total Burned Area (10³ha)				Total Biomass Burned (Million Tons)			
		East Asia	South Asia	Southeast Asia	Total Asia[b]	East Asia	South Asia	Southeast Asia	Total Asia[b]
2006	26.0	1,722	1,319	10,614	27,668	77.6	100.3	571.9	821.7
2007	13.0	2,470	1,754	12,523	24,483	88.1	112.2	363.2	611.6
2008	10.4	2,390	1,463	6,799	19,535	112.9	104.7	175.9	445.0
2009	23.5	1,350	3,200	11,413	20,864	93.1	157.0	486.1	771.8
2010	10.9	1,308	2,704	10,699	27,054	86.4	152.4	315.5	623.1
2011	12.2	2,335	1,566	7,531	14,390	86.3	117.5	272.3	502.5
2012	12.7	2,443	2,985	9,569	19,941	88.7	158.1	348.8	630.1
2013	12.7	1,412	1,790	7,747	12,409	82.5	128.1	319.3	550.8
2014	21.8	1,976	1,739	10,785	20,526	94.3	128.0	570.4	832.5
2015	27.0	2,479	1,083	12,143	23,462	99.1	100.9	605.3	855.4

a   Annual contribution of Asian emission to global emission, extracted from FAOSTAT (2020) biomass burning data.

b   Data include Central Asian and Western Asian countries; gray color indicates that the particular year is under El Niño condition.

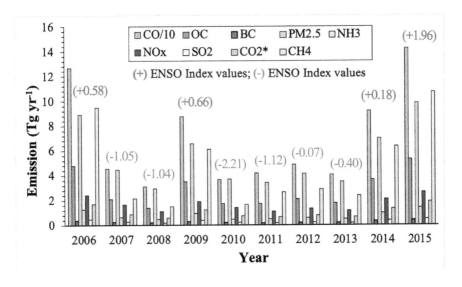

FIGURE 3.1 Annual BB emissions in Asia in the period 2006–2015 along with average Multivariate ENSO Index (MEI.v2) from June to December.

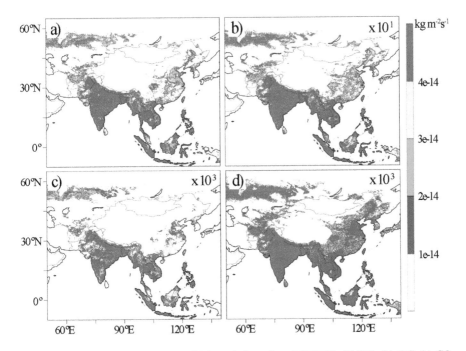

FIGURE 3.2 Spatial distribution of BB emissions from EDGAR: (a) BC; (b) OC; (c) CO; and (d) CO_2.

burning. It should be noted that tremendous amounts of emissions (in red) are shown up at multiple nations in SEA (i.e., Myanmar, Vietnam, Thailand, Laos, Cambodia, Indonesia, Malaysia, and the Philippines) and South Asia, which were attributed to forest fires and heavy biofuel use, respectively. Among the Asian countries, China has the highest agricultural emissions, followed by India (18%), Indonesia (13%), and Myanmar (8%) (Streets et al., 2003). This can be partially revealed by the high level of OC (Figure 3.2b) in the northern (i.e., Henan, Hebei, Shandong, Yangtze River Delta (YRD)), western (Guizhou), and southwestern parts of China (Guangxi (GX) and Yunnan), making China the leading nation in terms of anthropogenic BB emissions. For CO, the major part of emissions is centered in the northern part of the Yangtze River and South China (e.g., Guangdong (GD) and GX) resulting from residential heating in winter. It should be noted that these emissions illustrated in Figure 3.2 had huge seasonal variations and didn't occur at the same time. The figures only reflect the annual composite total.

BIOMASS BURNING IN SOUTH CHINA

South China is one of the fastest developing regions in China. It consists of five provinces/regions, including GD, GX, Hainan (HN), Hong Kong (HK), and Macau. Its development is centralized in Pearl River Delta (PRD), while it is less developed in GD, GX, and HN. It is observed that there are still several agricultural fields and underdeveloped rural areas in the surroundings (~59% rural population) with frequent crop residue burning events (He et al., 2011; Mestl et al., 2007). In regional BB inventory, burning is classified into forest fires, agricultural open straw burning (or frequently called infield straw burning), firewood burning, straw biofuel, and miscellaneous (e.g., biomass with municipal solid wastes or co-firing with coal) (Chen et al., 2017). Among these categories, agricultural residue burning (i.e., infield straw and straw biofuel) contributed the highest amounts to the total emissions, followed by domestic biofuel (i.e., firewood) burning. In general, infield and straw biofuel burning account for about 60%–80% of South China's overall emissions (He et al., 2011; Zhou et al., 2017). In PRD, Zhang et al. (2013) reported that biofuel burning (i.e., firewood and straws) could contribute up to 80% of overall emissions in winter, attributed to the burning in rural areas of Guangzhou, Dongguan, Jiangmen, Huizhou, and Zhaoqing. Figure 3.3 shows the distribution of burning types in South China in January, April, July, and October for $PM_{2.5}$ (Zhou et al., 2017). It is clear that BB varies largely with seasons and is subjected to geographic location, land characteristics, temperature, rainfall pattern, and type of agricultural activities (He et al., 2011). Throughout the year, routine agricultural activities result in substantial BB emissions in South China. These activities include infield straw burning and sowing season for beans during February–April, harvest seasons of first- and second-round crops in April and August, respectively, and straw biofuel for heating and cooking from October to February (Chen et al., 2017). As illustrated in Figure 3.3, infield straw (gray) and firewood burning (dark orange) are the major contributors of $PM_{2.5}$ in spring and winter from BB, respectively.

Table 3.2 summarizes the emissions in South China from 2000 to 2012. The results in Table 3.2 were synthesized from different studies covering different areas (e.g., PRD, GD, or South China) and objectives. In general, forest fires were estimated

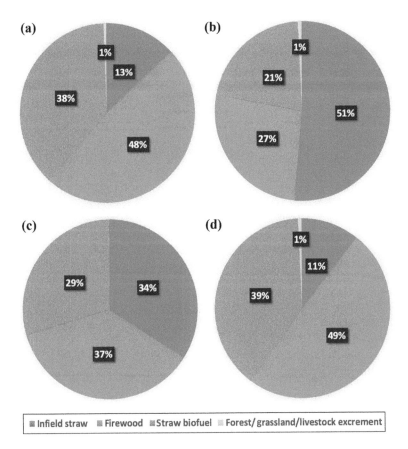

FIGURE 3.3 PM$_{2.5}$ distribution from different biomass sources for selected months: (a) January; (b) April; (c) July; and (d) October. (Adapted from Zhou et al., 2017.)

using satellite-derived fire spots to locate and estimate the amount of emissions. As the approach has its limitation to detect small fires, a supplementary approach based on population, land survey, and emission factors (EFs) was applied to estimate residential and localized agricultural burning in the region (Cao et al., 2005, 2008; He et al., 2011; HG-JWGSDEP, 2005; Zheng et al., 2009; Zhang et al., 2013). In the table, significant variations of OC, CO, NO$_x$, and NMVOCs are observed among the studies. For example, CO had the largest variability ranging from 1,017 to 3,694 kt/year for South China, 726 to 1,414 kt/year for GD, and 186 to 490 kt/year for PRD. The large variabilities are attributed to disparities in the activity data from different base years and the uncertainties in EFs (Li et al., 2016; Yan et al., 2006).

It is estimated that the influence of using different EFs could result in the change of straw biofuel, infield straw, and firewood emissions by up to 48%, 31%, and 28%, respectively. In general, infield straw burning releases much higher OC, NMVOCs, and PM$_{2.5}$ due to the abundant organic substances in the ground and soils, and its EF is 1–3 times higher than that of the straw burned as biofuel. For firewood burning, it tends to release more elemental carbon (EC) and CO than the other categories of

TABLE 3.2

Summary of Biomass Emission Inventories in South China (kt/year)

Year	2000[a]	2003[b,c]	2004[b]	2005[b]	2006[b]	2007[b]	2008[d]	2012[e]
Species/ Area of Cover	South China (GD)	PRD	PRD	PRD	PRD	PRD	PRD	South China (GD)
CO	3694 (1414)	489.0	490.0	481.0	427.0	408.0	186.0	1017 (726.0)
EC	27.5 (10.8)	4.9	4.9	4.8	6.5	6.0	2.3	28.0 (12.5)
OC	104.3 (42.3)	12.0	12.1	11.8	9.8	9.3	7.1	57.9 (21.0)
NO_x	104.2 (41.3)	5.7–9.3	9.2	9.0	9.5	8.3	4.9	55.2 (21.1)
NMVOCs	820.0 (330.0)	12.8–46.0	45.9	45.0	34.0	32.4	15.9	195.7 (70.0)
NH_3	76.0 (28.4)	5.7–7.7	7.7	7.6	7.2	6.9	–	31.0 (12.8)
SO_2	13.5 (5.3)	3.0	3.0	3.0	2.8	2.7	–	15.4 (5.9)
CH_4	177.3 (61.9)	31.1	31.2	30.6	21.2	20.2	–	115.9 (43.4)
$PM_{2.5}$	–	31.8	31.7	31.0	27.7	25.4	13.6	192.1 (74.2)
PM_{10}	–	–	–	–	–	–	–	203.6 (78.9)
CO_2	83,741 (31,518)	–	–	–	–	–	–	42,951 (17,619)

GD, Guangdong; PRD, Pearl River Delta; EC, elemental carbon.

[a] Cao et al. (2005), [b]He et al. (2011), [c]HG-JWGSDEP (2005), [d]Zhang et al. (2013), and [e]Zhou et al. (2017).

burning. From 2003 to 2007, it is observed that there was about 10%–35% emission reduction in the PRD region, specifically OC and VOC of about 22.5% and 29.4%, respectively. The reduction in emissions is attributed to the decrease in the infield straw burning where farmlands were converted to PRD's urban lands (He et al., 2011). Among the provinces in South China, GD, GX, and HN contribute around 35%–40% (37%–45%), 35%–46% (49.0%–55%), and 14%–27% (9%–11%) in 2000 (2012), respectively (Table 3.3). As all three regions were undergoing urbanization in the past 12 years, the result here only indicates the relative change of emission contribution within these three provinces.

As shown, in 2000, the relative contribution between GD and GX was nearly the same (only 1–6% difference). However, in 2012, the difference shifted from 14% to 25% (except EC), and GX has become the highest emitter of BB emissions in the region. It is observed that there was a considerable reduction in CO, OC, and NO_x and NMVOCs found between the years in GD, which were attributed to the reduction of infield straw burning. It echoes the earlier finding in the decrease of infield straw burning in PRD. A drop is also observed in HN with values of 10%–15% (e.g., 14%–23% in 2000 and 9%–10% in 2012). Overall, BB in South China is in a reduction trend. However, due to insufficient data from a single study or lack of consistent

TABLE 3.3

Contribution of BB Emissions for Selected Pollutants

Pollutant	CO		EC		OC		NO$_x$		NMVOCs	
Province/year	2000[a]	2012[b]	2000[a]	2012[b]	2000[a]	2012[b]	2000[a]	2012[b]	2000[a]	2012[b]
Guangdong (GD)	38	38	39	45	41	36	40	38	40	36
Guangxi (GX)	41	52	39	45	36	55	41	53	46	55
Hainan (HN)	21	10	22	10	23	9	20	9	14	9

[a] Cao et al. (2005) and [b]Zhou et al. (2017).

methodology from different studies, it is difficult to determine the reduction in emissions. A rough estimate of emission reduction from multiple studies varies from 30% to 70% in the last 12 years. This value is subjected to high uncertainty and should be used with caution.

SOURCES OF BIOMASS BURNING AND ITS RELATIVE INFLUENCE ON AIR QUALITY

To better understand the impact of BB on air quality, experimentalists and modelers have developed various techniques to identify BB sources and provide estimates for the contribution to air quality (Zhao et al., 2017; Chan, 2017). In general, studies that take the experimental approach have adopted the concept of source and receptor relationship, where the source region is referred to as the location of the burning and the receptor region is the place where air quality is being affected. The presence of distinct tracers (e.g., carbon isotopes, K+, and levoglucosan) and signature tracers (e.g., carbon isotopes, CO, BC, ratios of OC and BC, PM$_{2.5}$/PM$_{10}$) is used as proof of biomass influence, and the chemical source apportionment (SA) coupled with backward trajectory modeling is applied to evaluate the contribution of BB (Liu et al., 2014; Zhang et al., 2014). This approach is commonly used on short-term measurement campaigns for investigating the effect of long-range transport of BB from open burning and/or large forest fires. Other studies, such as investigation of the background influence of BB in remote/regional monitoring stations, also adopt a similar approach (Zhang et al., 2012).

It should be noted that the approach mentioned above can accurately confirm the presence of BB chemicals; however, due to the limitations of the trajectory approach where only dominant wind direction (probabilistically determined) is considered, the determined contribution by location may be subjected to considerable uncertainty, and its uncertainty increases with the number of sources and the distance between source and receptors. Moreover, the determined contribution can only reflect the tracers' influence and may not be directly applicable to other pollutants (e.g., SO$_2$, O$_3$, and NO$_x$), as the mixing ratios between BB and anthropogenic sources are different across different pollutants and locations. In contrast, modelers take advantage of numerical simulation models to determine the source of pollutants (both primary and secondary) and BB's relative contributions from different locations (Chan, 2017). However, models are also subjected to various uncertainties (e.g., grid selection, meteorological and emission inputs), requiring inputs from experimental/ measurement studies to validate and improve its performance. Therefore, combining experimental techniques with modeling approaches is a better way to address BB influence on local air quality.

Several studies have evaluated the impact of BB on South China's air quality using the techniques mentioned above. In general, these studies measured the concentrations of various carbonaceous species, as well as CO and O_3, to determine the source contribution and its influence on local air quality. For instance, Wang et al. (2007) showed, using their organic aerosol measurement, that BB in northern China has contributed around 3.0%–19.0% of $PM_{2.5}$. They also found that BB influences more than 58% of air pollution episodes in Guangzhou during October 2004. Similar results were reported by Zhang et al. (2010) on the enhancement of carbonaceous aerosols in South China from various BB sources (i.e., northern China and SEA) throughout the year. For SEA, Deng et al. (2008) have identified a distinct transport pathway from SEA to South China and confirmed that the extensive BB activities in SEA have a significant impact on aerosols and O_3 concentrations over the PRD region, while Chan (2017) has used the GEOS-Chem model to evaluate the influence of BB and found that SEA has the greatest air quality impact on South China with the contributions of 40% and 30% on column BC and CO during the springtime (March–April). To systematically evaluate BB's impact on South China, a conceptual model of BB sources has been developed with a thorough summary of individual sources that contribute to South China's air pollution. All results are synthesized and discussed as follows.

Conceptual Model of Biomass Burning Sources to South China

In South China, BB pollutants (particulates and gaseous) are generally released or contributed from a variety of sources, including local burning activities, regional transport from other parts of China, and global sources through long-range transport. Figure 3.4 describes the potential sources of BB.

The local sources cover activities within South China. These include burning of crop straw (mainly rice and wheat straw) as domestic biofuel, open burning of crop residues (e.g., rice–wheat, maize straws, sugarcane leaves, and other agricultural residues) in autumn (mainly in October), burning of firewood for household cooking and heating, and forest fires or savanna fires (although the contribution is minor). The regional sources consider the transport of biomass-related pollutants within China, mainly covering the intense crop burning in the rural area of northern and southwestern China. At last, the global sources refer to the direct transport of air pollutants from SEA through distinct meteorological conditions in spring and autumn, and the background pollutant enhancement from the rest of the world through the westerly wind in the free troposphere. Figure 3.5 shows the common pathways of long-range transport of BB in regional and global sources. In general, the regional sources are transported from northern China to South China through winter monsoon, where predominantly northeasterly occurs between October and February.

For global sources, three distinct pathways have been identified. The first two pathways (orange and red colors) take the surface marine boundary either from the north of the Philippines or from the south of the Indochina Peninsula to South China through

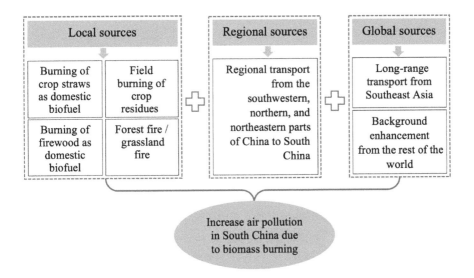

FIGURE 3.4 Sources of BB in South China.

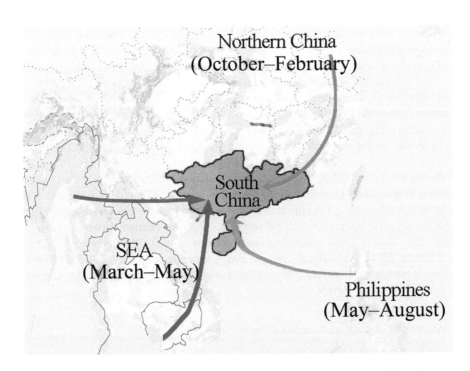

FIGURE 3.5 Major pathways of Asian BB to South China (Zhang et al., 2010; Ho et al., 2014).

the influence of summer monsoon occurring during May–August. It should be noted that the influence on air quality in these two pathways tends to be "not strong," as the pollutants travel thousands of kilometers before reaching South China, where they undergo major physical and chemical sinks during the transport. As a result, these two pathways' air quality effect is often treated as background pollution enhancements. The third pathway, which occurs from March to May, has been identified as the major pathway of Southeast Asian BB transport. In the process, BB pollutants are first transported to the free troposphere through the fires' buoyancy and then carried by westerly wind to South China. As the air mass reaches South China within 2 days, it doesn't give sufficient time for chemical sinks. Hence, these pollutants tend to be concentrated.

INFLUENCE OF AIR QUALITY FROM LOCAL AND REGIONAL SOURCES

In terms of local air quality effect, there are only limited studies that have looked at the impact of BB in South China from local sources, as anthropogenic sources have dominated the influence of air quality (Zheng et al., 2009; He et al., 2011). Some studies have shown that localized air quality influence could be substantial when the receptor is downwind of the sources. For example, Ho et al. (2014) reported that the relative contribution of BB emissions to OC could be as much as 33% in Zhaoqing from the nearby sources, while the contribution in Guangzhou is only 12%. It is clear that BB's influence has a huge variability and is highly affected by the size of the fire, type of burning, and the distance traveled by the pollutants. If entire South China is considered, BB's effect from local emissions could be small as South China seldom has widespread fires.

For regional sources, many studies have investigated the influence of Chinese BB on local air quality. Various experimental studies have made tracer measurements to understand their contribution. For example, Zhang et al. (2014) took aerosol filter samples from a background station in South China and evaluated the component of radiocarbon (^{14}C) in OC, EC, and water-insoluble OC (WINSOC). They found that EC's annual average contribution is around 7% (0.68 µg/m^3) using cluster analysis. Similar results were found by Ho et al. (2014), where 24-hour integrated PM$_{2.5}$ was collected from four different locations in four seasons. They found that the concentrations of levoglucosan and mannosan were on average 82.5 and 5.9 ng/m^3, and the relative contributions of BB emission to OC were estimated between 4% and 12%, in which winter had the greatest BB influence (1.7–2.8 times higher than other seasons) than other seasons. From a modeling study done by Li et al. (2017), the contributions of Chinese BB to the surface (column) PM$_{2.5}$ were reported as 2.7%–7.3% (0.8%–2.6%). It translates to surface PM$_{2.5}$ enhancements of 3.6 µg/m^3 in GD, 4.0 µg/m^3 in GX, and 0.51 µg/m^3 in HN.

INFLUENCE OF AIR QUALITY FROM GLOBAL SOURCES

Southeast Asia–Indochina Peninsula

Apart from the local and regional BB sources, intense BB in Southeast Asian countries has been recognized as one of the key sources of BB emissions in South China.

Zhang et al. (2010) have described an efficient long-range transport mechanism where BB was uplifted into the mid-altitude in the Indochina Peninsula and transported eastward to South China (see Figure 3.5), and caused degradation of air quality in South China from March to May. Li et al. (2017) reported that the average contributions of BB to surface (column) $PM_{2.5}$ were 1.7%–4.8% (42.7%–50.3%). It translates to enhancements of the surface (column) $PM_{2.5}$ of 0.85 $\mu g/m^3$ (67.2 mg/m^3) in GD, 2.66 $\mu g/m^3$ (89.0 mg/m^3) in GX, and 1.28 $\mu g/m^3$ (46.3 mg/m^3) in HN. Similar results were reported by Zhang et al. (2014) where the surface contribution of BC is about 9% and 0.72 $\mu g/m^3$, and the upper-level enhancement was around 30%–60% at higher altitudes of 2000–4000 m, and 10%–30% below 2000 m. A widespread enhancement of Aerosol Optical Depth (AOD) (i.e., 26%–62%) was also reported by Huang et al. (2013). These results have clearly shown that a large amount of BB-related emissions was transported at the free troposphere, and only a small portion of BB emissions was reentrant to the surface, impacting the surface air quality. Recently, Lam (2018) investigated this process and reported that the reentrant process could enhance the surface ozone and $PM_{2.5}$ concentrations by up to 20% in HK. Therefore, BB from the Indochina Peninsula could be an important source of pollution impacting South China's local air quality in the spring.

Southeast Asia–Philippines

Another Southeast Asian pathway that contributes to the degradation of air quality in South China is the oceanic pathway from the Philippines to South China. Zhang et al. (2012) and (2014) reported that biomass aerosols from the Philippines have resulted in 16%–28% and 4.9% of OC increase at the remote stations of HN (i.e., Jianfengling (JFL)) and HK (i.e., Hok Tsui (HT)), respectively. Moreover, the concentrations of levoglucosan (170%–340% or 57–106 ng/m^3) and BC (7% or 0.16 $\mu g/m^3$) were also increased. As these stations are located either in the mountain or on a remote island, low background concentrations may have resulted in high percentage increases in the measurement. In fact, the absolute increases in values are small.

The Rest of the World

Pollution released from burning activities (e.g., burning of African savanna) in western and central Africa can also be transported to South China. However, compared to all other sources mentioned above, the contribution of African BB is minor, and it only produces a background enhancement. For instance, Chan (2017) reported an annual contribution of 3% on CO from BB activities in Africa, while Lee et al. (2013) asserted that December and January are the most active periods of savanna burning in Africa, releasing a huge amount of carbonaceous particles as well as pyrogenic pollutants to the atmosphere. These pollutants are transported to South China until the springtime (April) due to the cross-equatorial wind stream's influence along the East African coast.

SUMMARY

BB releases different trace gases and aerosols in the atmosphere resulting in local, regional, and global impacts. Several studies (Crutzen and Andreae, 1990; Yan et al., 2006; Schultz et al., 2008; Taylor, 2010; Sigsgaard et al., 2015; Belkin, 2018; Yadav and Devi, 2018) have identified the impacts of BB on air quality, atmospheric composition, visibility, human health, and climate system on local, regional, and global scales. According to the World Health Organization (WHO) (2006), more than three billion people worldwide rely on solid fuels, including biofuel for cooking, heating, or other domestic purposes. In China, air quality degradation due to BB is also a major concern as many people (0.5 billion) depend on biofuel for cooking (Belkin, 2018). China is also one of the largest agricultural countries around the world, where massive agricultural residues are burned each year. In South China, the open burning of agricultural crop residues in spring and biofuel burning in winter increased the pollutant load intensely in the atmosphere. In general, local sources can impact nearby areas and a short period of high pollution and low visibility during those moments. Interestingly, the influence of Southeast Asian BB can be as big as the influence of local burning in South China, as reported in many studies (Zhang et al., 2014). This is mainly attributed to the efficient long-range transport of pollutants from SEA to South China during springtime with the extensive burning in SEA during El Niño year. Large surface and column enhancements of $PM_{2.5}$ and BC in remote stations in South China have confirmed such a phenomenon.

ACKNOWLEDGMENTS

This study was partially supported by the Research Grants Council of Hong Kong (CityU 21300214), the Environment and Conservation Fund (ECF Project 16/2015), and the RGC Research Postgraduate studentships.

REFERENCES

Belkin, H.E., 2018. Environmental human health issues related to indoor air pollution from domestic biomass use in rural China: A review. In: B. De Vivo, A. Lima, and H. Belkin (Eds.). *Environmental Geochemistry – Site characterization, Data analysis and Case histories*, (pp. 417–434). Elsevier, Amsterdam.

Bi, X., Zhang, G., Li, L., Wang, X., Li, M., Sheng, G., Fu, J. and Zhou, Z., 2011. Mixing state of biomass burning particles by single particle aerosol mass spectrometer in the urban area of PRD, China. *Atmospheric Environment.* 45(20), 3447–3453.

Biswas, S., Lasko, K.D. and Vadrevu, K.P. 2015a. Fire disturbance in tropical forests of Myanmar—Analysis using MODIS satellite datasets. *IEEE Journal of Selected Topics in Applied Earth Observations and Remote Sensing.* 8(5), 2273–2281.

Biswas, S., Vadrevu, K.P., Lwin, Z.M., Lasko, K. and Justice, C.O. 2015b. Factors controlling vegetation fires in protected and non-protected areas of Myanmar. *PLoS one.* 10(4), e0124346.

Cao, G., Zhang, X., Wang, Y. and Zheng, F. 2008. Estimation of emissions from field burning of crop straw in China. *Chinese Science Bulletin.* 53(5), 784–790.

Cao, G.L., Zhang, X.Y., Wang, D. and Zheng, F.C. 2005. Inventory of atmospheric pollutants discharged from biomass burning in China continent. *China Environmental Science,* 25(4), 389–393 (in Chinese).

Chan, K.L., 2017. Biomass burning sources and their contributions to the local air quality in Hong Kong. *Science of the Total Environment.* 596, 212–221.

Chan, C.Y., Chan, L.Y., Harris, J.M., Oltmans, S.J., Blake, D.R., Qin, Y., Zheng, Y.G. and Zheng, X.D. 2003. Characteristics of biomass burning emission sources, transport, and chemical speciation in enhanced springtime tropospheric ozone profile over Hong Kong. *Journal of geophysical research: atmospheres.* 108(D1), ACH–3.

Chan, L.Y., Chan, C.Y., Liu, H.Y., Christopher, S., Oltmans, S.J. and Harris, J.M., 2000. A case study on the biomass burning in Southeast Asia and enhancement of tropospheric ozone over Hong Kong. *Geophysical Research Letters*. 27(10), 1479–1482.

Chen, J., Li, C., Ristovski, Z., Milic, A., Gu, Y., Islam, M.S., Wang, S., Hao, J., Zhang, H., He, C. and Guo, H., Fu, H., Miljevic, B., Morawska, L., Thai, P., Lam, Y.F., Pereira, G., Ding, A., Huang, X., and Dumka, U. 2016. A review of biomass burning: Emissions and impacts on air quality, health and climate in China. *Science of the Total Environment*. 579, 1000–1034.

Chang, D. and Song, Y. 2010. Estimates of biomass burning emissions in tropical Asia based on satellite-derived data. *Atmospheric Chemistry and Physics*. 10(5), 2335–2351.

Cheng, Z., Wang, S., Fu, X., Watson, J.G., Jiang, J., Fu, Q., Chen, C., Xu, B., Yu, J., Chow, J.C. and Hao, J. 2014. Impact of biomass burning on haze pollution in the Yangtze River delta, China: a case study in summer 2011. *Atmospheric Chemistry and Physics*. 14(9), 4573–4585.

Crutzen, P.J. and Andreae, M.O. 1990. Biomass burning in the tropics: Impact on atmospheric chemistry and biogeochemical cycles. *Science*. 250(4988), 1669–1678.

Deng, X., Tie, X., Zhou, X., Wu, D., Zhong, L., Tan, H., Li, F., Huang, X., Bi, X. and Deng, T. 2008. Effects of Southeast Asia biomass burning on aerosols and ozone concentrations over the Pearl River Delta (PRD) region. *Atmospheric Environment*. 42(36), 8493–8501.

GFED (Global Fire Emissions Database). 2018. GFED-v4.1. Available at: http://www.falw.vu/~gwerf/GFED/GFED4/ (accessed 6 May 2019).

EDGAR (Emission Database for Global Atmospheric Research). 2017. EDGAR v4.3.2. Available at: https://edgar.jrc.ec.europa.eu/overvi (accessed 6 May 2019).

FAOSTAT. 2020. http://www.fao.org/faostat/en/.

Hayasaka, H., Noguchi, I., Putra, E.I., Yulianti, N. and Vadrevu, K. 2014. Peat-fire-related air pollution in Central Kalimantan, Indonesia. *Environmental Pollution*. 195, 257–266. doi: 10.1016/j.envpol.2014.06.031.

Hayasaka, H. and A. Sepriando. 2018. Severe air pollution due to peat fires during 2015 super El Niño in Central Kalimantan, Indonesia. In: Vadrevu, K.P., Ohara, T., and Justice, C. (Eds.). *Land-Atmospheric Research Applications in South/Southeast Asia*. Springer, Cham, pp. 129–142.

He, M., Zheng, J., Yin, S. and Zhang, Y. 2011. Trends, temporal and spatial characteristics, and uncertainties in biomass burning emissions in the Pearl River Delta, China. *Atmospheric Environment*. 45(24), 4051–4059.

Ho, K.F., Engling, G., Ho, S.S.H., Huang, R., Lai, S., Cao, J., and Lee, S.C. 2014. Seasonal variations of anhydrosugars in PM2.5 in the Pearl River Delta Region, China. *Tellus B: Chemical and Physical Meteorology*. 66(1). doi:10.3402/tellusb.v66.22577.

Huang, K., Fu, J.S., Hsu, N.C., Gao, Y., Dong, X., Tsay, S.C. and Lam, Y.F. 2013. Impact assessment of biomass burning on air quality in Southeast and East Asia during BASE-ASIA. *Atmospheric Environment*. 78, 291–302.

HG-JWGSDEP (Hong Kong-Guangdong Joint Working Group on Sustainable Development and Environmental Protection). 2005. Air Emission Inventory Handbook for Pearl River Delta Region. CH2M-IDC Hong Kong Ltd., Hong Kong.

Inoue, Y. 2018. Ecosystem carbon stock, Atmosphere and Food security in Slash-and-Burn Land Use: A Geospatial Study in Mountainous Region of Laos. In: Vadrevu, K.P., Ohara, T., and Justice, C. (Eds.). *Land-Atmospheric Research applications in South/Southeast Asia*. Springer, Cham.

Israr, I., Jaya, S.N.I, Saharjo, H.S., Kuncahyo, B., and Vadrevu, K.P. 2018. Spatio-temporal analysis of land and forest fires in Indonesia using MODIS active fire dataset. In: Vadrevu, K.P., Ohara, T., and Justice, C. (Eds.) *Land-Atmospheric Research Applications in South/Southeast Asia*. Springer, Cham, pp. 105–128.

Kim, K., Jahan, S.A., Kabir, E., and Brown, R.J. 2013. A review of airborne polycyclic aromatic hydrocarbons (PAHs) and their human health effects. *Environment International* 60, 71–80.

Lam, Y.F. 2018. Climate change and air quality in Southeastern China: Hong Kong study. In Akhtar, R. and Palagiano, C. (Eds.) *Climate Change and Air Pollution*. Springer, Cham, pp. 181–196.

Lasko, K. and Vadrevu, K.P. 2018. Improved rice residue burning emissions estimates: Accounting for practice-specific emission factors in air pollution assessments of Vietnam. *Environmental Pollution*. 236(5), 795–806.

Lasko, K., Vadrevu, K.P. and Nguyen, T.T.N. 2018a. Analysis of air pollution over Hanoi, Vietnam using multi-satellite and MERRA reanalysis datasets. *PloS One*. 13(5), e0196629.

Lasko, K., Vadrevu, K.P., Tran, V.T., Ellicott, E., Nguyen, T.T., Bui, H.Q. and Justice, C. 2017. Satellites may underestimate rice residue and associated burning emissions in Vietnam. *Environmental Research Letters*. 12(8), 085006.

Lasko, K., Vadrevu, K.P., Tran, V.T. and Justice, C. 2018b. Mapping double and single crop paddy rice with Sentinel-1A at varying spatial scales and polarizations in Hanoi, Vietnam. *IEEE Journal of Selected Topics in Applied Earth Observations and Remote Sensing*. 11(2), 498–512.

Lee, Y.C., Lam, Y.F., Kuhlmann, G., Wenig, M.O., Chan, K.L., Hartl, A. and Ning, Z. 2013. An integrated approach to identify the biomass burning sources contributing to black carbon episodes in Hong Kong. *Atmospheric Environment*. 80, 478–487.

Li, J., Zhang, Y., Wang, Z., Sun, Y., Fu, P., Yang, Y., Huang, H., Li, J., Zhang, Q., Lin, C., Lin, N., 2017. Regional impact of biomass burning in Southeast Asia on atmospheric aerosols during the 2013 Seven South-East Asian studies project, 2017. *Aerosol and Air Quality Research*. 17, 2924–2941. doi: 10.4209/aaqr.2016.09.0422.

Li, J., Bo, Y. and Xie, S. 2016. Estimating emissions from crop residue open burning in China based on statistics and MODIS fire products. *Journal of Environmental Sciences*. 44, 158–170.

Liu, J., Li, J., Zhang, Y., Liu, D., Ding, P., Shen, C., Shen, K., He, Q., Ding, X., Wang, X., Chen, D., Szidat, S., Zhang, G. 2014. Source apportionment using radiocarbon and organic tracers for PM2.5 carbonaceous aerosols in Guangzhou, South China: Contrasting local- and regional-scale haze events. *Environmental Science & Technology* 20, 12002–12011. doi: 10.1021/es503102w.

Liu, Y., Shao, M., Lu, S., Chang, C.C., Wang, J.L. and Chen, G. 2008. Volatile organic compound (VOC) measurements in the Pearl River Delta (PRD) region, China. *Atmospheric Chemistry and Physics*. 8(6), 1531–1545.

MEI (MEI dataset from NOAA Earth System Research Laboratory's physical sciences division, 2019. MEI. v2. Available at https://www.esrl.noaa.gov/psd/enso/mei/ (accessed 6 May 2019).

Mestl, H.E.S., Aunan, K., Seip, H.M., Wang, S., Zhao, Y., and Zhang, D., 2007. Urban and rural exposure to indoor air pollution from domestic biomass and coal burning across China. *Science of the Total Environment*. 377, 12–26.

Oanh, N.T.K., Permadi, D.A., Dong, N.P., and Nguyet, D.A. 2018. Emission of toxic air pollutants and greenhouse gases from crop residue open burning in Southeast Asia. In: Vadrevu, K.P., Ohara, T., and Justice, C. (Eds). *Land-Atmospheric Research applications in South/Southeast Asia*. Springer, Cham, pp. 47–68.

Prasad, V.K., Gupta, P.K., Sharma, C., Sarkar, A.K., Kant, Y., Badarinath, K.V.S., Rajagopal, T. and Mitra, A.P. 2000. NO_x emissions from biomass burning of shifting cultivation areas from tropical deciduous forests of India–estimates from ground-based measurements. *Atmospheric Environment*. 34(20), 3271–3280.

Prasad, V.K., Kant, Y., Gupta, P.K., Elvidge, C. and Badarinath, K.V.S. 2002. Biomass burning and related trace gas emissions from tropical dry deciduous forests of India: A study using DMSP-OLS data and ground-based measurements. *International Journal of Remote Sensing*. 23(14), 2837–2851.

Prasad, V.K., Lata, M. and Badarinath, K.V.S. 2003. Trace gas emissions from biomass burning from northeast region in India—estimates from satellite remote sensing data and GIS. *Environmentalist*. 23(3), 229–236.

Ramachandran, S. 2018. Aerosols and climate change: Present understanding, challenges and future outlook. In: *Land-Atmospheric Research Applications in South/Southeast Asia*. Vadrevu, K.P., Ohara, T., and Justice, C. (Eds.) Springer, Cham, pp. 341–378.

Saharjo, B.H., and Yungan, A. 2018. Forest and land fires in Riau province; A Case study in fire prevention, policy implementation with local concession holders. In: *Land-Atmospheric Research Applications in South/Southeast Asia*. Vadrevu, K.P., Ohara, T., and Justice, C. (Eds). Springer, Cham, pp. 143–170.

Streets, D.G., Yarber, K.F., Woo, J.H. and Carmichael, G.R. 2003. Biomass burning in Asia: Annual and seasonal estimates and atmospheric emissions. *Global Biogeochemical Cycles*. 17(4). doi:10.1029/2003GB002040

Sigsgaard, T., Forsberg, B., Annesi-Maesano, I., Blomberg, A., Bølling, A., Boman, C., Bønløkke, J., Brauer, M., Bruce, N., Héroux, M.E. and Hirvonen, M.R., 2015. Health impacts of anthropogenic biomass burning in the developed world. *European Respiratory Journal*. 46(6), 1577–1588.

Schultz, M.G., Heil, A., Hoelzemann, J.J., Spessa, A., Thonicke, K., Goldammer, J.G., Held, A.C., Pereira, J.M. and van Het Bolscher, M. 2008. Global wildland fire emissions from 1960 to 2000. *Global Biogeochemical Cycles*. 22(2). doi:10.1029/2007GB003031.

Taylor, D., 2010. Biomass burning, humans and climate change in Southeast Asia. *Biodiversity and Conservation*. 19(4), 1025–1042.

Vadrevu, K.P., Ohara, T., and Justice, C. 2014a. Air pollution in Asia. *Environmental Pollution*. 12, 233–235.

Vadrevu, K.P., Lasko, K., Giglio, L. and Justice, C. 2014b. Analysis of Southeast Asian pollution episode during June 2013 using satellite remote sensing datasets. *Environmental Pollution*. 12, 245–256.

Vadrevu, K.P., Lasko, K., Giglio, L. and Justice, C. 2015. Vegetation fires, absorbing aerosols and smoke plume characteristics in diverse biomass burning regions of Asia. *Environmental Research Letters*. 10(10), 105003.

Vadrevu, K.P., Ohara, T. and Justice, C. 2017. Land cover, land use changes and air pollution in Asia: A synthesis. *Environmental Research Letters*, 12(12), 120201.

Vadrevu, K.P., Ohara, T. and Justice, C. (Eds.) 2018. *Land-Atmospheric Research Applications in South and Southeast Asia*. Springer, Cham.

Vadrevu, K.P., Lasko, K., Giglio, L., Schroeder, W., Biswas, S. and Justice, C. 2019. Trends in vegetation fires in South and Southeast Asian countries. *Scientific Reports*. 9(1), 7422. doi:10.1038/s41598-019-43940-x.

Van Der Werf, G.R., Randerson, J.T., Giglio, L., Collatz, G.J., Kasibhatla, P.S. and Arellano Jr., A.F. 2006. Interannual variability in global biomass burning emissions from 1997 to 2004. *Atmospheric Chemistry and Physics*. 6(11), 3423–3441.

Wang, Q. Q., Shao, M., Liu, Y., William, K., Paul, G., Li, X., Liu, Y., & Lu, S. (2007). Impact of biomass burning on urban air quality estimated by organic tracers: Guangzhou and Beijing as cases. *Atmospheric Environment*. 41, 8380–8390.

Wang, T., Wei, X.L., Ding, A.J., Poon, S.C., Lam, K.S., Li, Y.S., Chan, L.Y. and Anson, M., 2009. Increasing surface ozone concentrations in the background atmosphere of Southern China, 1994–2007. *Atmospheric Chemistry and Physics*. 9, 6217–6227. doi:10.5194/acp-9-6217-2009.

WHO (World Health Organization), 2006. Fuel for Life: Household Energy and Health. Available at: https://www.who.int/airpollution/publications/fuelforlife.pdf (accessed 7 July 2019).

Wu, D., Tie, X., Li, C., Ying, Z., Lau, A.K.H., Huang, J., Deng, X. and Bi, X. 2005. An extremely low visibility event over the Guangzhou region: A case study. *Atmospheric Environment*. 39(35), 6568–6577.

Wu, D., Tie, X. and Deng, X., 2006. Chemical characterizations of soluble aerosols in southern China. *Chemosphere*. 64(5), 749–757.

Yadav, I.C. and Devi, N.L., 2018. Biomass Burning, Regional Air Quality, and Climate Change. In: J. Nriagu (Eds.) *Earth Systems and Environmental Sciences*. Edition: Encyclopedia of Environmental Health. Elsevier, Amsterdam.

Yan, X., Ohara, T. and Akimoto, H., 2006. Bottom-up estimate of biomass burning in mainland China. *Atmospheric Environment*. 40(27), 5262–5273.

Yan, C., Yu, J., Zhao, Y. and Zheng, M., 2017. Biomass burning sources in China. In: Guy P. Brasseur, I. Bouarar, X. Wang (Eds.) *Air Pollution in Eastern Asia: An Integrated Perspective* (pp. 135–166). Springer, Cham.

Zhang, J.L., Zhang, G., Zotter, P., Huang, R., Tang, J., Wacker, L., Prévôt, A.S.H., Szidat, S. 2014. Radiocarbon-based source apportionment of carbonaceous aerosols at a regional background site on Hainan Island, South China. *Environmental Science & Technology*, 48(5) 2651–2659. doi:10.1021/es4050852.

Zhang, Y., Shao, M., Lin, Y., Luan, S., Mao, N., Chen, W. and Wang, M., 2013. Emission inventory of carbonaceous pollutants from biomass burning in the Pearl River Delta Region, China. *Atmospheric Environment*. 76, 189–199.

Zhang, Y., Zhang, Z., Chan, C., Engling, G., Sang, X., Shi, S., and Wang, X. 2012. Levoglucosan and carbonaceous species in the background aerosol of coastal southeast China: case study on transport of biomass burning smoke from the Philippines. *Environmental Science and Pollution Research*. 19(1), 244–255.

Zhang, G., Li, J., Li, X.D., Xu, Y., Guo, L.L., Tang, J.H., Lee, C.S., Liu, X. and Chen, Y.J. 2010. Impact of anthropogenic emissions and open biomass burning on regional carbonaceous aerosols in South China. *Environmental Pollution*. 158(11), 3392–3400.

Zheng, J., Zhang, L., Che, W., Zheng, Z. and Yin, S. 2009. A highly resolved temporal and spatial air pollutant emission inventory for the Pearl River Delta region, China and its uncertainty assessment. *Atmospheric Environment*. 43(32), 5112–5122.

Zheng, M., Wang, F., Hagler, G.S.W., Hou, X., Bergin, M., Cheng, Y., Salmon, L.G., Schauer, J.J., Louie, P.K., Zeng, L. and Zhang, Y., 2011. Sources of excess urban carbonaceous aerosol in the Pearl River Delta Region, China. *Atmospheric Environment*, 45(5), 1175–1182.

Zhao, H., Zhang, X., Zhang, S., Chen, W., Tong, D.Q. and Xiu, A., 2017. Effects of agricultural biomass burning on regional haze in China: a review. Atmosphere, 8(5), 88.

Zheng, J., Zheng, Z., Yu, Y. and Zhong, L. 2010. Temporal, spatial characteristics and uncertainty of biogenic VOC emissions in the Pearl River Delta region, China. *Atmospheric Environment*. 44(16), 1960–1969.

Zhou, Y., Xing, X., Lang, J., Chen, D., Cheng, S., Wei, L., Wei, X., Liu, C., 2017. A comprehensive biomass burning emission inventory with high spatial and temporal resolution in China. *Atmospheric Chemistry and Physics*. 17, 2839–2864. doi:10.5194/acp-17-2839-2017.

4 Impact of Biomass Burning on Surface-Level Carbon Monoxide over Lahore and Karachi and Their Comparison with South Asian Megacities

Zia ul Haq and Salman Tariq
University of the Punjab, Pakistan

CONTENTS

INTRODUCTION

Carbon monoxide (CO) is a trace atmospheric constituent that adversely impacts human life and contributes to climate change at different spatial scales. Locally, it is poisonous to humans, and there is an ecological association between increased levels of ambient air CO and adverse cardiovascular disease (CVD), stroke, and birth outcomes (Graber et al., 2007, and references therein). At a regional scale, CO is responsible for photochemical smog and changes the troposphere's oxidizing capacity, thereby affecting the levels of important trace gases such as tropospheric ozone and methane. On a global scale, it forms hazardous surface-level ozone (Crutzen and Zimmermann, 1991; Daniel et al., 2007; Srivastava and Sheel, 2013). CO is mainly produced by methane oxidation, incomplete fossil fuel combustion (emitted from factories and cars), and biomass burning (from forest fires and agricultural burning) (WHO, 2000; Tahir et al., 2010). CO reaction with OH is responsible for ~90%–95% of its removal (Novelli et al., 1998, and references therein). The seasonality of atmospheric CO is also linked with meteorological parameters such as temperature, wind speed, and humidity (Elminir, 2005). The atmospheric concentrations of CO and aerosols may be associated in two important ways: Firstly, CO and aerosols have commonality in major emission sources such as fossil fuel combustion, large areas of anthropogenic activities, and biomass and crop residue burning; and secondly, aerosols change the Earth's radiation budget and affect cloud formation processes, thus fluctuating the lifetime of CO.

The identification of anthropogenic sources of CO and assessments of its trends, seasonality, and spatiotemporal patterns are of great importance to understand CO's role in the atmospheric chemistry over South Asia and climate change issues. Of the several sources, biomass burning is an important CO source, including the other greenhouse gas (GHG) emissions. Important causative factors of biomass burning in South/Southeast Asia include slash-and-burn agriculture, clearing of forests for oil palm plantations, and burning of crop residues for planting the next crop (Badarinath et al., 2008a, b; Biswas et al., 2015a, b; Justice et al., 2015; Lasko et al., 2017; Lasko and Vadrevu, 2018; Lasko et al., 2018a,b). Most of these practices are related to land clearing, and they result in GHG emissions which can have both climate change and radiative effects (Kant et al., 2000; Prasad et al., 2000, 2002, 2003, 2004, 2005, 2008, 2008a; Gupta et al., 2001). Also, biomass burning can result in land degradation and alter the biogeochemical cycles in the terrestrial ecosystems, including loss of biomass and bioenergy (Prasad et al., 2001, 2004, 2005; Vadrevu, 2008; Prasad and Badarinth, 2004; Vadrevu and Choi, 2011; Vadrevu et al., 2008, 2012, 2013, 2014a, b; Vadrevu and Lasko, 2015). Thus, mapping and monitoring of biomass burning events, including quantifying the emissions and controlling factors, gain significance in both space and time (Vadrevu et al., 2006; Vadrevu and Justice, 2011; Hayasaka et al., 2014; Hayasaka and Sepriando, 2018; Vadrevu et al., 2015, 2017, 2018, 2019; Vadrevu and Lasko, 2018; Israr et al., 2018; Saharjo and Yungan, 2018; Ramachandran, 2018; Tariq and Ul-Haq, 2019).

In this study, we review some of the studies that focused on CO mapping and monitoring as the species is of our interest. In the recent past, some studies have focused on this topic by using satellite remote sensing (e.g., Ghude et al., 2011; Kumar et al., 2013; Worden et al., 2013; Girach and Nair, 2014; Ul-Haq et al., 2015, 2016; Ul-Haq, 2018). For example, Ghude et al. (2011) examined MOPITT data over the Indian subcontinent to show large-scale vertical movement of surface-level CO into to the upper troposphere due to Asian summer monsoon. Kumar et al. (2013) used MOPITT CO retrievals over the South Asian region during January–February 2008 to describe its sources, and spatiotemporal and vertical distribution. They showed that wintertime CO enhancement was mostly due to man-made emissions, and CO inflow from the lateral boundaries. Kumar et al. (2013) indicated the modulation of CO over the Arabian Sea (ArS) and the Bay of Bengal (BoB) by CO transported from the Indian-subcontinent landmass and regions outside South Asia. Worden et al. (2013) used multi-satellite data during 2000–2012 to report modest-to-slight decreasing trends in the total column of CO over the Northern Hemisphere and Southern Hemisphere and some other selected regions including India. They attributed these decreasing trends to changes in source and sink chemistry, and decreases in economic activities to global financial crisis in the last decade. Girach and Nair (2014) examined trends and seasonality of lower and upper tropospheric CO and total column CO over Indian regions using MOPITT observations during 2000–2014 and reported decreasing trends in lower tropospheric CO and CTC and an increasing trend for upper tropospheric CO. Ul-Haq et al. (2015) used Atmospheric Infrared Sounder (AIRS)/Advanced Microwave Sounding Unit (AMSU; Susskind et al., 2003) data for Pakistan and adjoining regions during 2003–2012 to describe seasonal and spatial patterns of CO and identified high-concentration areas. In another study, Ul-Haq et al. (2016) reported a slight decrease in CO over South Asia by using AIRS CO retrievals performed at 618 hPa during 2003–2015. They found a strong seasonality in AIRS CO concentration with spring-season maxima mainly due to crop residue burning. They also performed an analysis of SCanning Imaging Absorption spectroMeter for Atmospheric CartograpHY (SCIAMACHY; Bovensmann at al., 1999) total column CO tendencies which indicated minor rising trends over some parts of South Asia. They also analyzed background CO, recent emissions (RE), and spatial anomalies in RE over high-anthropogenic-activity zones of the Indus Basin, Ganges Basin, and Eastern Region using AIRS and SCIAMACHY CO data.

In this study, we analyze spatiotemporal distribution, seasonality, and trends in MOPITT-derived surface-level CO over Lahore and Karachi, and their comparison with other megacities of South Asia, i.e., Delhi, Mumbai, Kolkata, and Dhaka, during 2001–2017. We also explored spatial patterns in MERRA-2 CO emissions and the main drivers of CO seasonality over the study region.

STUDY AREA AND METEOROLOGY

South Asia is the most densely populated region on the Earth and consists of Afghanistan, Pakistan, India, Nepal, Sri Lanka, Maldives, Bhutan, and Bangladesh, with a surface extent of 5,134,613 km². This region is located between Afghanistan in Southwest Asia and Myanmar in Southeast Asia. This broad region extends from Afghanistan in the west to Myanmar. There are large urban areas (megacities), namely Lahore, Karachi, Delhi, Mumbai, Kolkata, and Dhaka (Figure 4.1).

South Asia is under serious threat from sea-level rise, increasing incidences of extreme weather events, including shifting monsoon patterns. Anthropogenic activities such as increased use of wood and fossil fuel burning for energy generation together with large-scale pre-monsoon and post-monsoon crop residue fires in Punjab (in Pakistan and India) and eastern regions of Bangladesh result in the formation of Asian brown cloud (ABC) with a thickness of about 3.2 km. The other causes of pollution in the region include unplanned and rapid urbanization, deforestation, and emissions from animal manure, solid waste, etc. (Badarinath et al., 2009; Ghude et al., 2011; Ul-Haq et al., 2015, 2016; Ul-Haq, 2018).

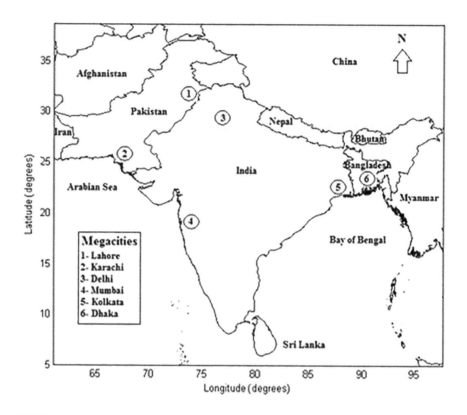

FIGURE 4.1   Location map of Lahore and Karachi and other megacities of South Asia.

South Asia has huge variations in climatic conditions ranging from subzero temperatures in the northern Himalayan Mountains to semiarid zone and desert zones covering the Deccan Plateau and the lower Indus Valley, western India, and southern Pakistan. A moderate climate is found in the Himalayan foothills and northern Indo Gangetic Plains (IGP). The tropical conditions are observed in the central Indian Deccan Plateau. This region is heavily affected by wet (summer, June through September) and dry (winter, October through February) monsoon systems causing alternating periods of wet and dry weather (UNEP, 2009; Joshi, 2015). The variations of some important meteorological factors in the selected megacities are shown in Figure 4.2.

DATASETS AND METHODOLOGY

The monthly mean daytime surface mixing ratio of CO (product ID: MOP03JM, product level 3, product version 007) was obtained from the MOPITT sensor flying on the Terra spacecraft and used to assess the spatiotemporal patterns, seasonality, and trends, and to identify large source areas and activities over South Asia. MOPITT makes use of near-infrared (NIR) and thermal infrared (TIR) radiances to retrieve CO on a global scale. The sensor has a spatial resolution of 22 km at nadir and a resolution of 2 km (Drummond, 1992; Deeter et al., 2003). MOPITT data and description of the

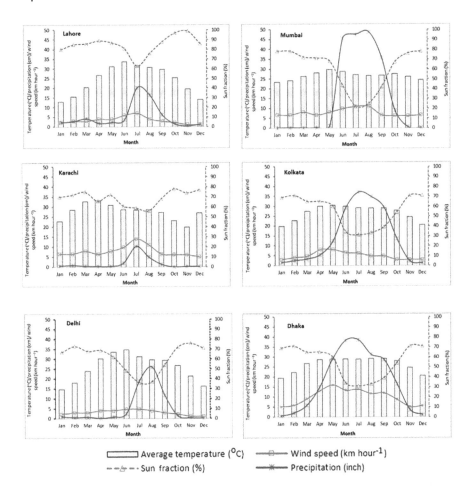

FIGURE 4.2 Climographs showing averaged variations of some important metrological factors over Karachi and Lahore and other megacities of South Asia. Dotted line corresponds to secondary vertical axis. Karachi has a maximum temperature of ~27.6°C and minimum precipitation of ~1.9 inches, and Lahore has the highest sun fraction of ~84%. Dhaka has the highest precipitation and wind speed averaged at 17.4 inches and 9.8 km/hr, respectively.

level-3 products and the retrieval algorithms' details may be obtained from NASA's web portal at https://earthdata.nasa.gov/. Several studies have shown the usefulness of MOPITT data for spatial–temporal distributions, transport, sources, and sinks of CO in the troposphere (e.g., Ghude et al., 2011; Kumar et al., 2013; Girach and Nair, 2014).

In this study, we discuss the spatial and temporal patterns of CO emissions using the MERRA-2 dataset (The Modern-Era Retrospective analysis for Research and Applications version 2; Gelaro et al., 2017). MERRA-2 data are an improved, advanced data assimilation system combining hyperspectral radiance and microwave data, GPS radio occultation data, ozone profile observations, and several other data-sets (Gelaro et al., 2017; Lasko et al., 2018a). These data are gridded at 0.5×0.625 (latitude × longitude) degrees and 72 vertical levels, from the surface to 0.01 hPa.

Some important meteorological parameters have been obtained from AIRS (Pagano, et al., 2003) on board NASA's Aqua satellite. These AIRS-derived meteorological parameters include surface-level air temperature and relative humidity. In addition to the datasets described above, aerosol optical depth (AOD) and Normalized Difference Vegetation Index (NDVI) were obtained from Moderate Resolution Imaging Spectroradiometer (MODIS; Kaufman, et al., 1997) on board the Terra satellite. Also, near-surface wind speed dataset is obtained from Tropical Rainfall Measuring Mission (TRMM; Liu et al., 2012) and Global Land Data Assimilation System (GLDAS; Fang et al., 2009), respectively. Table 4.1 highlights the datasets used in the study.

To analyze the concentrations of surface-level CO and its variability at monthly and annual time steps, we have plotted the CO concentration maps and performed spatial and temporal correlations of CO with meteorological parameters, the NDVI, and aerosols. For reliable results, spatial correlation mapping of monthly mean CO with other factors was obtained for only those grid points with at least 102 values of monthly mean data (i.e., 50% of the total available months from 2001 to 2017).

ANALYSIS AND DISCUSSION

MERRA-2 ENSEMBLE CO EMISSIONS

Area-and time-averaged MERRA-2 CO emissions are found be $9.53 \pm 0.76 \times 10^{-11}$ kg/m²/s with elevating values of 12.46% (*slope*: 0.0699; *y*-intercept: 8.903×10^{-11}

TABLE 4.1
Datasets Used in the Study

Product Name (Identifier)	Sensor/ Model	Retrieval Time (Day/ Night)	Spatial Resolution (Degrees)	Product Name/Version/ Level	Units	Level Description
Atmospheric CO (MOP03JM)	MOPITT	Daytime	$1° \times 1°$	Version 007/ level 3	ppbv	Surface
CO emissions (M2TMNXCHM)	MERRA-2	--	$0.5° \times 0.625°$	Version 5.12.4/level 3	kg/m²/s	Surface
AOD (MYD08_M3)	MODIS	Daytime	$1° \times 1°$	Version 6/ level 3	Unitless	Total column
Temperature (AIRX3STM)	AIRS	Daytime	$1° \times 1°$	Version 6/level 3	Kelvin	Surface
Relative humidity (AIRX3STM)	AIRS	Daytime	$1° \times 1°$	Version 6/ level 3	%	Surface
Wind speed (GLDAS_ NOAH025_M)	GLDAS	--	$0.25° \times 0.25°$	GLDAS_ NOAH025_M/- version 2.1/level 3	m/s	Near-surface level
NDVI (MODIS13C2)	MODIS	Daytime	$0.05° \times 0.05°$	Version 6/ level 3	Unitless	Surface

kg/m²/s; $R^2 = 0.22$) in South Asia during 2001–2017. The highest value is observed in March ($17.9 \pm 5.2 \times 10^{-11}$ kg/m²/s), followed by April ($14 \pm 2.3 \times 10^{-11}$ kg/m²/s), whereas the lowest value is observed in December ($6.9 \pm 0.3 \times 10^{-11}$ kg/m²/s), followed by January ($7 \pm 0.4 \times 10^{-11}$ kg/m²/s). Lahore, Karachi, and other megacities can be clearly identified in the maps (Figure 4.3). MERRA-2 ensemble CO emissions depict very different monthly mean patterns for Lahore and Karachi and other South Asian megacities. Lahore exhibited the highest seasonality pattern, mostly linked to pre-monsoon biomass burning. During the pre-monsoon periods, MERRA-2

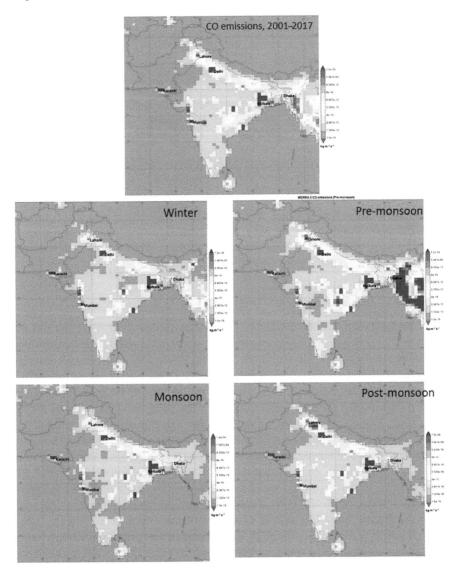

FIGURE 4.3 Seasonal spatial distributions of MERRA-2 ensemble CO emissions in Lahore and Karachi and adjoining megacities of South Asia from January 2001 to December 2017.

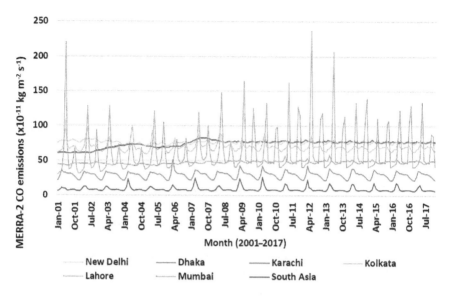

FIGURE 4.4   Temporal distributions of MERRA-2 ensemble CO emissions in Lahore and Karachi and adjoining megacities of South Asia from January 2001 to December 2017.

ensemble CO emissions reached up to 236×10^{-11} kg/m²/s in May 2012 and the least, 220×10^{-11} kg/m²/s, during May 2001. Karachi has the second-highest value (74.4×10^{-11} kg/m²/s) after Kolkata (77×10^{-11} kg/m²/s). Also, Karachi has the most consistent seasonal pattern since 2008 (Figure 4.4).

Temporal Variability of Surface-Level CO

Large amplitude in spatiotemporal and seasonal variations of tropospheric CO has been noted over megacities of South Asia. The main factors that contribute to these variations are anthropogenic emissions and local meteorology. A notable interannual variability in the annual mean values of CO is observed, ranging from 125.5 ppb (in 2017) to 167.5 ppb (in 2001), mostly due to the CO fluctuations linked to biomass burning and transported CO and partly driven by precipitation rates (Szopa et al. 2007; Van der Werf et al. 2004). The observed relativity high peaks of CO during 2005 and 2007 agree with findings by Turquety et al. (2008). They also attributed the occurrence of these peaks to the intense biomass burning events from South Asia. From 2001 to 2017, the averaged MOPITT-derived surface-level CO value of 138 ± 10 ppbv with an overall decrease of -19.6% (*slope*: -1.88 ± 0.3 ppbv; *y-intercept*: 155 ± 3 ppbv; R^2: 0.79) is noticed for satellite descending-mode (daytime) retrievals (Figure 4.5a). Our results agree with some recent studies for the South Asian region (e.g., Girach and Nair, 2014). These decreasing trends of CO are linked to carbon-containing fuels, waste treatment, and reduction in agriculture emissions, as inferred by Zhang and Huixiang (2012).

Monthly mean variations in surface-level CO over South Asia from January 2001 to December 2017 are presented in Table 4.1; the values varied from 79.1 ppbv in July

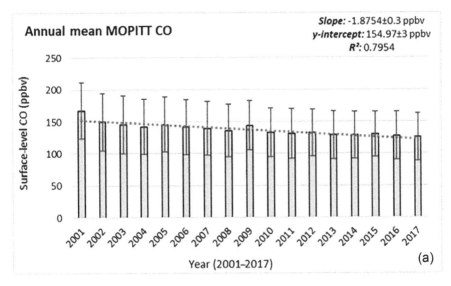

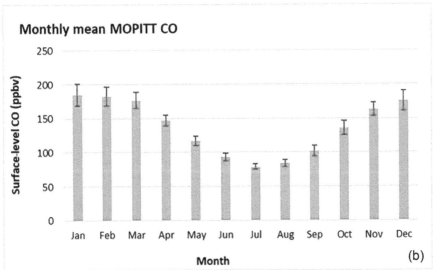

FIGURE 4.5 Distribution of (a) annual and (b) monthly mean MOPITT-retrieved surface-level CO (ppbv) over South Asia from January 2001 to December 2017. The vertical error bars represent standard deviations of observations.

to 184.9 ppbv in January during the study period. January showed the highest change in CO concentration of −22.6%, followed by December with −21.8%, whereas April showed the lowest declining trend of −12.6%, followed by July with −12.8%, during the study period (Figure 4.5b and Table 4.2).

It is observed that among all CO sources and sinks, loss due to CO reaction with OH and emissions from crop residue and biomass burning appears to be the main cause for seasonal fluctuations in surface-level CO concentrations. Further, surface-level

TABLE 4.2

Statistics of Monthly Mean MOPITT-Retrieved Surface-Level CO (ppbv) over South Asia Including Megacities from January 2001 to December 2017

Month	Mean (ppbv)	Median (ppbv)	Maximum Value (ppbv)	Minimum Value (ppbv)	Change (%)
January	184.9 ± 16.0	181.61	224.15	165.91	−22.6
February	182.4 ± 13.7	182.29	215.74	165.86	−19.8
March	176.8 ± 11.7	175.49	203.46	159.62	−16.6
April	147 ± 7.9	145.42	164.60	135.05	−12.6
May	116.8 ± 6.9	115.36	135.88	108.03	−13.2
June	93.2 ± 5.7	92.28	106.01	82.57	−15.4
July	79.1 ± 3.9	79.18	89.45	73.14	−12.8
August	84.3 ± 5.2	83.08	96.09	75.68	−15.3
September	102.0 ± 7.6	101.36	118.58	88.50	−16.2
October	136.1 ± 10.4	133.99	153.86	119.98	−18.5
November	163.2 ± 10.3	161.76	178.58	151.01	−16.2
December	175.8 ± 14.8	173.67	203.99	159.78	−21.8

CO is negatively related to temperature and wind speed, and it is positively related to humidity (Elminir, 2005; Barrero et al., 2006; Ocak and Turalioglu, 2008). CO washout and precipitation scavenging are also observed during the rainy season from June to September (Figure 4.5). Also, during the wet monsoon, large-scale deep convective activity is responsible for polluted surface air uplift, mainly in the BoB and central parts of India (Ul-Haq et al., 2015; Ul-Haq, 2018). CO concentration at the surface level reaches its peak in winter (November–January) mainly because of low atmospheric oxidation of CH_4, which is an important source of tropospheric CO, and by the photolysis of CO_2 (Lyma and Jensen, 2001; Jae et al., 2013; Ruzmaikin et al., 2014), weak winds, low temperatures, large-scale crop residue burning in rice fields near Lahore, and more wintertime domestic fuel burning for heating (Deeter et al., 2007; IPCC, 2007, Zhang et al., 2012; Sahu et al., 2013). Also, a shallower boundary layer results in lower vertical dispersion, which reduces the dilution and removal rates of CO.

Spatial correlation maps have been generated to explore the association of surface-level CO with meteorological factors (air temperature at the surface, near-surface wind speed, and relative humidity at the surface), aerosols (AOD), and vegetation cover (NDVI) (Figure 4.6).

LARGE-SCALE PRE-MONSOON AND POST-MONSOON
BIOMASS BURNING EVENTS AND CO

With respect to air pollutant emissions, pre-monsoon (mostly from wheat fields) and post-monsoon (mostly from rice fields), crop residue burning is a prominent South Asian source. To investigate the linkage between CO and crop residue burning, fire and thermal anomalies (day and night) acquired from MODIS/Terra have been

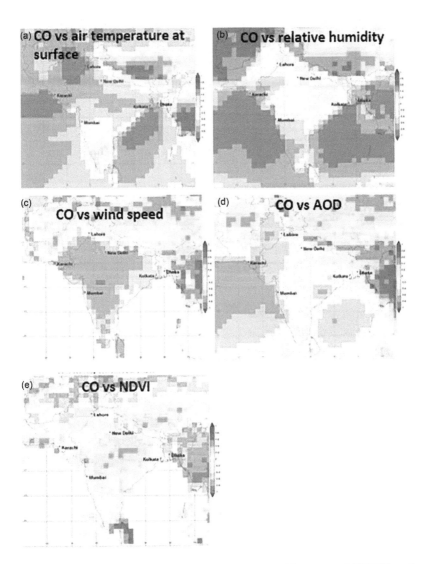

FIGURE 4.6 Spatial correlation maps between monthly mean MOPITT-retrieved surface-level CO and (a) air temperature at surface, (b) relative humidity at surface, (c) near-surface wind speed, (d) aerosol burden (AOD), and (e) vegetation cover (NDVI) over megacities of South Asia from January 2001 to December 2017.

superimposed on MOPITT CO retrievals for two days of pre-monsoon (March 2, 2017, and May 7, 2017) and one day of post-monsoon (October 25, 2016) (Figure 4.7).

SPATIAL VARIABILITY OF CO

We observed that high columns of CO exist over the eastern areas of the study region covering the Indo Gangetic Plains (IGP) (Pakistan and India) and large

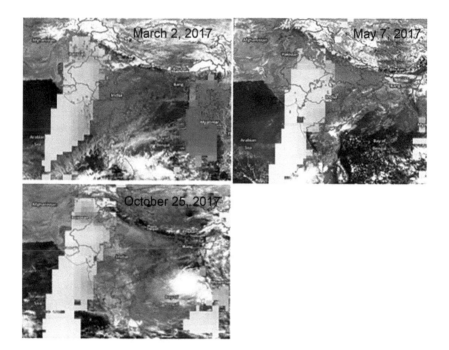

FIGURE 4.7   Maps showing variability of MOPITT-retrieved CO and MODIS/Terra-sensed fire thermal anomalies for different dates in the eastern and western regions of South Asia.

vegetative areas (Pakistani and Indian Punjab, and Indian states of Haryana and Utter Pradesh) attributed to high population density and high industrial and agricultural activities. A significant contribution of anthropogenic CO emissions related to crop residue burning has been observed (Figure not shown) over South Asia during post-monsoon (September–November). The western areas of the study region that consist of the Balochistan Plateau and the Sulaiman Mountains had relatively lower values of tropospheric CO than other areas in all months due to the combined effect of less population, minor agricultural activity, elevated temperatures, less humidity, and a smaller number of biomass and crop residue burning events. In wintertime (December–February), CO concentration rises due to increasing wintertime biomass burning for home heating and declining wind speeds and temperatures.

CO OVER LAHORE AND KARACHI AND OTHER MEGACITIES

Due to the geographic locations and climatic conditions, the six megacities of the study area can be paired into three groups, i.e., Lahore and Delhi, which are located in the semiarid monsoon region of the Indo-Gangetic Plain (IGP) down to the foothills of the Himalayan region; Karachi and Mumbai – Pakistan's and India's financial hub, respectively – which are coastal megacities situated in the southeastern part of South Asia; and Kolkata and Dhaka, which are located in the tropical wet zone near the western and eastern coasts of the BoB in India and Bangladesh, respectively (Table 4.3 and Figure 4.8). CO column over Lahore

TABLE 4.3

Statistics of MOPITT-Derived Surface-Level CO (ppbv) over Lahore and Karachi and Other Megacities of South Asia During January 2001–December 2017

Name of Megacity (State/Province, Country)	Location	Population (Million)	Area (km²)	Average CO (ppbv)	Change (%) (Equation of Straight Line of Best Fit, Coefficient of Determination)
Megacities located in the Indo-Gangetic Plains (IGP)					
Lahore (Punjab, Pakistan)	31.32°N, 74.22°E	10.23	1172	186 ± 12	-20.8 ($y = -2.1285x + 205.74$, $R^2 = 0.77$)
New Delhi (National Capital Region of India)	28.67°N, 77.22°E	16.31	1484	195 ± 12	-16 ($y = -2.1138x + 213.69$, $R^2 = 0.79$)
Coastal megacities					
Karachi (Sindh, Pakistan)	24.51°N, 67.72°E	16.05	3527	131 ± 14	-26.9 ($y = -2.545x + 154.1$, $R^2 = 0.89$)
Mumbai (Maharashtra, India)	19.04°N, 72.52°E	12.47	4355	164 ± 14	-17.7 ($y = -1.992x + 181.58$, $R^2 = 0.49$)
Megacities located near the Bay of Bengal (BoB)					
Kolkata (West Bengal, India)	22.55°N, 88.31 °E	14.11	1886	278 ± 17	-15.8 ($y = -2.9809x + 304.93$, $R^2 = 0.77$)
Dhaka (Dhaka Division, Bangladesh)	23.48°N, 90.24°E	14.39	815	286 ± 22	-17.6 ($y = -3.4492x + 317.14$, $R^2 = 0.62$)

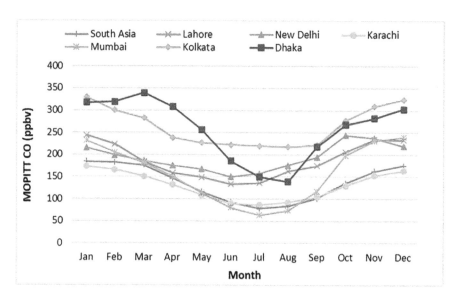

FIGURE 4.8   Distribution of monthly mean MOPITT-retrieved surface-level CO (ppbv) over Lahore and Karachi and other megacities of South Asia from January 2001 to December 2017.

is greatly influenced by extensive pre (wheat fields)- and post-monsoon (rice fields) crop residue burning in Pakistan's adjoining areas consisting of Narowal, Hafizabad, Sheikhupura, Kasur, and Faisalabad, and neighboring Uttar Pradesh, Punjab, and Haryana in India.

CONCLUSION

This study used several datasets of MOPITT and MERRA-2 CO emissions, AOD, NDVI, fire and thermal anomalies, and meteorological parameters to establish the linkage between CO and biomass burning and other parameters over Lahore and Karachi, including other megacities of South Asia, i.e., Delhi, Mumbai, Kolkata, and Dhaka, during 2001–2017. The significant impact of fire and thermal anomalies recorded during the pre-monsoon seasons of 2016 and 2017 has been observed over Lahore and adjoining regions of South Asia. In these regions, MERRA-2 showed CO emissions up to 236×10^{-11} kg/m^2/s in pre-monsoon in 2012 and 220×10^{-11} kg/m^2/s by May 2001 in Lahore. Karachi has the second-highest averaged value (74.4×10^{-11} kg/m^2/s) after Kolkata (77×10^{-11} kg/m^2/s). MOPITT CO retrievals revealed that Karachi had the highest decreasing trend of −26.9%, mostly linked to meteorology and decreasing emissions from anthropogenic activities, followed by Lahore (−20.8%), during 2001–2017.

ACKNOWLEDGMENTS

We greatly acknowledge the MERRA-2, Worldview (https://worldview.earthdata.nasa.gov), and NASA teams (http://earthdata.com) for the datasets.

REFERENCES

Badarinath, K.V.S., Kharol, S.K., Krishna Prasad, V., Kaskaoutis, D.G. and Kambezidis, H.D. 2008a. Variation in aerosol properties over Hyderabad, India during intense cyclonic conditions. *International Journal of Remote Sensing*. 29(15), 4575–4597.

Badarinath, K.V.S., Kharol, S.K., Prasad, V.K., Sharma, A.R., Reddi, E.U.B., Kambezidis, H.D. and Kaskaoutis, D.G. 2008b. Influence of natural and anthropogenic activities on UV Index variations–a study over tropical urban region using ground based observations and satellite data. *Journal of Atmospheric Chemistry*. 59(3), 219–236.

Badarinath, K.V.S., Shailesh, K.K., Anu, R.S. and Krishna, V.P. 2009. Analysis of aerosol and carbon monoxide characteristics over Arabian Sea during crop residue burning period in the Indo-Gangetic Plains using multi-satellite remote sensing datasets. *Journal of Atmospheric and Solar-Terrestrial Physics*, 71(12), 1267–1276. doi:10.1016/j.jastp.2009.04.004.

Biswas, S., Lasko, K.D. and Vadrevu, K.P. 2015a. Fire disturbance in tropical forests of Myanmar—Analysis using MODIS satellite datasets. *IEEE Journal of Selected Topics in Applied Earth Observations and Remote Sensing*. 8(5), 2273–2281.

Biswas, S., Vadrevu, K.P., Lwin, Z.M., Lasko, K. and Justice, C.O. 2015b. Factors controlling vegetation fires in protected and non-protected areas of Myanmar. *PLoS One*. 10(4) e0124346.

Barrero, M.A., Grimalt, J.O. and Cantón, L. 2006. Prediction of daily ozone concentration maxima in urban atmosphere. *Chemometrics & Intelligent Laboratory System*, 80, 67–76.

Bovensmann, H., Burrows, J.P., Buchwitz, M., Frerick, J., Noël, S., Rozanov, V.V., Chance, K.V. and Goede, A.P.H. 1999. SCIAMACHY: Mission objectives and measurement modes. *Journal of the Atmospheric Sciences*. 56(2), 127–150.

Crutzen, P.J. and Zimmermann, P.H. 1991. The changing photochemistry of the troposphere. *Tellus*. 43AB, 136–151.

Daniel, J.S., Velders, G.J.M., Douglass, A.R., Forster, P.M.D., Hauglustaine, D.A., Isaksen, I.S.A., Kuijpers, L.J.M., McCulloch, A., Wallington, T.J., Ashford, P., Montzka, S.A., Newman, P.A. and Waugh, D.W. 2007. Halocarbon scenarios, ozone depletion potentials, and global warming potentials, in Scientific Assessment of Ozone Depletion: 2006. Global Ozone Res. Monit. Proj. Rep. 50, pp. 8.1–8.39, World Meteorol. Organ. Geneva, Switzerland.

Deeter, M.N., Edwards, D.P., Gille, J.C. and Drummond, J.R. 2007. Sensitivity of MOPITT observations to carbon monoxide in the lower troposphere. *Journal of Geophysical Research*. 112, D24306. doi:10.1029/2007JD008929.

Deeter, M.N., Emmons, L.K., Francis, G.L., Edwards, D.P., Gille, J.C., Warner, J.X., Khattatov, B., Ziskin, D., Lamarque, J.-F., Ho, S.-P., Yudin, V., Attié, J.-L., Packman, D., Chen, J., Mao D. and Drummond, J.R. 2003. Operational carbon monoxide retrieval algorithm and selected results for the MOPITT instrument. *Journal of Geophysical Research*. 108(D14), 4399. doi:10.1029/2002JD003186.

Drummond, J.R. 1992. Measurements of Pollution in the Troposphere (MOPITT). In: *The Use of EOS for Studies of Atmospheric Physics*. Gille, J.C. and Visconti, G. (Eds.). North-Holland, New York, pp. 77–101.

Elminir, H.K. 2005. Dependence of urban air pollutants on meteorology. *The Science of the Total Environment*. 350, 225–237.

Elminir, H.K. 2005. Dependence of urban air pollutants on meteorology. *Science of the Total Environment* 350, 225–237.

Fang, H., Beaudoing, H.K., Rodell, M., Teng, W.L. and Vollmer, B.E. 2009. Global land data assimilation system (GLDAS) products, services and application from NASA hydrology data and information services center (HDISC). In: ASPRS 2009 Annual Conference Baltimore, Maryland, March 8–13, 2009.

Gelaro, R., McCarty, W., Suárez, M.J., Todling, R., Molod, A., Takacs, L., Randles, C.A., Darmenov, A., Bosilovich, M.G., Reichle, R. and Wargan, K., 2017. The modern-era retrospective analysis for research and applications, version 2 (MERRA-2). *Journal of Climate*, 30(14), pp. 5419–5454.

Ghude, S.D., Beig, G., Kulkarni, P.S., Kanawade, V.P., Fadnavis, S., Remedios, J.J. and Kulkarni, S.H. 2011. Regional CO pollution over the Indian subcontinent and various transport pathways as observed by MOPITT. *International Journal of Remote Sensing*. 32(21), 6133–6148. doi:10.1080/01431161.2010.507796.

Girach, I.A. and Nair, P.R. 2014. Carbon monoxide over Indian region as observed by MOPITT. *Atmospheric Environment*, 99, 599–609.

Graber, J.M., Macdonald, S.C., Kass, D.E., Smith, A.E. and Anderson, H.A. 2007. Carbon monoxide: The case for environmental public health surveillance. *Public Health Reports*. 122(2), 138–144.

Gupta, P.K., Prasad, V.K., Sharma, C., Sarkar, A.K., Kant, Y., Badarinath, K.V.S. and Mitra, A.P. 2001. CH_4 emissions from biomass burning of shifting cultivation areas of tropical deciduous forests–experimental results from ground-based measurements. *Chemosphere-Global Change Science*. 3(2), 133–143.

Hayasaka, H., Noguchi, I., Putra, E.I., Yulianti, N. and Vadrevu, K. 2014. Peat-fire-related air pollution in Central Kalimantan, Indonesia. *Environmental Pollution*. 195, 257–266. doi:10.1016/j.envpol.2014.06.031.

Hayasaka, H. and A. Sepriando. 2018. Severe air pollution due to peat fires during 2015 Super El Niño in Central Kalimantan, Indonesia. In: *Land-Atmospheric Research Applications in South/Southeast Asia*. Vadrevu, K.P., Ohara, T. and Justice, C. (Eds.). Springer, Cham, pp. 129–142.

IPCC. 2007. Climate Change 2007: The physical science basis. In: *Contribution of Working Group I to the Fourth Assessment Report of the Intergovernmental Panel on Climate Change*. Solomon, S., Qin, D., Manning, M., Chen, Z., Marquis, M., Averyt, K.B., Tignor, M., Miller, H.L. (Eds.). Cambridge University Press, Cambridge, United Kingdom; New York, USA, p. 996.

Israr, I., Jaya, S.N.I, Saharjo, H.S., Kuncahyo, B. and Vadrevu, K.P. 2018. Spatio-temporal analysis of land and forest fires in Indonesia using MODIS active fire dataset. In: *Land-Atmospheric Research Applications in South/Southeast Asia*. Vadrevu, K.P., Ohara, T., and Justice, C. (Eds.). Springer, Cham, pp. 105–128.

Jae, N.L., Wu, D.L. and Ruzmaikin, A. 2013. Interannual variations of MLS carbon monoxide induced by solar cycle. *Journal of Atmospheric and Solar-Terrestrial Physics*. 102, 99–104.

Joshi, R.M. 2015. Education in South Asia. In: *International Encyclopedia of the Social & Behavioral Sciences*. Second ed., pp. 194–197.

Justice, C., Gutman, G. and Vadrevu, K.P. 2015. NASA land cover and land use change (LCLUC): An interdisciplinary research program. *Journal of Environmental Management*. 148(15), 4–9.

Kant, Y., Ghosh, A.B., Sharma, M.C., Gupta, P.K., Prasad, V.K., Badarinath, K.V.S. and Mitra, A.P. 2000. Studies on aerosol optical depth in biomass burning areas using satellite and ground-based observations. *Infrared Physics & Technology*. 41(1), 21–28.

Kaufman, Y.J., Tanre, D., Remer, L.A., Vermote, E.F., Che, A. and Holben, B.N. 1997. Operational remote sensing of tropospheric aerosol over land from EOS moderate resolution imaging spectroradiometer. *Journal of Geophysical Research*. 102, 17051–17067.

Kumar, R., Naja, M., Pfister, G.G., Barth, M.C. and Brasseur, G.P. 2013. Source attribution of carbon monoxide in India and surrounding regions during wintertime. *Journal of Geophysical Research: Atmospheres*. 118, 1981–1995. doi:10.1002/jgrd.50134.

Lasko, K., Vadrevu, K.P., Tran, V.T., Ellicott, E., Nguyen, T.T., Bui, H.Q. and Justice, C. 2017. Satellites may underestimate rice residue and associated burning emissions in Vietnam. *Environmental Research Letters*. 12(8), 085006.

Lasko, K. and Vadrevu, K.P. 2018. Improved rice residue burning emissions estimates: Accounting for practice-specific emission factors in air pollution assessments of Vietnam. *Environmental Pollution*. 236(5), 795–806.

Lasko, K., Vadrevu, K.P. and Nguyen, T.T.N. 2018a. Analysis of air pollution over Hanoi, Vietnam using multi-satellite and MERRA reanalysis datasets. *PloS One*. 13(5), e0196629.

Lasko, K., Vadrevu, K.P., Tran, V.T. and Justice, C. 2018b. Mapping double and single crop paddy rice with Sentinel-1A at varying spatial scales and polarizations in Hanoi, Vietnam. *IEEE Journal of Selected Topics in Applied Earth Observations and Remote Sensing*. 11(2), 498–512.

Liu, Z., Ostrenga, D., Teng, W. and Kempler, S., 2012. Tropical Rainfall measuring mission (TRMM) precipitation data and services for research and applications. *American Meteorological Society BAMS*. 93, 1317–1325.

Lyma, J.L. and Jensen, R.J. 2001. Chemical reactions occurring during direct solar reduction of CO_2. *Science of the Total Environment*. 277, 7–14.

Novelli, P.C., Masarie, K.A. and Lang, P.M., 1998. Distributions and recent changes of carbon monoxide in the lower troposphere. *Journal of Geophysical Research*. 103, 19015–19033. doi:10.1029/98JD01366.

Ocak, S. and Turalioglu, F.S. 2008. Effect of meteorology on the atmospheric concentrations of traffic-related pollutants in Erzurum, Turkey. *Journal of International Environmental Application & Science*. 3(5), 325–335.

Pagano, T.S., Aumann, H.H., Hagan, D.E. and Overoye, K. 2003. Prelaunch and in-flight radiometric calibration of the atmospheric infrared sounder (AIRS). *IEEE Transactions on Geoscience and Remote Sensing*. 41(2), 265–273.

Prasad, V.K., Gupta, P.K., Sharma, C., Sarkar, A.K., Kant, Y., Badarinath, K.V.S., Rajagopal, T. and Mitra, A.P. 2000. NOx emissions from biomass burning of shifting cultivation areas from tropical deciduous forests of India–estimates from ground-based measurements. *Atmospheric Environment*. 34(20), 3271–3280.

Prasad, V.K., Kant, Y. and Badarinath, K.V.S., 2001. CENTURY ecosystem model application for quantifying vegetation dynamics in shifting cultivation areas: A case study from Rampa Forests, Eastern Ghats (India). *Ecological Research*. 16(3), 497–507.

Prasad, V.K., Kant, Y., Gupta, P.K., Elvidge, C. and Badarinath, K.V.S. 2002. Biomass burning and related trace gas emissions from tropical dry deciduous forests of India: A study using DMSP-OLS data and ground-based measurements. *International Journal of Remote Sensing*. 23(14), 2837–2851.

Prasad, V.K., Lata, M. and Badarinath, K.V.S. 2003. Trace gas emissions from biomass burning from northeast region in India—estimates from satellite remote sensing data and GIS. *Environmentalist*. 23(3), 229–236.

Prasad, V.K. and Badarinth, K.V.S., 2004. Land use changes and trends in human appropriation of above ground net primary production (HANPP) in India (1961–98). *Geographical Journal*. 170(1), 51–63.

Prasad, V.K., Badarinath, K.V.S., Yonemura, S. and Tsuruta, H. 2004. Regional inventory of soil surface nitrogen balances in Indian agriculture (2000–2001). *Journal of Environmental Management*. 73(3), 209–218.

Prasad, V.K., Anuradha, E. and Badarinath, K.V.S. 2005. Climatic controls of vegetation vigor in four contrasting forest types of India—evaluation from National Oceanic and Atmospheric Administration's advanced very high resolution radiometer datasets (1990–2000). *International Journal of Biometeorology*. 50(1), 6–16.

Prasad, V.K., Badarinath, K.V.S. and Eaturu, A. 2008a. Biophysical and anthropogenic controls of forest fires in the Deccan Plateau, India. *Journal of Environmental Management.* 86(1), 1–13.

Prasad, V.K., Badarinath, K.V.S. and Eaturu, A. 2008. Effects of precipitation, temperature and topographic parameters on evergreen vegetation greenery in the Western Ghats, India. *International Journal of Climatology: A Journal of the Royal Meteorological Society.* 28(13), 1807–1819.

Ramachandran, S. 2018. Aerosols and Climate change: Present Understanding, Challenges and Future Outlook. In: *Land-Atmospheric Research Applications in South/Southeast Asia.* Vadrevu, K.P., Ohara, T. and Justice, C. (Eds.). Springer, Cham, pp. 341–378.

Ruzmaikin, A., Lee, J.N. and Wu, D.L. (2014). Patterns of carbon monoxide in the middle atmosphere and effects of solar variability. *Advances in Space Research,* 54, 320–326.

Saharjo, B.H. and Yungan, A. 2018. Forest and land fires in Riau province; A Case study in fire prevention, policy implementation with local concession holders. In: *Land-Atmospheric Research Applications in South/Southeast Asia.* Vadrevu, K.P., Ohara, T., and Justice, C. (Eds.). Springer, Cham, pp. 143–170.

Sahu, L.K., Varun, S., Kajino, M. and Nedelec, P. 2013. Variability in tropospheric carbon monoxide over an urban site in Southeast Asia. *Atmospheric Environment.* 68, 243–255.

Srivastava, S., Sheel, V. 2013. Study of tropospheric CO and O_3 enhancement episode over Indonesia during autumn 2006 using the model for ozone and related chemical tracers (MOZART-4). *Atmospheric Environment.* 67, 53–62.

Susskind, J., Barnet, C.D. and Blaisdell, J.M. 2003. Retrieval of atmospheric and surface parameters from AIRS/AMSU/HSB data in the presence of clouds. *IEEE Transactions on Geoscience and Remote Sensing.* 41(2), 390–409.

Szopa, S., Hauglustaine, D. and Ciais, P. 2007. Relative contributions of biomass burning emissions and atmospheric transport to carbon monoxide interannual variability. *Geophysical Research Letters.* 34, L18810. doi:10.1029/L030231.

Tahir, S.N.A., Rafique, M. and Alaamer, A.S. 2010. Biomass fuel burning and its implications: deforestation and greenhouse gases emissions in Pakistan. *Environmental Pollution* 158, 2490–2495.

Tariq, S. and Ul-Haq, Z. 2019. Satellite remote sensing of aerosols and gaseous pollution over Pakistan. In: *Land-Atmospheric Research Applications in South/Southeast Asia.* Vadrevu, K.P., Ohara, T. and Justice, C. (Eds.). Springer, Cham, pp. 523–552.

Turquety, S., Clerbaux, C., Law, K., Coheur, P.-F., Cozic, A., Szopa, S., Hauglustaine, D.A., Hadji-Lazaro, J., Gloudemans, A.M.S., Schrijver, H., Boone, C.D., Bernath, P.F. and Edwards, D.P., 2008. CO emission and export from Asia: an analysis combining complementary satellite measurements (MOPITT, SCIAMACHY and ACE-FTS) with global modeling. *Atmospheric Chemistry and Physics.* 8, 5187–5204.

Ul-Haq, Z., Rana, A.D., Ali, M., Mahmood, K., Tariq, S. and Qayyum, Z. 2015. Carbon monoxide (CO) emissions and its tropospheric variability over Pakistan using satellite-sensed data. *Advances in Space Research.* 56, 583–595. doi:10.1016/j.asr.2015.04.026.

Ul-Haq, Z., Tariq, S. and Ali, M. 2016. Anthropogenic emissions and space-borne observations of carbon monoxide over South Asia. *Advances in Space Research.* 58, 1610–1626. doi:10.1016/j.asr.2016.06.033.

Ul-Haq, Z. 2018. PhD Thesis "An assessment of major Ozone Depleting Substances (ODSs) and their impacts on climate change using RS/GIS in Lahore, Pakistan. submitted to HEC, Pakistan repository.

UNEP. United Nations Environmental Program 2009. Handbook for the Montreal Protocol on substances that deplete the ozone layer. 6th ed. Nairobi, Kenya.

Vadrevu, K.P., Eaturu, A. and Badarinath, K.V.S. 2006. Spatial distribution of forest fires and controlling factors in Andhra Pradesh, India using spot satellite datasets. *Environmental Monitoring and Assessment.* 123(1–3), 75–96.

Vadrevu, K.P., 2008. Analysis of fire events and controlling factors in eastern India using spatial scan and multivariate statistics. *Geografiska Annaler: Series A, Physical Geography*. 90(4), 315–328.

Vadrevu, K.P., Badarinath, K.V.S. and Anuradha, E. 2008. Spatial patterns in vegetation fires in the Indian region. *Environmental Monitoring and Assessment*, 147(1–3), 1. doi:10.1007/s10661-007-0092-6.

Vadrevu, K.P. and Justice, C.O. 2011. Vegetation fires in the Asian region: Satellite observational needs and priorities. *Global Environmental Change*. 15(1), 65–76.

Vadrevu, K.P. and Choi, Y. 2011. Wavelet analysis of airborne CO_2 measurements and related meteorological parameters over heterogeneous landscapes. *Atmospheric Research*. 102(1–2), 77–90.

Vadrevu, K.P., Csiszar, I., Ellicott, E., Giglio, L., Badarinath, K.V.S., Vermote, E. and Justice, C. 2012. Hotspot analysis of vegetation fires and intensity in the Indian region. *IEEE Journal of Selected Topics in Applied Earth Observations and Remote Sensing*. 6(1)224–238.

Vadrevu, K.P., Giglio, L. and Justice, C. 2013. Satellite based analysis of fire–carbon monoxide relationships from forest and agricultural residue burning (2003–2011). *Atmospheric Environment*. 64, 179–191.

Vadrevu K.P., Ohara T. and Justice C. 2014a. Air pollution in Asia. *Environmental Pollution*. 12, 233–235.

Vadrevu, K.P., Lasko, K., Giglio, L. and Justice, C. 2014b. Analysis of Southeast Asian pollution episode during June 2013 using satellite remote sensing datasets. *Environmental Pollution*. 12, 245–256.

Vadrevu, K.P., Lasko, K., Giglio, L. and Justice, C. 2015. Vegetation fires, absorbing aerosols and smoke plume characteristics in diverse biomass burning regions of Asia. *Environmental Research Letters*. 10(10), 105003.

Vadrevu, K.P. and Lasko, K.P., 2015. Fire regimes and potential bioenergy loss from agricultural lands in the Indo-Gangetic Plains. *Journal of Environmental Management*. 148, 10–20.

Vadrevu, K.P., Ohara, T. and Justice, C. 2017. Land cover, land use changes and air pollution in Asia: A synthesis. *Environmental Research Letters*. 12(12), 120201.

Vadrevu, K.P., Ohara, T. and Justice, C. (Eds.). 2018. *Land-Atmospheric Research Applications in South and Southeast Asia*. Springer, Cham.

Vadrevu, K.P. and Lasko, K. 2018. Intercomparison of MODIS AQUA and VIIRS I-Band fires and emissions in an agricultural landscape—Implications for air pollution research. *Remote Sensing*. 10(7), 978. doi:10.3390/rs10070978.

Vadrevu, K.P., Lasko, K., Giglio, L., Schroeder, W., Biswas, S. and Justice, C. 2019. Trends in vegetation fires in South and Southeast Asian countries. *Scientific Reports*. 9(1), 7422. doi:10.1038/s41598-019-43940-x.

Van der Werf, G.R., Randerson, J.T., Collatz, G.J., Giglio, L., Kasibhatla, P.S., Arellano Jr., A.F., Olsen, S.C. and Kasischke, E.S. 2004. Continental-scale partitioning of fire emissions during the 1997 to 2001 El Nino/La Nina period. *Science*. 303, 73–76.

WHO, 2000. Air Quality Guidelines for Europe, European Series, second ed., vol. 91. WHO Regional Publications.

Worden, H.M., Deeter, M.N., Frankenberg, C., George, M., Nichitiu, F., Worden, J., Aben, I., Bowman, K.W., Clerbaux, C., Coheur, P.F., de Laat, A.T.J., Detweiler, R., Drummond, J.R., Edwards, D.P., Gille, J.C., Hurtmans, D., Luo, M., Mart´ınez-Alonso, S., Massie, S., Pfister, G. and Warner, J.X. 2013. Decadal record of satellite carbon monoxide observations. *Atmospheric Chemistry and Physics*. 13, 837–850.

Zhang, X., van Geffen, J., Liao, H., Zhang, P. and Lou, S. 2012. Spatiotemporal variations of tropospheric SO_2 over China by SCIAMACHY observations during 2004–2009. *Atmospheric Environment*. 60, 238–246.

Zhang, Y. and Huixiang, X. 2012. The sources and sinks of carbon monoxide in the St. Lawrence estuarine system. *Deep-Sea Research: II*. 81–84, 114–123.

Section II

Biomass Burning Emissions

5 Estimating Biomass Burning Emissions in South and Southeast Asia from 2001 to 2017 Based on Satellite Observations

Yusheng Shi
Aerospace Information Research Institute,
Chinese Academy of Sciences, China

Yu Tian
The University of Tokyo, Japan

Shuying Zang
Harbin Normal University, China

Yasushi Yamaguchi
Nagoya University, Japan

Tsuneo Matsunaga
National Institute for Environmental Studies, Japan

Zhengqiang Li and Xingfa Gu
Aerospace Information Research Institute,
Chinese Academy of Sciences, China

CONTENTS

INTRODUCTION

Biomass burning (BB) emissions significantly impact global atmospheric chemistry and climate change. It has become an important source of uncertainty in biogeochemical cycles and atmospheric transport simulations (Andreae and Merlet, 2001). Therefore, accurate estimates of BB emissions are important to understand the interactions between fire, vegetation, and climate, and the relationships between emissions and deposition cycles in terrestrial ecosystems and the atmosphere (Vadrevu and Justice, 2011; Justice et al., 2015; Shi et al., 2019).

south and southeast Asia (S/SEA) (including mainland southeast Asia (MSEA), equatorial southeast Asia (ESEA), and south Asia (SA)) have been experiencing rapid deforestation (Vadrevu et al., 2006; Vadrevu, 2008; Vadrevu et al., 2008; 2012; Biswas et al., 2015a, b; Vadrevu et al., 2015, 2017), slash-and-burn cultivation (Shi et al., 2014), crop residue burning (Vadrevu et al., 2013; Lasko et al., 2017, 2018a,b; Lasko and Vadrevu, 2018; Vadrevu and Lasko, 2018), peatland burning for oil palm plantations (Page et al., 2002; Hayasaka et al., 2014; Hayasaka and Sepriando, 2018; Israr et al., 2018; Saharjo and Yungan, 2018), etc. All of these are closely related to fires with extensive burning, resulting in serious emissions of trace gases and aerosols, including loss of biomass and bioenergy (Kant et al., 2000; Badarinath et al., 2008a, b; Prasad et al., 2000; Gupta et al., 2001; Prasad et al., 2001, 2002, 2003, Vadrevu et al., 2013, 2014a, b; Vadrevu and Lasko, 2015; Vadrevu et al., 2017, 2018, 2019). For example, during 2000–2006, forest fire-related BB in tropical Asia released approximately 58 Tg CO_2 emissions annually (Chang and Song, 2010).

Numerous studies estimated BB emissions using burned areas, fuels, combustion efficiency (CE), and emission factors (EFs) (Shi et al., 2015a). Earlier studies aiming to calculate biomass fuels for emission estimation used biome-averaged values from statistical data or observational data together with country-based or land-type-based biomass density. Meanwhile, CE was often allocated according to the land type. However, it was also strongly affected by moisture conditions. Therefore, such simple representation and allocation overlooked the spatial variations and heterogeneities over large areas, without reflecting the characteristics of variabilities in biomass and CE among pixels. Such estimation will result in large uncertainty in BB emissions inventory and atmospheric transport simulations.

This study developed a high-resolution ($0.1° \times 0.1°$), multi-year, monthly BB emissions inventory covering all land types in S/SEA during 2001–2017. We employed satellite burned area product MCD64A1, satellite and observational data of biomass density, and spatiotemporal variable CE to estimate the BB emissions. The BB emission species included black carbon (BC), methane (CH_4), carbon monoxide (CO), carbon dioxide (CO_2), ammonia (NH_3), nonmethane organic compounds (NMOC), nitrogen oxides (NO_x), organic carbon (OC), fine particulate matter ($PM_{2.5}$), and sulfur dioxide (SO_2).

MATERIALS AND METHODS

BB emissions are estimated based on burned area, biomass fuel, CE, and EFs (Shi et al., 2015b; Wiedinmyer et al., 2011):

$$E_{i,x,t} = \sum_{j=1}^{n} BA_{x,t} \times F_x \times CE_x \times EF_{i,j}$$

where j indicates different vegetation types; i indicates different emission species; $E_{i,x,t}$ is the emission amount of species i in location x and month t; $BA_{x,t}$ is the total burned area (km^2/month) of vegetation type in location x and month t; F_x is the available biomass fuel loading (kg/m^2) in location x; CE_x is the CE in location x, defined as the fraction of combusted fuel to the total amount (unitless); $EF_{i,j}$ is the EF of species i for vegetation type j, conveying the mass of species per mass of dry matter burned (g/kg).

Burned Area and Active Fire

Burned area datasets were derived from the monthly MODIS 500-m direct broadcast burned area product (MCD64A1) (http://modis-fire.umd.edu/). The MCD64A1 product has been validated by observational data from different fire types worldwide (Randerson et al., 2012; Giglio et al., 2013; Shi et al., 2014). Here, we used the monthly burned areas generated after resampling the original 500-m binary map into a 0.1° percentage grid map. Then, the active fire products (MOD14A2/MYD14A2) during 2001–2017 were also used as a supplementary tool to describe small fires. We quantified the fire density on a 0.1° grid as the percentage of pixels classified as fire pixels within each continent and land-cover units and calculated fire frequency as the return interval of fire (Shi and Yamaguchi, 2014).

Biomass Fuels

Forest

The forest aboveground biomass (AGB) was generated by combining two existing datasets (Saatchi et al., 2011; Baccini et al., 2012), which employed data from multiple satellites and thousands of ground plots to map the spatial variations of biomass. The forest AGB in S/SEA at 1 km resolution used an independent reference dataset of field observations and locally calibrated high-resolution biomass maps, harmonized and upscaled to 400 one-km AGB estimates (Saatchi et al., 2011; Baccini et al., 2012; Avitabile et al., 2016). The validation of the modeled AGB by a mixture of ground and Lidar observations showed good agreement with a relative error of 33.4% in S/SEA (Saatchi et al., 2011). The forest AGB density was highly variable in S/SEA with 335 and 211 Mg/ha in intact and non-intact forest ecozones (Avitabile et al., 2016).

In the peat forest, belowground biomass (BGB) will also burn because of the organic matter in the carbon-rich soil, especially in ESEA. Here, we assumed that BB consumes almost 50% of the BGB in peatland (van der Werf et al., 2010), which is close to the results from Chang and Song (2010) with a mean peat dry bulk density of $100\,\text{kg/m}^3$, carbon concentration of 60%, and burn depth of 51 cm in ESEA.

Shrub and Grass

The biomass of shrub and grass was estimated by the maximum standing biomass in savannas. The annual rainfall (P) was demonstrated to have a linear relationship with the measured biomass density (kg dry matter/m²) (Ito and Penner, 2004):

$$\text{Biomass density} = 4.9 \times 10^{-4} \times P - 0.58$$

Monthly precipitation data for the study period (2001–2017) were obtained from the Global Precipitation Climatology Project (GPCP) of the NASA Goddard Space Flight Center.

Crop Residues

The Food and Agriculture Organization Statistical Yearbook (FAOSTAT) calculated the amount of dry crop residues burned using harvested area data and the mean default crop values of fuel and combustion factors based on the IPCC Guidelines for National Greenhouse Gas Inventories (IPCC, 2006; Shi and Yamaguchi, 2014). Here, we derived the annually burned crop residues covering broad categories of crop types (maize, sugarcane, rice, and wheat) for each country in the S/SEA from 2001 to 2017, as provided by the FAOSTAT (http://faostat.fao.org). The burned crop residues in each pixel were determined by the active fire products MOD14A2/MYD14A2 and cropland map from MODIS Land Cover Type product MCD12Q1 fire counts, fire frequency, and spatial distribution of fires on a 0.1° grid cell.

COMBUSTION EFFICIENCY

The CE is primarily determined by fuel types and moisture conditions (Shi et al., 2014). Here, we calculated the CE for each kind of fuel type (forest, woodland, and grassland) based on the tree cover fraction (Tc; MOD44B Vegetation Continuous Fields). Forest, woodland, and grassland were categorized as regions with Tc above 60%, 40%–60%, and below 40%, respectively (Ito and Penner, 2004).

$$\text{CE}_{\text{forest}} = \left(1 - e^{-1}\right)^{mcf} \quad \text{Tc} > 60\%$$

Here, fuel moisture category factors (mcf) were introduced to reflect the monthly variations in moisture conditions (Zhang et al., 2008). The mcf value was estimated using the fuel moisture condition by MODIS Vegetation Condition Index (VCI) data, calculated by the Normalized Difference Vegetation Index (NDVI) from MOD13A3 Vegetation Indices. The *mcf* value decreases from wet to dry fuel condition (very dry: 0.33; dry: 0.5; moderate: 1; moist: 2; wet: 4; and very wet: 5).

$$CE_{woodland} = \exp(-0.013 \times Tc) \quad 40\% < Tc \leq 60\%$$

$$CE_{grassland} = \frac{1}{100} \times \left(-213 \times \frac{NDVI_t - NDVI_{min}}{NDVI_{max} - NDVI_{min}} + 138 \right) \quad Tc \leq 40\%$$

We limited the value of CE to the range from 0.44 to 0.98 to avoid extrapolation beyond measured values for grassland fires (Shi et al., 2019).

EMISSION FACTORS

Detailed EFs for various gases and aerosols in each biomass type for S/SEA were compiled (Table 5.1). The vegetation type in each fire pixel was determined by the MCD12Q1 for 2001–2017. We used the International Geosphere-Biosphere Programme (IGBP) land-cover classification to assign each fire pixel to one of 16 land-use/land-cover (LULC) classes. Finally, five land types were generated: forest, woody savanna/shrubland (WS), savanna/grassland (SG), cropland, and peatland.

RESULTS AND DISCUSSION

SPATIAL PATTERNS AND VARIATIONS

First, we found that all emissions species presented similar spatial distributions, even though their magnitudes varied significantly. Taking CO_2 as an example (Figure 5.1), in general, CO_2 emissions could be observed throughout most of MSEA, part of ESEA, and SA with obvious spatial variations. In MSEA, high emissions were observed in west and east Myanmar, north Thailand, north Laos, east Cambodia, and south Vietnam; in these regions, most CO2 emissions were greater than 200 Gg/year in each $0.1° \times 0.1°$ grid. In ESEA, CO_2 emissions were high in south Sumatra and south Kalimantan of Indonesia and south Papua New Guinea and relatively low in Malaysia and the Philippines. Most CO_2 emissions were from Indonesia with high amounts greater than 500 Gg/year. In SA, large CO_2 emissions originated from some of northwest India (Punjab and Haryana) and part of central India (Chhattisgarh).

Then, we quantified each country's contributions to the total CO_2 emissions in S/SEA during 2001–2017 (Table 5.2). Myanmar was determined to be the most significant contributor to total BB CO_2 emissions in S/SEA (285.3 Tg, 28%), followed by Cambodia (161.1 Tg, 16%), India (154.8 Tg, 15%), Thailand (134.6 Tg, 13%), and then Indonesia (98.4 Tg, 10%). After categorizing the LULC classes into five major land types based on vegetation and fire sources (Figure 5.2), we quantified each land type's contributions to the total BB CO_2 emissions in S/SEA during 2001–2017

TABLE 5.1

Land Use/Land Cover Classifications as Assigned by the MODIS Land Cover Type (LCT), Assigned Generic Land Cover Class, and Emission Factors (g/kg Dry Matter)

LCT classes	Types	CO_2	CO	CH_4	NO_x	NMOC	SO_2	NH_3	$PM_{2.5}$	OC	BC
Evergreen needleleaf forest	Forest	1514	118	6	1.8	28	1	3.5	13	7.8[a]	0.2[a]
Evergreen broadleaf forest	Forest	1643	92	5.1	2.6	24	0.5	0.8	9.7	4.7	0.5
Deciduous needleleaf forest	Forest	1514	118	6	3	28	1	3.5	13.6	7.8[a]	0.2[a]
Deciduous broadleaf forest	Forest	1630	102	5	1.3	11	1	1.5	13	9.2	0.6
Mixed forests	Forest	1630	102	5	1.3	14	1	1.5	13	9.2	0.6
Closed shrublands	WS	1716	68	2.6	3.9	4.8	0.7	1.2	9.3	6.6[a]	0.5[a]
Open shrublands	WS	1716	68	2.6	3.9	4.8	0.7	1.2	9.3	6.6[a]	0.5[a]
Woody savannas	WS	1716	68	2.6	3.9	4.8	0.7	1.2	9.3	6.6[a]	0.5[a]
Savannas	SG	1692	59	1.5	2.8	9.3	0.5	0.5	5.4[b]	2.6	0.4
Grasslands	SG	1692	59	1.5	2.8	9.3	0.5	0.5	5.4[b]	2.6	0.4
Permanent wetlands	SG	1692	59	1.5	2.8	9.3	0.5	0.5	5.4[b]	2.6	0.4
Croplands	Crop	1537	111	6	3.5	57	0.4[b]	2.3	5.8	3.3[b]	0.7[b]
Cropland/natural vegetation mosaic	SG	1692	59	1.5	2.8	9.3	0.5	0.5	5.4[b]	2.6	0.4
Barren or sparsely vegetated	SG	1692	59	1.5	2.8	9.3	0.5	0.5	5.4[b]	2.6	0.4
Peatland	Peat	1703[c]	210[c]	20.8[c]	2.26[d]	7.0[d]	0.71[d]	2.55[c]	9.1	4.3[c]	0.57[c]

[a] From McMeeking (2008); [b] from Andreae and Merlet (2001); [c] from Christian et al. (2003); and [d] from van der Werf et al. (2010). Others were from Akagi et al. (2011), Wiedinmyer et al. (2011), and GFED4 (http://www.falw.vu/~gwerf/GFED/GFED4/emission_factors/).

Forest, boreal, tropical, temperate forest; WS, woody savanna/shrubland; SG, savanna/grassland.

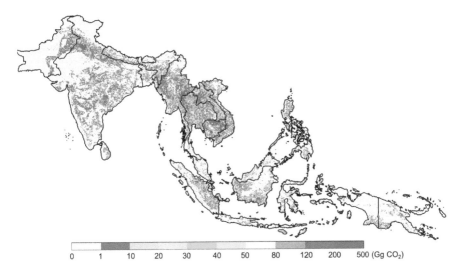

FIGURE 5.1 Annual total BB CO$_2$ emissions (0.1°×0.1°) in S/SEA during 2001–2017.

FIGURE 5.2 Land-cover map of S/SEA.

(Table 5.2). It was clear that forest fire was the largest contributor (363.2 Tg, 36%), followed by WS (341.4 Tg, 34%), SG (124.5 Tg, 12%), cropland (107.8 Tg, 11%), and peatland (73.4 Tg, 7%). Forest fires were observed extensively in Myanmar, Thailand, and Laos. WS fires were mostly concentrated in Myanmar and Cambodia. For SG fires, CO$_2$ emissions were mainly distributed in Thailand and Cambodia. Cropland residue burning was high in India, an agriculture-dominated country. Peatland burning in south Sumatra and south Kalimantan also exhibited high CO$_2$ emissions.

TABLE 5.2

BB CO$_2$ Emissions (Tg/year) in S/SEA Countries from All Land Types During 2001–2017

	Forest	WS	SG	Crop	Peat	Total
Myanmar	133.3	131.6	16.9	3.5	0.0	285.3
Cambodia	29.0	100.4	28.2	3.5	0.0	161.1
India	38.8	40.7	10.4	64.9	0.0	154.8
Thailand	56.9	40.7	30.0	7.0	0.0	134.6
Indonesia	9.1	1.7	10.5	6.6	70.5	98.4
Laos	52.1	8.7	6.7	0.5	0.0	68.0
Vietnam	18.5	14.5	11.9	10.0	0.0	54.9
Nepal	11.5	0.1	2.3	2.2	0.0	16.1
Papua New Guinea	6.2	0.4	4.0	0.3	2.9	13.8
Pakistan	0.5	0.4	0.2	5.1	0.0	6.2
Bangladesh	0.7	1.4	0.3	3.4	0.0	5.8
Philippines	2.2	0.5	1.9	0.6	0.0	5.2
Malaysia	3.4	0.1	0.6	0.1	0.0	4.2
Bhutan	0.8	0.1	0.1	0.0	0.0	1.0
Sri Lanka	0.2	0.1	0.5	0.1	0.0	0.9
Total	363.2	341.4	124.5	107.8	73.4	1010.3

However, the five land types' contributions to the CO$_2$ emissions in each country varied considerably (Table 5.2). In Myanmar, forest fires were the primary source (47%), followed by WS, SG, and cropland. In Cambodia, the primary fuel source was WS (62%), and forest fires were the second-largest emitter (18%). In India, the predominant source was crop residues (42%). In Indonesia, most CO$_2$ emissions were from peatland burning (72%). In other words, the dominant sources of BB emissions were forest fires in MSEA, peatland burning in ESEA, and crop residues in SA.

In general, the spatial patterns of high CO$_2$ emissions correlated well with the annual percentages of burned areas, a decisive parameter that controls the patterns of BB CO$_2$ emissions. However, a large fraction of the high CO$_2$ emissions were also affected by fuel loads. The actual amounts of BB CO$_2$ emissions were jointly determined by the available fuels, combustion factors, and EFs of all land types during 2001–2017. Forest fires typically release more CO$_2$ emissions than fires in WS, SG, and cropland due to their characteristically high fuel loads. Consequently, forest burning with high-density biomass in Myanmar resulted in higher CO$_2$ emissions than the burning of WS with low-density fuels in Cambodia, even though there were fewer burned areas in Myanmar. Although the burned peatland area in Indonesia was small, the CO$_2$ emissions from peatland burning were very large, which plays a dominant role in ESEA. This is due to the high density of BGB.

More specifically, the average annual BB emissions in S/SEA during 2001–2017 were 0.3 Tg BC, 2.7 Tg CH$_4$, 52.0 Tg CO, 1010.3 Tg CO$_2$, 0.7 Tg NH$_3$, 10.8 Tg NMOC,

TABLE 5.3

Average Annual BB Emissions (Tg) for Each Land Type in S/SEA During 2001–2017

Species	Forest (%)	WS (%)	SG (%)	Crop (%)	Peat (%)	Total (Tg)
BC	**35**	31	14	15	5	0.3
CH_4	**40**	19	7	15	21	2.7
CO	**37**	25	11	14	11	52.0
CO_2	**36**	34	12	11	7	1010.3
NH_3	26	**34**	8	22	10	0.7
NMOC	**45**	9	9	35	2	10.8
NO_x	29	**40**	16	12	3	1.9
OC	36	**43**	10	7	4	3.0
$PM_{2.5}$	**41**	34	12	8	5	5.1
SO_2	34	**39**	16	8	5	0.4

WS, woody savanna/shrubland; SG, savanna/grassland.
The bold part indicates the maximum value of each line.

1.9 Tg NO_x, 3.0 Tg OC, 5.1 Tg $PM_{2.5}$, and 0.4 Tg SO_2 (Table 5.3). The results showed that forest fires were the biggest contributor to the total emissions for most species (36% on average), followed by burning of WS (31%), cropland (14%), SG (12%), and peatland (7%). However, for different emission species, the contribution of each land type to the total amount varied moderately. For example, WS was the biggest source of NH_3, NO_x, OC, and SO_2 emissions, which was attributable to the high WS EFs. In SEAS, evergreen broadleaf forest was the dominant forest type. However, the EFs of evergreen broadleaf forest for NH_3, NO_x, OC, and SO_2 were lower than those of WS, resulting in higher emissions from WS for these trace gases and particles.

TEMPORAL PATTERNS AND VARIATIONS

Overall, BB CO_2 emissions showed strong interannual variations with an increasing trend during 2001–2017; the total BB CO_2 emissions increased sharply from 2001 to 2004, fluctuated between 2004 and 2017, and generally decreased until 2017 (Figure 5.3). The CO_2 emissions in 2004 were by far the highest (1561 Tg), followed by 2007 (1354 Tg), 2010 (1327 Tg), and 2015 (1248 Tg). The amounts of CO_2 emissions in the four peak years were almost two or three times larger than those in other years. Similarly, the 4 years exhibited the largest burned areas in MCD64A1 burned area products with $22.4 \times 10^4 km^2$ in 2004, $18.1 \times 10^4 km^2$ in 2007, $17.8 \times 10^4 km^2$ in 2010, and $18.5 \times 10^4 km^2$ in 2015, which were higher than the total annual average of $15.4 \times 10^4 km^2$ during 2001–2017. This indicated that the total burned area is one of the most important factors in determining BB CO_2 emissions.

Generally, the interannual variations of BB CO_2 emissions during 2001–2017 were predominantly determined by forest fires. Among the five land types, forest fires, the largest emitter of BB, presented consistent interannual variations with the

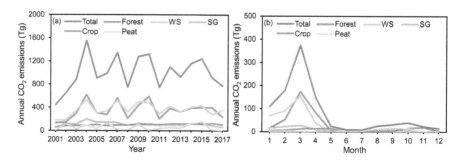

FIGURE 5.3   Interannual and intra-annual variations of BB CO_2 emissions in S/SEA from 2001 to 2017.

total amount from 2001 to 2017. WS fires displayed consistent variations with forest fires. SG fires, cropland fires, and peatland fires all experienced steady trends with minor changes from year to year. The highest peatland fire CO_2 emissions were detected in 2015 because of drought-induced fires in Indonesia as the result of El Niño (Figure 5.3; van der Werf et al., 2017); increased peatland fire CO_2 emissions during El Niño years were also observed in 2002, 2004, 2006, 2009, and 2015 (Figure 5.3; Shi et al., 2014), but the contribution from peatland fires was still small compared with the other land types.

BB CO_2 emissions also presented dramatic seasonal variations within 1 year (Figure 5.3). In total, S/SEA experienced the largest annual peak in March (376 Tg) each year, which is the typical fire season for the primary contributors, i.e., Myanmar, Cambodia, India, and Thailand. Intra-annual variations in CO_2 emissions revealed that they were highest during January–March, when deforestation and land clearing were under way (Shi and Matsunaga, 2017). Consequently, values in CO_2 emissions and burned areas were typically high during the dry period (January–March). Comparing with the land types, we found that the main contributors to the peak values were from the forest fires and WS fires because of the accessible forest and short vegetation burning for field preparation. Another peak occurred during the period September–November, when Indonesia experienced extensive burning practices in peatland for palm oil plantations by humans during the drought season.

COMPARISONS WITH OTHER STUDIES

Here, the widely available data source of GFEDv4.1s, a dataset of all emissions of trace gases and aerosols from BB, was selected for comparison with the results of this study (Figure 5.4). Generally, our estimates of BB emissions in S/SEA during 2010–2017 were comparable with GFEDv4.1s for most emission species even though there were some overestimations. The difference was primarily due to fuel loads and combustion factors. In GFEDv4.1s, the biomass density/fuel was simulated by the biogeochemical model. This study used a dataset that integrates site measurements and Lidar-based observations in forests from 400 one-km samples in S/SEA. Besides, in this study, each land type's combustion factor was linearly scaled by introducing the vegetation index, which can reflect the ratio of burned biomass to the total available biomass.

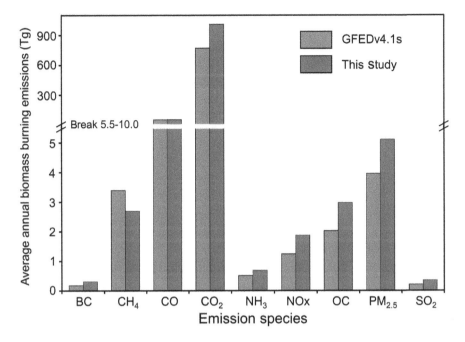

FIGURE 5.4 Comparison of average annual BB emissions in S/SEA during 2001–2017 between this study and GFEDv4.1s.

Fires in S/SEA were primarily caused by humans starting small fires for clearing croplands and deforestation, and then produced speckle-patterned fire scars. Compared to GFEDv4.1s ($0.25° \times 0.25°$), this study developed a $0.1° \times 0.1°$ spatial resolution and multi-year BB emissions inventory for the period from 2001 to 2017. Therefore, this inventory was more effective at capturing the spatial variations of small-sized fires, which were frequently smoothed and misinterpreted by the large pixels of coarse data.

EMISSION UNCERTAINTY

The employed MCD64A1 burned area product has been shown to be reliable in some big fires (Shi et al., 2014), but it usually misses small agricultural burning that is less than 40 ha, which is the minimum detectable burn size for MCD64A1 (Giglio et al., 2013). Besides, the cloud and aerosol contamination in S/SEA also affects the accuracy of interpreted burned areas. According to Avitabile et al. (2016), the fuel load was within an uncertainty range of approximately 50% around the mean value. The CE was calculated from empirical formulae with uncertainties ranging from 20% to 30% (Shi and Yamaguchi, 2014), and the typical uncertainty of the EF was on the order of 20%–30% (Hoelzemann et al., 2004). Then, we ran 20,000 Monte Carlo simulations to estimate the range of BB emission with a 90% confidence interval; the estimated annual emission ranges were 0.1–0.5 Tg BC, 1.9–3.9 Tg CH_4, 36.2–78.8 Tg CO, 725.9–1524.5 Tg CO_2, 0.4–1.5 Tg NH_3, 5.2–18.4 Tg NMOC, 1.0–3.2 Tg NO_x, 1.7–5.2 Tg OC, 2.9–12.1 Tg $PM_{2.5}$, and 0.2–0.7 Tg SO_2.

CONCLUSIONS

This study developed a multi-year and 0.1° spatial resolution BB emissions inventory in S/SEA during 2001–2017 and investigated the spatial and temporal variations of CO_2 emissions over the past 17 years. The average annual BB emissions in S/SEA during 2001–2017 were 0.3 Tg BC, 2.7 Tg CH_4, 52.0 Tg CO, 1010.3 Tg CO_2, 0.7 Tg NH_3, 10.8 Tg NMOC, 1.9 Tg NO_x, 3.0 Tg OC, 5.1 Tg $PM_{2.5}$, and 0.4 Tg SO_2. Forest fires were the largest contributor, accounting for 36% (363.2 Tg) of the total BB CO_2 emissions. Meanwhile, Myanmar released 285.3 Tg CO_2/year, being the largest emitter among all countries in S/SEA. The total BB CO_2 emissions showed strong inter-annual variations from 2001 to 2017 with peak emissions during January–March annually. Forest fires and WS fires were the primary land types controlling both the inter- and intra-annual emission characteristics.

ACKNOWLEDGMENT

This research was supported by the National Key R&D Program of China, 2018YFB0504800 (2018YFB0504801); the Key Deployment Projects of the Chinese Academy of Sciences (ZDRW-ZS-2019-1-3); the CAS Pioneer Hundred Talents Program, China (Y8YR2200QM, Y7S00100CX, and Y7S0010030); the National Natural Science Foundation of China (41701498); and the NIES GOSAT-2 Project (1RA-31), Japan.

REFERENCES

Andreae, M.O., Merlet, P. 2001. Emission of trace gases and aerosols from biomass burning. *Glob. Biogeochem. Cycles* 15, 955–966.

Akagi, S.K., Yokelson, R.J., Wiedinmyer, C., Alvarado, M.J., Reid, J.S., Karl, T., Crounse, J.D., Wennberg, P.O. 2011. Emission factors for open and domestic biomass burning for use in atmospheric models. *Atmos. Chem. Phys.* 11, 4039–4072.

Avitabile, V., Herold, M., Heuvelink, G.B., Lewis, S.L., Phillips, O.L., Asner, G.P., Armston, J., Ashton, P.S., Banin, L., Bayol, N., Berry, N.J., Boeckx, P., de Jong, B.H., DeVries, B., Girardin, C.A., Kearsley, E., Lindsell, J.A., Lopez-Gonzalez, G., Lucas, R., Malhi, Y., Morel, A., Mitchard, E.T., Nagy, L., Qie, L., Quinones, M.J., Ryan, C.M., Ferry, S.J., Sunderland, T., Laurin, G.V., Gatti, R.C., Valentini, R., Verbeeck, H., Wijaya, A., Willcock, S. 2016. An integrated pan-tropical biomass map using multiple reference datasets. *Glob. Chang. Biol.* 22, 1406–1420.

Baccini, A., Goetz, S.J., Walker, W.S., Laporte, N.T., Sun, M., Sulla-Menashe, D., Hackler, J., Beck, P.S.A., Dubayah, R., Friedl, M.A., Samanta, S., Houghton, R.A. 2012. Estimated carbon dioxide emissions from tropical deforestation improved by carbon-density maps. *Nat. Clim. Change* 2, 182–185.

Badarinath, K.V.S., Kharol, S.K., Krishna Prasad, V., Kaskaoutis, D.G. and Kambezidis, H.D. 2008a. Variation in aerosol properties over Hyderabad, India during intense cyclonic conditions. *Int. J. Remote Sens.* 29(15), 4575–4597.

Badarinath, K.V.S., Kharol, S.K., Prasad, V.K., Sharma, A.R., Reddi, E.U.B., Kambezidis, H.D. and Kaskaoutis, D.G. 2008b. Influence of natural and anthropogenic activities on UV Index variations–a study over tropical urban region using ground based observations and satellite data. *J. Atmos. Chem.* 59(3), 219–236.

Biswas, S., Lasko, K.D. and Vadrevu, K.P. 2015a. Fire disturbance in tropical forests of Myanmar—Analysis using MODIS satellite datasets. *IEEE J. Sel. Top. Appl. Earth Obs. Remote Sens.* 8(5), 2273–2281.

Biswas, S., Vadrevu, K.P., Lwin, Z.M., Lasko, K. and Justice, C.O. 2015b. Factors controlling vegetation fires in protected and non-protected areas of Myanmar. *PLoS One* 10(4), e0124346.

Chang, D., Song, Y. 2010. Estimates of biomass burning emissions in tropical Asia based on satellite-derived data. *Atmos. Chem. Phys.* 10, 2335–2351.

Christian, T.J., Kleiss, B., Yokelson, R.J., Holzinger, R., Crutzen, P.J., Hao, W.M., Saharjo, B.H., Ward, D.E. 2003. Comprehensive laboratory measurements of biomass burning emissions: 1. Emissions from Indonesian, African, and other fuels. *J. Geophys. Res.* 108, 4719.

Giglio, L., Randerson, J.T., van der Werf, G.R. 2013. Analysis of daily, monthly, and annual burned area using the fourth-generation global fire emissions database (GFED4). *J. Geophys. Res. Biogeosci.* 118, 317–328.

Gupta, P.K., Prasad, V.K., Sharma, C., Sarkar, A.K., Kant, Y., Badarinath, K.V.S., Mitra, A.P. 2001. CH_4 emissions from biomass burning of shifting cultivation areas of tropical deciduous forests–experimental results from ground-based measurements. *Chemosphere- Global Change Sci.* 3(2), 133–143.

Hayasaka, H., Noguchi, I., Putra, E.I., Yulianti, N., Vadrevu, K. 2014. Peat-fire-related air pollution in Central Kalimantan, Indonesia. *Environmental Pollution.* 195, 257–266. doi: 10.1016/j.envpol.2014.06.031

Hayasaka, H., Sepriando, A. 2018. Severe Air Pollution Due to Peat Fires During 2015 Super El Niño in Central Kalimantan, Indonesia. In: Vadrevu, K.P., Ohara, T., and Justice, C. (Eds). Land-Atmospheric Research applications in South/Southeast Asia. Springer, Cham, pp. 129–142.

Hoelzemann, J.J., Schultz, M.G., Brasseur, G.P., Granier, C. 2004. Global wildland fire emission model (GWEM): evaluating the use of global area burnt satellite data. *J. Geophys. Res.* 109, D14S04.

IPCC, 2006. Guidelines for National Greenhouse Gas Inventories. In: Agriculture, Forestry and Other Land Use, vol. 4. Chapter 2 Generic methodologies applicable to multiple land-use categories. 2006, Tab. 2.4.

Israr, I., Jaya, S.N.I., Saharjo, H.S., Kuncahyo, B., and Vadrevu, K.P. 2018. Spatio-temporal analysis of land and forest fires in Indonesia using MODIS active fire dataset. In: Vadrevu, K.P., Ohara, T., and Justice, C. (Eds). *Land-Atmospheric Research Applications in South/Southeast Asia.* Springer, Cham, pp. 105–128.

Ito, A., Penner, J.E., 2004. Global estimates of biomass burning emissions based on satellite imagery for the year 2000. *J. Geophys. Res.* 109, D14S05.

Justice, C., Gutman, G., Vadrevu, K.P. 2015. NASA land cover and land use change (LCLUC): An interdisciplinary research program. *J. Environ. Manag.* 148(15), 4–9.

Kant, Y., Ghosh, A.B., Sharma, M.C., Gupta, P.K., Prasad, V.K., Badarinath, K.V.S., Mitra, A.P. 2000. Studies on aerosol optical depth in biomass burning areas using satellite and ground-based observations. *Infrared Phys. Technol.* 41(1), 21–28.

Lasko, K., Vadrevu, K.P., Tran, V.T., Ellicott, E., Nguyen, T.T., Bui, H.Q., Justice, C. 2017. Satellites may underestimate rice residue and associated burning emissions in Vietnam. *Environ. Res. Lett.* 12(8), 085006.

Lasko, K., Vadrevu, K.P. 2018. Improved rice residue burning emissions estimates: Accounting for practice-specific emission factors in air pollution assessments of Vietnam. *Environ. Pollut.* 236 (5), 795–806.

Lasko, K., Vadrevu, K.P., Nguyen, T.T.N. 2018a. Analysis of air pollution over Hanoi, Vietnam using multi-satellite and MERRA reanalysis datasets. *PLoS One.* 13(5), e0196629.

Lasko, K., Vadrevu, K.P., Tran, V.T., Justice, C. 2018b. Mapping double and single crop Paddy rice with Sentinel-1A at varying spatial scales and polarizations in Hanoi, Vietnam. *IEEE J. Sel. Top. Appl. Earth Obs. Remote Sens.* 11(2), 498–512.

McMeeking, G.R. 2008. The optical, chemical, and physical properties of aerosols and gases emitted by the laboratory combustion of wildland fuels, Ph.D. Dissertation, Department of Atmospheric Sciences, Colorado State University, 109–113, Fall 2008.

Page, S.E., Siegert, F., Rieley, J.O., Boehm, H.-D.V., Jaya, A., Limin, S. 2002. The amount of carbon released from peat and forest fires in Indonesia in 1997. *Nature* 420, 61–65.

Prasad, V.K., Gupta, P.K., Sharma, C., Sarkar, A.K., Kant, Y., Badarinath, K.V.S., Rajagopal, T., Mitra, A.P. 2000. NO_x emissions from biomass burning of shifting cultivation areas from tropical deciduous forests of India–estimates from ground-based measurements. *Atmos. Environ.* 34(20), 3271–3280.

Prasad, V.K., Kant, Y., Badarinath, K.V.S. 2001. CENTURY ecosystem model application for quantifying vegetation dynamics in shifting cultivation areas: A case study from Rampa Forests, Eastern Ghats (India). *Ecol. Res.* 16(3), 497–507.

Prasad, V.K., Kant, Y., Gupta, P.K., Elvidge, C., Badarinath, K.V.S. 2002. Biomass burning and related trace gas emissions from tropical dry deciduous forests of India: A study using DMSP-OLS data and ground-based measurements. *Int. J. Remote Sens.* 23(14), 2837–2851.

Prasad, V.K., Lata, M., Badarinath, K.V.S. 2003. Trace gas emissions from biomass burning from northeast region in India—estimates from satellite remote sensing data and GIS. *Environmentalist.* 23(3), 229–236.

Randerson, J.T., Chen, Y., van der Werf, G.R., Rogers, B.M. Morton, D.C. 2012. Global burned area and biomass burning emissions from small fires. *J. Geophys. Res.* 117, G04012.

Saatchi, S.S., Harris, N.L., Brown, S., Lefsky, M., Mitchard, E.T.A., Salas, W., Zutta, B.R., Buermann, W., Lewis, S.L., Hagen, S., Petrova, S., White, L., Silman, M., Morel, A. 2011. Benchmark map of forest carbon stocks in tropical regions across three continents. *Proc. Natl. Acad. Sci. U. S. A.* 108, 9899–9904.

Saharjo, B.H., and Yungan, A. 2018. Forest and land fires in Riau province; A Case study in fire prevention, policy implementation with local concession holders. In: Vadrevu, K.P., Ohara, T., and Justice, C. (Eds). *Land-Atmospheric Research Applications in South/Southeast Asia.* Springer, Cham, pp. 143–170.

Shi, Y., Matsunaga, T. 2017. Temporal comparison of global inventories of CO_2 emissions from biomass burning during 2002–2011 derived from remotely sensed data. *Environ. Sci. Pollut. Res.* 24, 16905–16916.

Shi, Y., Matsunaga, T., Saito, M., Yamaguchi, Y., Chen, X. 2015a. Comparison of global inventories of CO_2 emissions from biomass burning during 2002–2011 derived from multiple satellite products. *Environ. Pollut.* 206, 479–487.

Shi, Y., Matsunaga, T., Yamaguchi, Y. 2015b. High-resolution mapping of biomass burning emissions in three tropical regions. *Environ. Sci. Technol.* 49, 10806–10814.

Shi, Y., Sasai, T., Yamaguchi, Y. 2014. Spatio-temporal evaluation of carbon emissions from biomass burning in Southeast Asia during the period 2001–2010. *Ecol. Model.* 272, 98–115.

Shi, Y., Yamaguchi, Y. 2014. A high-resolution and multi-year emissions inventory for biomass burning in Southeast Asia during 2001–2010. *Atmos. Environ.* 98, 8–16.

Shi, Y., Zhao, A., Matsunaga, T., Yamaguchi, Y., Zang, S., Li, Z., Yu, T., Gu, X. 2019. High-resolution inventory of mercury emissions from biomass burning in tropical continents during 2001–2017. *Sci. Total Environ.* 653, 638–648.

Vadrevu, K.P. 2008. Analysis of fire events and controlling factors in eastern India using spatial scan and multivariate statistics. *Geogr Ann A, Phys. Geogr.* 90(4), 315–328.

Vadrevu, K.P., Badarinath, K.V.S. and Anuradha, E. 2008. Spatial patterns in vegetation fires in the Indian region. *Environ. Monit. Assess.*, 147(1–3), 1. doi: 10.1007/s10661-007-0092-6

Vadrevu, K.P. and Justice, C.O. 2011. Vegetation fires in the Asian region: Satellite observational needs and priorities. *Glob Environ Res.* 15(1), 65–76.

Vadrevu, K., Ohara, T., Justice, C. 2017. Land cover, land use changes and air pollution in Asia: A synthesis. *Environ. Res. Lett.* 12, 120201.

Vadrevu, K., Giglio, L., Justice, C. 2013. Satellite based analysis of fire–carbon monoxide relationships from forest and agricultural residue burning (2003–2011). *Atmos. Environ.* 64, 179–191.

Vadrevu, K.P., Csiszar, I., Ellicott, E., Giglio, L., Badarinath, K.V.S., Vermote, E., Justice, C. 2012. Hotspot analysis of vegetation fires and intensity in the Indian region. *IEEE J. Sel. Top. Appl. Earth Obs. Remote Sens.* 6(1), 224–238.

Vadrevu, K.P., Eaturu, A. and Badarinath, K.V.S. 2006. Spatial distribution of forest fires and controlling factors in Andhra Pradesh, India using spot satellite datasets. *Environmental Monitoring and Assessment.* 123(1–3), 75–96.

Vadrevu, K.P., Giglio, L., Justice, C. 2013. Satellite based analysis of fire–carbon monoxide relationships from forest and agricultural residue burning (2003–2011). *Atmos. Environ.* 64, 179–191.

Vadrevu K.P., Ohara T, Justice C. 2014a. Air pollution in Asia. *Environ. Pollut.* 12, 233–235.

Vadrevu, K.P., Lasko, K., Giglio, L., Justice, C. 2014b. Analysis of Southeast Asian pollution episode during June 2013 using satellite remote sensing datasets. *Environ. Pollut.* 12, 245–256.

Vadrevu, K.P., Lasko, K., Giglio, L., Justice, C. 2015. Vegetation fires, absorbing aerosols and smoke plume characteristics in diverse biomass burning regions of Asia. *Environ. Res. Lett.* 10(10), 105003.

Vadrevu, K.P., Lasko, K.P. 2015. Fire regimes and potential bioenergy loss from agricultural lands in the Indo-Gangetic Plains. *J. Environ. Manag.* 148:10–20.

Vadrevu, K.P., Ohara, T., Justice, C. 2017. Land cover, land use changes and air pollution In Asia: A synthesis. *Environ. Res. Lett.* 12(12), 120201.

Vadrevu, K.P., Ohara, T., Justice, C. eds., 2018. *Land-Atmospheric Research Applications in South and Southeast Asia.* Springer, Cham.

Vadrevu, K.P., Lasko, K. 2018. Intercomparison of MODIS AQUA and VIIRS I-Band fires and emissions in an agricultural landscape—Implications for air pollution research. *Remote Sens.* 10(7), 978. doi:10.3390/rs10070978.

Vadrevu, K.P., Lasko, K., Giglio, L., Schroeder, W., Biswas, S., Justice, C. 2019. Trends in vegetation fires in south and southeast Asian countries. *Sci. Rep.* 9(1), 7422. doi:10.1038/s41598-019-43940-x.

van der Werf, G.R., Randerson, J.T., Giglio, L., Collatz, G.J., Mu, M., Kasibhatla, P.S., Morton, D.C., DeFries, R.S., Jin, Y., van Leeuwen, T.T. 2010. Global fire emissions and the contribution of deforestation, savanna, forest, agricultural, and peat fires (1997–2009). *Atmos. Chem. Phys.* 10, 11707–11735.

van der Werf, G.R., Randerson, J.T., Giglio, L., van Leeuwen, T.T., Chen, Y., Rogers, B.M., Mu, M., van Marle, M.J.E., Morton, D.C., Collatz, G.J., Yokelson, R.J., Kasibhatla, P.S. 2017. Global fire emissions estimates during 1997–2016. *Earth Syst. Sci. Data* 9, 697–720.

Wiedinmyer, C., Akagi, S.K., Yokelson, R.J., Emmons, L.K., Al-Saadi, J.A., Orlando, J.J., Soja, A.J. 2011. The Fire INventory from NCAR (FINN): A high resolution global model to estimate the emissions from open burning. *Geosci. Model Dev.* 4, 625–641.

Zhang, X., Kondragunta, S., Schmidt, C., Kogan, F. 2008. Near real time monitoring of biomass burning particulate emissions ($PM_{2.5}$) across contiguous United States using multiple satellite instruments. *Atmos. Environ.* 42, 6959–6972.

6 Black Carbon Emissions from Biomass Burning in Southeast Asia – A Review

Max Gerrit Adam and Rajasekhar Balasubramanian
National University of Singapore, Singapore

CONTENTS

INTRODUCTION

Aerosols (also known as particulate matter (PM)) in the atmosphere can impact the climate through absorption and scattering of radiation. While some components of PM, such as organic carbon (OC) and sulfates, predominantly scatter radiation leading to a cooling effect of the atmosphere, others such as black carbon (BC) tend to warm the atmosphere (Bond et al. 2013, Menon et al. 2002). Thus, the proportion of BC to OC and sulfates determines whether a region experiences a net warming or cooling effect. The radiative impacts of BC extend beyond direct absorption of radiation by influencing cloud cover (semi-direct impact) and reflectivity of snow and ice surfaces (indirect), which in turn can change precipitation and melt rates (Hadley and Kirchstetter 2012, Koch and Del Genio 2010). However, unlike greenhouse gases such as carbon dioxide (CO_2), which is the most important climate forcer with a long lifetime of several years, BC has a short lifetime of days to weeks and hence is known as a short-lived climate forcer (SLCF) (UNEP 2011). As a result, BC concentrations are highly inhomogeneous and tend to vary according to regions and the magnitude of emission sources. The other concern about BC is that its inhalation is associated with cardiovascular and respiratory diseases (WHO 2012). Due to their small size

(<100 nm), BC particles and co-emitted harmful species can penetrate deeply into the lungs leading to higher morbidity and mortality (Janssen et al. 2011, WHO 2012). Given that SEA is densely populated, many people are at increased risk of adverse health effects from emissions of BC (Shindell et al. 2012, Permadi et al. 2018).

The primary sources of BC include the incomplete combustion of fossil fuel (FF) and biofuels and biomass burning (BB) (Bond et al. 2013). However, uncertainties concerning the extent of BC emissions from its main sources, including FF combustion and BB, and the resulting climate impacts remain (IPCC 2013). In Southeast Asia (SEA), BB through vegetation fires, Crop Residue Open Burning (CROB), and peat fires leads to transboundary air pollution (TAP) and contributes significantly to the local and global BC budget, but is poorly quantified (Streets et al. 2003, Chen and Taylor 2018, Bond et al. 2013; Vadrevu et al. 2019). As a result, uncertainties exist as to how much BC is emitted from BB in SEA. In recent years, an increasing number of (ground-based) observational, remote sensing, and model simulation studies have focused on understanding the role of aerosols and BC in the context of Southeast Asian BB emissions (Reid et al. 2013, Permadi et al. 2018).

This chapter provides an overview of BC's emissions and characteristics from BB in SEA by integrating recent research findings from the scientific literature. We examine the emission of BC from BB in SEA and its impacts on climate and human health. We also point out critical knowledge gaps and provide recommendations for further studies, including BC mitigation measures. Reducing BC emissions from BB can lead to co-benefits in terms of improved human health and ecosystem services.

BC EMISSION FROM BB IN SEA

BC is emitted from various BB fires, including trees, cultivated plants, bushes, grass, peat, and lignite coal (van Leeuwen et al. 2014). Forest fires in SEA generally take place in regions of peat swamps (especially Indonesia) and underground lignite coal deposits, often during the dry season when the conditions for fires (deliberate or accidental) are conducive (Liu et al. 2014; Hayasaka et al. 2014). The sequence of a forest fire is characterized by the following four stages: ignition, flaming, smoldering, and extinction. When heat, oxygen, and the fuel source combine and have a chemical reaction, a fire will ensue. The central part of the emissions occurs during flaming (characterized by high temperatures up to 1800 K) and smoldering (low temperatures). During the flaming stage, more BC is emitted, while during the smoldering stage, more incompletely oxidized products (e.g., CO, CH_4, OC such as polycyclic aromatic hydrocarbons (PAHs)) are emitted. Factors that influence the BC emission from BB are the type of vegetation, its water content, density and structure, weather conditions (lightning can serve as ignition, while heavy rains can extinguish a fire), and topography (Cochrane 2010, Radojevic 2003). BC particles are emitted from BB in the fine particle range (PM with less than 2.5 µm in diameter) (Schwarz et al. 2008, Bond et al. 2013). Compared to BC emissions from FF combustion, the BC from BB is generally accompanied by higher emissions of OC. A typical OC/BC ratio in urban traffic environments is two, while the OC/BC ratio can be five or higher in regions with more prevalent BB emissions and where smoldering dominates (Novakov et al. 2005).

To assess BB emissions' and various impacts, it is crucial to understand the specific sources from which BC can be emitted. In this regard, emission inventories (EIs) can quantify local emissions of BC based upon the characteristics of sources for a particular location over a specific time period (Kanabkaew and Oanh 2011). To obtain accurate EIs, one needs to determine all possible emission sources and the corresponding activity data, such as production. The information on source activity is gathered from statistical records made available by the government or satellite-based remote sensing. This is often achieved by monitoring fires, including land cover/land use from sensors on board satellites (Vadrevu and Justice 2011, Justice et al. 2003; 2015). To help identify possible pollution source locations, researchers also frequently use air (mass) trajectory information. With the help of meteorological data, these trajectories give information about an air parcel's origin, which can be traced back in time to its source by up to several days. The two most commonly used trajectory models are Hybrid Single-Particle Lagrangian Integrated Trajectory model (HYSPLIT) (Stein et al. 2015) and FLEXible PARTicle dispersion model (FLEXPART) (Stohl et al. 2005), both of which are free and computationally easy to use. In the following sections, the sources of BC from BB are described further, while emission factors for BC from BB sources are shown in Table 6.1.

CROP RESIDUE OPEN BIOMASS BURNING

In many countries in SEA, the crops (e.g., rice, maize, sugarcane) after harvesting are burned, which act as a source of BC. CROB is practiced to control pests and weeds, improve soil fertility through ash, and prepare the fields for the next cycle of harvesting of which, for example, there are 2–3 cycles per year for paddy (rice). CROB emissions result from the low-temperature combustion of biomass. Particularly in agriculture-based economies such as Thailand and, to some extent, Vietnam, the CROB emissions contribute significantly to BC emissions (Oanh et al. 2018, Kanokkanjana, Cheewaphongphan, and Garivait 2011).

A comprehensive EI study by Oanh et al. (2018) reported that total Southeast Asian BC emissions from CROB were dominated by those from Indonesia and

---

TABLE 6.1

Emission Factors of BC for Biomass Burning (in g/kg)

Tropical Forest	Peatland Fires	CROB	Savanna	Extratropical Forest	References
0.52	0.20[a]	0.75	0.37	0.56	Akagi et al. (2011)
0.66 ± 0.31	–	0.69 ± 0.13	0.48 ± 0.18	0.56 ± 0.19	Andreae and Merlet (2001)
0.56–0.61[b]	–	0.69	0.48	–	Bond et al. (2004)

[a] EF includes an assumed tropical forest overstory.

[b] Total forest emission factor.

Vietnam and showed an increasing trend in annual emissions for the study period 2010–2015. Emissions of BC from CROB varied daily, with higher values observed during the dry season harvesting time. In Indonesia, BC emissions occur during August–October, but during February–March in northern and central Vietnam and the May–June period in southern Vietnam (Oanh et al. 2018). A recent study focusing on air pollutant emissions from CROB reported that only a small fraction of the total PM (2 kilotons out of 4 megatons) could be attributed to BC (Junpen et al. 2018).

FOREST FIRES

Due to agricultural management considerations such as planting palm oil trees, major land areas with peat swamp forests are cleared. Often, forest fires are at the surface level since the fuel is composed of biomass present on the forest's ground surface, including leaf litter, twig, grass, undergrowth, shrub, climber, and seedling (Radojevic 2003). Also, ineffective implementation of legislation to control fire practices and poor fire management, particularly during fire-prone periods, exacerbate the challenges to curb the occurrences of fires. Kalimantan in Indonesia is the region most affected by the aforementioned issues, and fires in this part have strongly affected air quality in neighboring countries (Jayarathne et al. 2018, Stockwell et al. 2016; Vadrevu et al. 2018). In a study from Thailand focusing on BC emissions from dry dipterocarp forest fires (DDF) in Thailand, the authors found a high fraction of BC (8.25%) in the total mass of $PM_{2.5}$ and calculated a global warming potential of BC from DDF which equaled that of CO_2 over a 20-year timescale (Chaiyo and Garivait 2014).

PEAT FIRES

While the bulk of global peatland area is distributed in northern latitudes, tropical peatlands account for ~11% of the global share and an estimated mean peat thickness varying from 1.3 to 11 m (Page et al. 2011). Peat is a carbon-rich soil composed of decomposed organic matter which has accumulated in water-rich, anaerobic, and nutrient-deficient conditions over timescales from years to as long as hundreds of years (Turetsky et al. 2011). Environmental changes, including drainage, forest clearing, and increasing drought periods, threaten to destabilize the low-lying peat areas in SEA. The characteristics of smoldering are the slow and low-temperature burning of biomass, which leads to long-lasting fires (weeks, months up to years) underground over extensive regions as they are difficult to extinguish even during strong periods of precipitation (Rein 2013). The effects of peat smoke on climate are incompletely understood thus far. As a consequence, there are little data available on emission factors for BC for tropical peat fires. Recently, Jayarathne et al. (2018) calculated BC emissions from $PM_{2.5}$ collected *in situ* during the 2015 El Niño peat fire episode in Central Kalimantan, Indonesia, with only minor emissions of BC. Another study by Chakrabarty et al. (2016) showed that aerosols emitted from peatland fires in Alaska and Siberia are predominantly brown carbon (BrC, a class of OC which is optically active), also with only small amounts of BC.

EFFECTS OF BC ON RADIATIVE FORCING (RF) AND CLIMATE

The most commonly used metric in climate science for assessing BB's impact on Earth's climate is RF, which pertains to a net change in the energy balance of the Earth system, expressed in W/m^2. BC strongly absorbs solar radiation due to its black color, thereby warming the atmosphere (Kirchstetter et al. 2017). Estimates of the RF for BC vary from 0.2 to 1.1 W/m^2 or up to 55% of that of CO_2, making it the second most potent atmospheric warming agent (Ramanathan and Carmichael 2008, IPCC 2013). BC can impact Earth's radiative balance in a variety of ways, and the emissions of BC are inevitably linked with degradation in air quality, which in turn affects climate (Myhre et al. 2013). The impact of BC on climate can occur through the following: (1) Changes in temperature, rainfall, humidity, wind patterns, and radiation can affect the chemistry and transport of air pollutants; and (2) vice versa; i.e., air pollutants can directly or indirectly exert RF which subsequently impacts temperatures, circulation patterns, and the hydrological cycle (Myhre et al. 2013). The most important climate impact of BC is the absorption of solar radiation leading to a warming effect of the atmosphere (direct effect). Furthermore, BC impacts cloud cover and thus reflectivity by evaporation of clouds when suspended near them, leading to a reduction of cloud cover and lifetime (semi-direct effect). BC can promote cloud formation when it is coated with hygroscopic materials (organic compounds); i.e., "aged" BC can serve as cloud condensation nuclei, which can ultimately change the reflectivity of the cloud (indirect effect) (Koch and Del Genio 2010). Also, when BC is deposited on snow and ice surfaces, it changes the albedo and may lead to a faster rate of melting (Flanner et al. 2007). Wang et al. (2015), based on ice-core analysis from a Tibetan glacier and with the help of air mass trajectories, showed an increasing BC contribution in recent years from BB in SEA on the snow surfaces. Similarly, from measurements of BC coupled with air trajectory analysis, Zhang et al. (2018) have shown that BC from BB in SEA is deposited on glaciers in the Himalayas, which may lead to a faster rate of melting. Certain light-absorbing organic compounds, known as BrC, have been found in BB-influenced emissions and can enhance BC's radiative effects (Laskin et al. 2015). Furthermore, recent research has shown that the warming effect can also be enhanced by coating the BC particle surface with OC, which focuses photons on the BC core, also known as the lensing effect (Lack and Cappa 2010). This is relevant since research by Schwarz et al. (2008) has shown that BC from BB sources tends to be larger in size and thicker coating than BC particles generated in urban environments from incomplete combustion of FF. Balasubramanian and Wang (2006) showed that the fraction of BC in PM samples taken on hazy days is lower during non-haze days, implying that BC's RF is potentially less on hazy days. However, due to the aforementioned research, organic compounds' presence can alter the radiative properties, introducing further uncertainty as to the magnitude of the RF of BC from BB.

The BC measurements are often carried out by light absorption measurements whereby the change in the PM sample absorption collected on a filter is proportional to the amount of BC deposited in the filter. This method is frequently used for reliable BC measurements in ambient air (Hansen et al. 1984). Dedicated field campaigns such as the Atmospheric Brown Cloud (ABC) (Ramanathan et al. 2007)

TABLE 6.2

Ambient Concentrations of BC Reported in Various Regions of SEA

Region	Site	Method	BC ($\mu g/m^3$)	References
Singapore	Urban	Filter	2.76	See, Balasubramanian, and Wang (2006)
Sonla, Vietnam	Urban	Filter	3.3–3.5	Lee et al. (2015)
Bandung, Indonesia	Urban	Smoke Stain Reflectometer	3.05	Santoso, Dwiana Lestiani, and Hopke (2013)
Jakarta, Indonesia	Urban	Smoke Stain Reflectometer	3.37	Santoso, Dwiana Lestiani, and Hopke (2013)
Bangkok, Thailand	Urban/suburban	Filter	5.43–8.64	Wimolwattanapun, Hopke, and Pongkiatkul (2011)
Kuala Lumpur, Malaysia	Urban	Reflectometer	3.85	Salako et al. (2016)
Makassar, Indonesia	Urban	Smoke Stain Reflectometer	2.01	Rashid et al. (2014)
Chiang Mai, Thailand	Urban	Filter	2.85	Thepnuan et al. (2019)
Riau, Indonesia	Background	Filter	3.23	Fujii et al. (2014)
Petaling Jaya, Malaysia	Urban	Filter	3.1	Fujii et al. (2015)
Manila, Philippines	Urban	MAAP	7.0	Alas et al. (2018)

and 7-SEAS (Reid et al. 2013) reported the significance of BC emissions from BB. Table 6.2 provides a summary of BC measurements reported in the literature for SEA. However, many of these measurements did not provide an extensive database to understand the extent of BB's influence on atmospheric BC levels. This is because BC is not included as a criterion air pollutant in national ambient air quality standards, and hence, there is no dedicated network of long-term measurements of BC in SEA (Chen and Taylor 2018). Reid et al. (2013) highlighted a lack of observational data for aerosols in SEA.

Nonetheless, model simulation studies provide another way to study the effects of BC on the climate. They simulate the physical and chemical processes occurring in Earth's atmosphere, including aerosols, in a mathematical framework. Model simulations can be applied to study past, present, and future scenarios of BB in SEA. Comparison with measurement data from ground-based observations, remote sensing, or satellite data tests the model's performance. With advances in computational resources and a more detailed understanding of the concentrations' underlying processes, model simulations have found a wider application in scientific studies. By integrating over several time steps in a mathematical model framework, the meteorological and chemical evolution of the climate-relevant parameters is simulated. The following differential equation illustrates how an air pollutant concentration is computed in a computer model (Chen and Taylor 2018):

$$\frac{\delta q}{\delta t} = -\Delta \cdot qV + E - R \qquad (6.1)$$

where q represents the concentration of air pollutants, t represents the time step, V represents the vector of wind direction, E represents the air pollutant emission, and R represents the removal of the air pollutant, which includes deposition and transformation of the pollutant of interest.

An example of a model to simulate BC concentrations is the Weather Research and Forecasting model with a chemical component (WRF-CHIMERE), used for daily forecasts of air pollutants in the atmosphere and long-term simulations with a spatial resolution of a few kilometers (Menut et al. 2013). WRF-CHIMERE simulations by Permadi et al. (2018) investigated the BC direct effects under two emission scenarios for the year 2030, business as usual (BAU2030) with no interventions to BC emissions in the Southeast Asian region and reduced emissions (RED2030), which featured measures to cut down BC emissions in four sectors (open BB, road transport, residential cooking, and industry). The BC direct radiative forcing (DRF) showed an increase in BAU2030 to 2.0 W/m² from 0.98 in the base year 2007. Under RED2030, however, the DRF value increased to just 1.4 W/m². Other studies have simulated BC emissions in Asia as part of global model simulations and found that BC from BB in SEA may contribute significantly to RF and thus warming. However, there are uncertainties in BC's emission (Shindell et al. 2012, Ramanathan and Carmichael 2008). This is, in part, attributable to the lack of systematic EIs that do not account for the full extent of BB emissions. In agreement with observational findings, Xu et al. (2018) modeled the impact of BC from BB emissions in SEA on glacier surface deposits of BC. During the wintertime, the BB signature on BC was negligible; however, during the springtime (April), a marked enhancement in BC from BB was observed in the Tibetan Plateau (Xu et al. 2018).

Estimating BC concentrations in the atmosphere can also be carried out via remote sensing. This is achieved by measuring the receiving electromagnetic radiation from the Earth's surface (passive systems) or determining the changes in an electromagnetic signal sent out and reflected or scattered back. The Aerosol Robotic Network (AERONET) has been used to investigate BC levels in SEA by making use of the aerosol optical depth (AOD), which is a measure of the amount of sunlight absorbed or scattered in a vertical column (Cohen and Wang 2014). Cohen and Wang (2014) used the AERONET data to simulate global BC concentrations. They found that BC emissions, especially in the Southeast Asian region, are underpredicted against emission values published in the literature.

IMPACT OF BC ON HUMAN HEALTH

The emissions of BC from BB have wide-ranging effects on surface air quality, which can, in turn, lead to adverse impacts on public health, such as chronic obstructive pulmonary disease (COPD), lung cancer, and cardiovascular diseases (Reid et al. 2016, Marlier et al. 2013). Globally, it is estimated that 6.5 million deaths were associated with air pollution in 2015 (Landrigan et al. 2018). In recent decades, recurring BB events have posed serious regional air pollution problems in SEA. Uncontrolled forest and peat fires in SEA, particularly in Indonesia, are a major source of aerosols which can be transported by transboundary winds to neighboring countries, most notably Malaysia, Thailand, and Singapore, resulting in smoke haze episodes lasting days up to months (See et al. 2007, Aouizerats et al. 2015, Heil and Goldammer 2001). As a consequence of regional smoke haze, atmospheric visibility is reduced with increased concentrations of particulate-bound toxic pollutants, resulting in elevated health risks in both indoor and outdoor environments (Sharma and Balasubramanian 2017; Betha, Behera, and Balasubramanian 2014, Pavagadhi et al. 2013, Huang et al. 2013, Lee, Bar-Or, and Wang 2017). Furthermore, biofuels' combustion (wood, agricultural waste, and dried manure in cooking stoves) releases significant BC in Asia, which can exacerbate adverse health effects (Venkataraman et al. 2005).

Since SEA has a large population density, several studies have quantified the health risks resulting from exposure to smoke haze particles in SEA (Betha et al. 2014, Marlier et al. 2013, Koplitz et al. 2016). However, these studies have not focused on BC, but rather $PM_{2.5}$, and its toxic (e.g., Al and Mn) and carcinogenic (e.g., Cr, Ni, Cd, and As) components. While BC is not directly carcinogenic, it is implicated in negative health effects as it can act as a carrier of toxic or carcinogenic organic compounds (e.g., PAHs; nitrated PAHs; quinones; hydroquinones) (WHO 2012).

Fine particulate matter, $PM_{2.5}$, is the commonly used metric in health impact assessment (HIA) studies to estimate air pollution's effects on mortality and morbidity. However, since BC is a component of $PM_{2.5}$ and a tracer of combustion particles, it can be argued that BC is a better indicator of harmful health effects from combustion-driven sources when compared to unspeciated $PM_{2.5}$ (Janssen et al. 2011). Both the short-term (daily) and long-term personal exposure to BC can lead to higher health risks. HIA studies, which explore mortality with respect to BC, often use $PM_{2.5}$ as the determinant parameter for health outcomes. The following equation is used to estimate the number of excess deaths (Wang and Mauzerall 2006, Permadi et al. 2018):

$$\Delta_{cases} = I_{ref} \cdot POP \cdot CR \cdot \Delta C \qquad (6.2)$$

where Δ_{cases} is the change in number of cases of mortality per year due to changes in the annual ambient $PM_{2.5}$ concentration; I_{ref} is the baseline mortality rate (%); POP is the population, i.e., the number of exposed persons; CR is the concentration–response coefficient for mortality rate (the unit is % change in mortality and morbidity as a result of a 1 $\mu g/m^3$ change in annual average $PM_{2.5}$ concentration); and ΔC is the change in annual ambient $PM_{2.5}$ concentration under a given emission scenario.

Model simulation studies by Shindell et al. (2012) and Permadi et al. (2018) have investigated the health and climate change impacts of BC emission reductions until 2030. Shindell et al. (2012) found that reducing BC emissions until the year 2030 resulted in a significant number of avoided premature deaths, particularly for Indonesia (89,100 avoided deaths) and Vietnam (58,700). Permadi et al. (2018), based on their reduced emissions scenario for 2030, estimated that 401,000 premature deaths could be avoided relative to the scenario of the base year 2007. However, if BC emissions continued to rise as indicated by current trends, the number of additional premature deaths for the Southeast Asian region would be as much as 201,000. In all emission scenarios, Indonesia and Thailand were the most severely affected countries or could benefit the most from lowered BC emissions. Marlier et al. (2013) combined satellite observations and atmospheric model simulations for the years 1997–2006 to show that enhanced PM concentrations from BB during El Niño occurrences were responsible for an estimated 10,800-person annual increase in cardiovascular mortality. Based on satellite-derived emissions of $PM_{2.5}$ (calculated as $BC + OC$) during the 2006 (September–October) and 2015 (September–October) smoke haze episodes, Koplitz et al. (2016) quantified the public health impacts in terms of the number of excess deaths across Indonesia, Malaysia, and Singapore as 100,300. This number is twice as high as the 2006 smoke haze episode due to larger burned land areas, particularly in Indonesia. More realistic health risk assessment studies are warranted at a network of sampling locations at a ground level based on the comprehensive chemical speciation of $PM_{2.5}$ during BB episodes.

SUMMARY AND OUTLOOK

BC has significant impacts on human well-being by influencing climate and air quality. While uncertainties exist on the global scale for BC's emission sources in general, those linked to BB are of particular concern. SEA represents a significant source region of BC from BB. However, the influence of BB in SEA on BC levels is not well quantified due to the lack of observational data despite its implications on regional and global climate and adverse health impacts. In this context, it is desired to integrate BC measurements through (1) dedicated field campaigns in strategic locations during BB seasons, (2) continuous monitoring of BC at fixed sites in countries within SEA, (3) personal monitoring of human exposure to PM and its toxic chemical components, and (4) model simulations of the long-range transport of BC emitted from BB. The developing economies of the countries within SEA are projected to rapidly grow in the near future with further increases in emissions of airborne pollutants, including BC, which emphasizes the need to quantify BC from BB.

The challenges involved in improving the observational infrastructure in conjunction with BC mitigation measures in Southeast Asian countries are to find the political will to cooperate across countries. Competing interests among the stakeholders involved (e.g., from governments, corporate, civil society) can be counterproductive toward decarbonizing economies and/or mitigating aerosol emissions from BB. The transboundary nature of smoke haze pollution means that governments in the Southeast Asian region must develop a collaborative framework to implement effective measures to reduce BC emissions from BB. The outcome of these BC mitigation measures would lead to co-benefits to improve public health, stabilize regional climate, and sustain ecosystem services (UNEP 2011).

REFERENCES

Akagi, S.K., Yokelson, R.J., Wiedinmyer, C., Alvarado, M.J., Reid, J.S., Karl, T., Crounse, J.D., and Wennberg, P.O. 2011. Emission factors for open and domestic biomass burning for use in atmospheric models. *Atmospheric Chemistry and Physics*. 11 (9), 4039–4072.

Alas, H.D., Müller, T., Birmili, W., Kecorius, S., Cambaliza, M.O., Simpas, J.B.B., Cayetano, M., Weinhold, K., Vallar, E., Galvez, M.C., and Wiedensohler, A. 2018. Spatial characterization of black carbon mass concentration in the atmosphere of a Southeast Asian megacity: An air quality case study for Metro Manila, Philippines. *Aerosol and Air Quality Research.* 18, 2301–2317.

Andreae, M.O., and Merlet, P. 2001. Emission of trace gases and aerosols from biomass burning. *Global Biogeochemical Cycles.* 15 (4), 955–966.

Aouizerats, B., Van Der Werf, G.R., Balasubramanian, R., and Betha, R. 2015. Importance of transboundary transport of biomass burning emissions to regional air quality in Southeast Asia during a high fire event. *Atmospheric Chemistry & Physics.* 15(1), 363–373.

Balasubramanian, R., Wang, W., and Murray, R.W. 2006. Redox ionic liquid phases: Ferrocenated imidazoliums. Journal of the American Chemical Society, 128(31), pp. 9994–9995.

Betha, R., Behera, S.N., and Balasubramanian, R. 2014. 2013 Southeast Asian smoke haze: fractionation of particulate-bound elements and associated health risk. *Environmental Science & Technology.* 48(8), 4327–4335.

Bond, T.C., Doherty, S.J., Fahey, D.W., Forster, P.M., Berntsen, T., DeAngelo, B.J., Flanner, M.G., Ghan, S., Kärcher, B., Koch, D., and Kinne, S. 2013. Bounding the role of black carbon in the climate system: A scientific assessment. *Journal of Geophysical Research: Atmospheres.* 118(11), 5380–5552.

Bond, T.C., Streets, D.G., Yarber, K.F., Nelson, S.M., Woo, J.H., and Klimont, Z. 2004. A technology-based global inventory of black and organic carbon emissions from combustion. *Journal of Geophysical Research: Atmospheres.* 109(D14). doi:10.1029/2003JD003697.

Chaiyo, U., and Garivait, S. 2014. Estimation of black carbon emissions from dry dipterocarp forest fires in Thailand. *Atmosphere.* 5(4), 1002–1019.

Chakrabarty, R.K., Gyawali, M., Yatavelli, R.L., Pandey, A., Watts, A.C., Knue, J., Chen, L.W.A., Pattison, R.R., Tsibart, A., Samburova, V., and Moosmüller, H. 2016. Brown carbon aerosols from burning of boreal peatlands: Microphysical properties, emission factors, and implications for direct radiative forcing. *Atmospheric Chemistry and Physics.* 16(5), 3033–3040.

Chen, Q., and Taylor, D. 2018. Transboundary atmospheric pollution in Southeast Asia: current methods, limitations and future developments. *Critical Reviews in Environmental Science and Technology.* 48(16–18), 997–1029.

Cochrane, M. 2010. *Tropical Fire Ecology: Climate Change, Land Use and Ecosystem Dynamics.* Springer Science & Business Media, Berlin.

Cohen, J.B., and Wang, C. 2014. Estimating global black carbon emissions using a top-down Kalman Filter approach. *Journal of Geophysical Research: Atmospheres.* 119(1), 307–323.

Flanner, M.G., Zender, C.S., Randerson, J.T., and Rasch, P.J. 2007. Present-day climate forcing and response from black carbon in snow. *Journal of Geophysical Research: Atmospheres.* 112(D11). doi:10.1029/2006JD008003.

Fujii, Y., Tohno, S., Amil, N., Latif, M.T., Oda, M., Matsumoto, J., and Mizohata, A. 2015. Annual variations of carbonaceous PM2. 5 in Malaysia: influence by Indonesian peatland fires. *Atmospheric Chemistry and Physics.* 15(23), 13319–13329.

Fujii, Y., Iriana, W., Oda, M., Puriwigati, A., Tohno, S., Lestari, P., Mizohata, A., and Huboyo, H.S. 2014. Characteristics of carbonaceous aerosols emitted from peatland fire in Riau, Sumatra, Indonesia. *Atmospheric Environment.* 87, 164–169.

Hadley, O.L., and Kirchstetter, T.W. 2012. Black-carbon reduction of snow albedo. *Nature Climate Change.* 2(6), 437–440.

Hansen, A D A, Rosen, H., and Novakov, T. 1984. The aethalometer - an instrument for the real-time measurement of optical absorption by aerosol particles. *Science of the Total Environment.* 36, 191–196.

Hayasaka, H., Noguchi, I., Putra, E.I., Yulianti, N., and Vadrevu, K., 2014. Peat-fire-related air pollution in Central Kalimantan, Indonesia. *Environmental Pollution*, 195, 257–266.

Heil, A., and Goldammer, J. 2001. Smoke-haze pollution: A review of the 1997 episode in Southeast Asia. *Regional Environmental Change*, 2 (1):24–37.

Huang, K., Fu, J.S., Hsu, N.C., Gao, Y., Dong, X., Tsay, S.C., and Lam, Y.F. 2013. Impact assessment of biomass burning on air quality in Southeast and East Asia during BASE-ASIA. *Atmospheric Environment*. 78, 291–302.

IPCC. 2013. Climate change 2013: The physical science basis. Intergovernmental panel on climate change, working group I contribution to the IPCC fifth assessment report (AR5). New York.

Janssen, N.A., Hoek, G., Simic-Lawson, M., Fischer, P., Van Bree, L., Ten Brink, H., Keuken, M., Atkinson, R.W., Anderson, H.R., Brunekreef, B., and Cassee, F.R. 2011. Black carbon as an additional indicator of the adverse health effects of airborne particles compared with PM10 and PM2. 5. *Environmental Health Perspectives*. 119(12), 691–1699.

Jayarathne, T., Stockwell, C.E., Gilbert, A.A., Daugherty, K., Cochrane, M.A., Ryan, K.C., Putra, E.I., Saharjo, B.H., Nurhayati, A.D., Albar, I., and Yokelson, R.J. 2018. Chemical characterization of fine particulate matter emitted by peat fires in Central Kalimantan, Indonesia, during the 2015 El Niño. *Atmospheric Chemistry and Physics*. 18(4), 2585.

Junpen, A., Pansuk, J., Kamnoet, O., Cheewaphongphan, P., and Garivait, S. 2018. Emission of air pollutants from rice residue open burning in Thailand, 2018. *Atmosphere*. 9(11), 449.

Justice, C.O., Smith, R., Gill, M., Csiszar, I., Justice, C.O., Smith, R., Gill, M., and Csiszar, I. 2003. Satellite-based fire monitoring: Current capabilities and future directions. *International Journal of Wildland Fire*. 12(4), 247–258.

Justice, C., Gutman, G., and Vadrevu, K.P. 2015. NASA land cover and land use change (LCLUC): An interdisciplinary research program. *Journal of Environmental Management*. 148(15), 4–9.

Kanabkaew, T., and Oanh, N.T.K., 2011. Development of spatial and temporal emission inventory for crop residue field burning. Environmental Modeling & Assessment. 16(5), 453–464.

Kanokkanjana, K., Cheewaphongphan, P., and Garivait, S., 2011. Black carbon emission from paddy field open burning in Thailand. *IPCBEE Proc*. 6, 88–92.

Kirchstetter, T.W., Preble, C.V., Hadley, O.L., Bond, T.C., and Apte, J.S. 2017. Large reductions in urban black carbon concentrations in the United States between 1965 and 2000. *Atmospheric Environment*. 151, 17–23.

Koch, D., and Del Genio, A.D., 2010. Black carbon semi-direct effects on cloud cover: review and synthesis. *Atmospheric Chemistry and Physics*. 10(16), 7685–7696.

Koplitz, S.N., Mickley, L.J., Marlier, M.E., Buonocore, J.J., Kim, P.S., Liu, T., Sulprizio, M.P., DeFries, R.S., Jacob, D.J., Schwartz, J., and Pongsiri, M. 2016. Public health impacts of the severe haze in Equatorial Asia in September–October 2015: Demonstration of a new framework for informing fire management strategies to reduce downwind smoke exposure. *Environmental Research Letters*. 11(9), 094023.

Lack, D.A., and Cappa, C.D. 2010. Impact of brown and clear carbon on light absorption enhancement, single scatter albedo and absorption wavelength dependence of black carbon. *Atmospheric Chemistry and Physics*. 10(9), 4207–4220.

Landrigan, P.J., Fuller, R., Acosta, N.J., Adeyi, O., Arnold, R., Baldé, A.B., Bertollini, R., Bose-O'Reilly, S., Boufford, J.I., Breysse, P.N., and Chiles, T. 2018. The Lancet Commission on pollution and health. *The Lancet*. 391(10119), 462–512.

Laskin, A., Laskin, J., and Nizkorodov, S.A. 2015. Chemistry of atmospheric brown carbon. *Chemical Reviews*. 115(10), 4335–4382.

Lee, C.T., Ram, S.S., Nguyen, D.L., Chou, C.C., Chang, S.Y., Lin, N.H., Chang, S.C., Hsiao, T.C., Sheu, G.R., Ou-Yang, C.F., and Chi, K.H. 2015. Aerosol chemical profile of near-source biomass burning smoke in Sonla, Vietnam during 7-SEAS campaigns in 2012 and 2013. *Aerosol and Air Quality Research*. 16(11), 2603–2617.

Lee, H.H., Bar Or, R.Z., and Wang, C. 2017. Biomass burning aerosols and the low-visibility events in Southeast Asia. *Atmospheric Chemistry and Physics.* 17(2), 965–980.

Liu, Y., Goodrick, S., and Heilman, W. 2014. Wildland fire emissions, carbon, and climate: Wildfire–climate interactions. *Forest Ecology and Management.* 317, 80–96.

Marlier, M.E., DeFries, R.S., Voulgarakis, A., Kinney, P.L., Randerson, J.T., Shindell, D.T., Chen, Y., and Faluvegi, G. 2013. El Niño and health risks from landscape fire emissions in southeast Asia. *Nature Climate Change.* 3(2), 131–136.

Menon, S., Hansen, J., Nazarenko, L., and Luo, Y. 2002. Climate effects of black carbon aerosols in China and India. *Science.* 297(5590), 2250–2253.

Menut, L., Bessagnet, B., Khvorostyanov, D., Beekmann, M., Blond, N., Colette, A., Coll, I., Curci, G., Foret, G., Hodzic, A., and Mailler, S., 2013. CHIMERE 2013: A model for regional atmospheric composition modelling. *Geoscientific Model Development.* 6(4), 981–1028.

Myhre, G., Samset, B.H., Schulz, M., Balkanski, Y., Bauer, S., Berntsen, T.K., Bian, H., Bellouin, N., Chin, M., Diehl, T. and Easter, R.C., 2013. Radiative forcing of the direct aerosol effect from AeroCom Phase II simulations. Atmospheric Chemistry and Physics, 13(4), 1853–1877.

Novakov, T., Menon, S., Kirchstetter, T.W., Koch, D., and Hansen, J.E. 2005. Aerosol organic carbon to black carbon ratios: Analysis of published data and implications for climate forcing. *Journal of Geophysical Research: Atmospheres.* 110(D21). doi:10.1029/2005JD005977.

Oanh, N.T.K., Permadi, D.A., Hopke, P.K., Smith, K.R., Dong, N.P., and Dang, A.N., 2018. Annual emissions of air toxics emitted from crop residue open burning in Southeast Asia over the period of 2010–2015. *Atmospheric Environment.* 187, 163–173.

Page, S.E., Rieley, J.O., and Banks, C.J. 2011. Global and regional importance of the tropical peatland carbon pool. *Global Change Biology.* 17(2), 798–818.

Pavagadhi, S., Betha, R., Venkatesan, S., Balasubramanian, R., and Hande, M.P. 2013. Physicochemical and toxicological characteristics of urban aerosols during a recent Indonesian biomass burning episode. *Environmental Science and Pollution Research.* 20(4), 2569–2578.

Permadi, D.A., Oanh, N.T.K., and Vautard, R. 2018. Assessment of emission scenarios for 2030 and impacts of black carbon emission reduction measures on air quality and radiative forcing in Southeast Asia. *Atmospheric Chemistry and Physics*, 18(5), 3321–3334.

Radojevic, M. 2003. Chemistry of forest fires and regional haze with emphasis on Southeast Asia. *Pure and Applied Geophysics.* 160(1–2), 157–187.

Ramanathan, V., Li, F., Ramana, M.V., Praveen, P.S., Kim, D., Corrigan, C.E., Nguyen, H., Stone, E.A., Schauer, J.J., Carmichael, G.R., and Adhikary, B. 2007. Atmospheric brown clouds: Hemispherical and regional variations in long-range transport, absorption, and radiative forcing. *Journal of Geophysical Research: Atmospheres.* 112(D22). doi:10.1029/2006JD008124.

Ramanathan, V., and Carmichael, G. 2008. Global and regional climate changes due to black carbon. *Nature Geoscience.* 1(4), 221.

Rashid, M., Yunus, S., Mat, R., Baharun, S., and Lestari, P., 2014. PM_{10} black carbon and ionic species concentration of urban atmosphere in Makassar of South Sulawesi Province, Indonesia. *Atmospheric Pollution Research.* 5(4), 610–615.

Reid, C.E., Brauer, M., Johnston, F.H., Jerrett, M., Balmes, J.R., and Elliott, C.T. 2016. Critical review of health impacts of wildfire smoke exposure. *Environmental Health Perspectives.* 124(9), 1334–1343.

Reid, J.S., Hyer, E.J., Johnson, R.S., Holben, B.N., Yokelson, R.J., Zhang, J., Campbell, J.R., Christopher, S.A., Di Girolamo, L., Giglio, L., and Holz, R.E. 2013. Observing and understanding the Southeast Asian aerosol system by remote sensing: An initial review and analysis for the Seven Southeast Asian Studies (7SEAS) program. *Atmospheric Research.* 122, 403–468.

Rein, G. 2013. Smouldering fires and natural fuels. In *Fire Phenomena in the Earth System – An Interdisciplinary Approach to Fire Science.* Edited by C.M. Belcher, pp. 15–34. John Wiley & Sons, Hoboken, NJ.

Salako, G.O., Hopke, P.K., Cohen, D.D., Begum, B.A., Biswas, S.K., Pandit, G.G., Chung, Y.S., Abd Rahman, S., Hamzah, M.S., Davy, P., and Markwitz, A. 2016. Exploring the variation between EC and BC in a variety of locations. *Aerosol and Air Quality Research.* 12(1), 1–7.

Santoso, M., Dwiana Lestiani, D., and Hopke, P.K., 2013. Atmospheric black carbon in PM2. 5 in Indonesian cities. *Journal of the Air & Waste Management Association.* 63(9) 1022–1025.

Schwarz, J.P., Gao, R.S., Spackman, J.R., Watts, L.A., Thomson, D.S., Fahey, D.W., Ryerson, T.B., Peischl, J., Holloway, J.S., Trainer, M., and Frost, G.J. 2008. Measurement of the mixing state, mass, and optical size of individual black carbon particles in urban and biomass burning emissions. *Geophysical Research Letters.* 35(13). doi:10.1029/2008GL033968.

See, S.W., Balasubramanian, R., and Wang, W. 2006. A study of the physical, chemical, and optical properties of ambient aerosol particles in Southeast Asia during hazy and nonhazy days. *Journal of Geophysical Research: Atmospheres.* 111(D10). doi: 10.1029/2005JD006180.

See, S.W., Balasubramanian, R., Rianawati, E., Karthikeyan, S., and Streets, D.G. 2007. Characterization and source apportionment of particulate matter$\leq$ 2.5 μm in Sumatra, Indonesia, during a recent peat fire episode. *Environmental Science & Technology.* 41(-10) 3488–3494.

Sharma, R., and Balasubramanian, R. 2017. Indoor human exposure to size-fractionated aerosols during the 2015 Southeast Asian smoke haze and assessment of exposure mitigation strategies. *Environmental Research Letters.* 12(11), 114026.

Shindell, D., Kuylenstierna, J.C., Vignati, E., van Dingenen, R., Amann, M., Klimont, Z., Anenberg, S.C., Muller, N., Janssens-Maenhout, G., Raes, F., and Schwartz, J. 2012. Simultaneously mitigating near-term climate change and improving human health and food security. *Science.* 335(6065), 183–189.

Stein, A.F., Draxler, R.R., Rolph, G.D., Stunder, B.J., Cohen, M.D., and Ngan, F. 2015. NOAA's HYSPLIT atmospheric transport and dispersion modeling system. *Bulletin of the American Meteorological Society.* 96(12), 2059–2077.

Stockwell, C.E., Jayarathne, T., Cochrane, M.A., Ryan, K.C., Putra, E.I., Saharjo, B.H., Nurhayati, A.D., Albar, I., Blake, D.R., Simpson, I.J., and Stone, E.A. 2016. Field measurements of trace gases and aerosols emitted by peat fires in Central Kalimantan, Indonesia, during the 2015 El Niño. *Atmospheric Chemistry and Physics.* 16(18), 11711–11732.

Stohl, A., Forster, C., Frank, A., Seibert, P., and Wotawa, G. 2005. The Lagrangian particle dispersion model FLEXPART version 6.2. *Atmospheric Chemistry and Physics.* 5 (9), 2461–2474.

Streets, D.G., Yarber, K.F., Woo, J.H., and Carmichael, G.R. 2003. Biomass burning in Asia: Annual and seasonal estimates and atmospheric emissions. *Global Biogeochemical Cycles.* 17(4), doi:10.1029/2003GB002040.

Thepnuan, D., Chantara, S., Lee, C.T., Lin, N.H., and Tsai, Y.I. 2019. Molecular markers for biomass burning associated with the characterization of $PM_{2.5}$ and component sources during dry season haze episodes in Upper South East Asia. *Science of the Total Environment.* 658, 708–722.

Turetsky, M.R., Donahue, W., and Benscoter, B.W. 2011. Experimental drying intensifies burning and carbon losses in a northern peatland. *Nature Communications.* 2(1)1–5.

UNEP, WMO. 2011. *Integrated Assessment of Black Carbon and Tropospheric Ozone: Summary for Decision Makers.* Nairobi: United Nations Environment Programme.

Vadrevu, K.P., and Justice, C.O. 2011. Vegetation fires in the Asian region: satellite observational needs and priorities. *Global Environmental Research*. 15(1), 65–76.

Vadrevu, K.P., Ohara, T., and Justice, C. 2018. *Land-Atmospheric Research Applications in South and Southeast Asia*. Springer, Cham, 725 pages.

Vadrevu, K.P., Lasko, K., Giglio, L., Schroeder, W., Biswas, S., and Justice, C. 2019. Trends in vegetation fires in south and southeast Asian countries. *Scientific Reports*. 9(1), 7422. doi:10.1038/s41598-019-43940-x.

Van Leeuwen, T.T., Van der Werf, G.R., Hoffmann, A.A., Detmers, R.G., Rücker, G., French, N.H., Archibald, S., Carvalho Jr., J.A., Cook, G.D., De Groot, W.J., and Hély, C. 2014. Biomass burning fuel consumption rates: A field measurement database. *Biogeosciences*. 11(24):7305–7329. doi:10.5194/bg-11-7305-2014.

Venkataraman, C., Habib, G., Eiguren-Fernandez, A., Miguel, A.H., and Friedlander, S.K. 2005. Residential biofuels in South Asia: carbonaceous aerosol emissions and climate impacts. *Science*. 307(5714), 1454–1456.

Wang, M., Xu, B., Cao, J., Tie, X., Wang, H., Zhang, R., Qian, Y., Rasch, P.J., Zhao, S., Wu, G., and Zhao, H. 2015. Carbonaceous aerosols recorded in a southeastern Tibetan glacier: analysis of temporal variations and model estimates of sources and radiative forcing. *Atmospheric Chemistry Physics*. 15(3), 1191–1204.

Wang, X., and Mauzerall, D.L. 2006. Evaluating impacts of air pollution in China on public health: Implications for future air pollution and energy policies. *Atmospheric Environment*. 40(9), 1706–1721.

WHO. 2012. Health effects of black carbon: WHO Regional Office for Europe Copenhagen.

Wimolwattanapun, W., Hopke, P.K., and Pongkiatkul, P. 2011. Source apportionment and potential source locations of PM2.5 and PM2.5–10 at residential sites in metropolitan Bangkok. *Atmospheric Pollution Research*. 2 (2), 172–181.

Xu, R., Tie, X., Li, G., Zhao, S., Cao, J., Feng, T., and Long, X. 2018. Effect of biomass burning on black carbon (BC) in South Asia and Tibetan Plateau: The analysis of WRF-Chem modeling. *Science of the Total Environment*. 645, 901–912.

Zhang, X., Xu, J., Kang, S., Liu, Y., and Zhang, Q. 2018. Chemical characterization of long-range transport biomass burning emissions to the Himalayas: Insights from high-resolution aerosol mass spectrometry. *Atmospheric Chemistry & Physics*, 18(7), 4617–4638.

7 Satellite-Based Estimation of Global CO_2 Emissions from Biomass Burning

Haemi Park and Wataru Takeuchi
The University of Tokyo, Japan

CONTENTS

INTRODUCTION

Wildfires are a major source of environmental disturbance in terrestrial ecosystems. Huge amounts of greenhouse gases (GHGs), which can affect climate change, are released during the wildfire events, which include carbon dioxide (CO_2), carbon monoxide (CO), and methane (CH_4), as well as aerosols (Reddington et al., 2015). CO_2 accounts for the largest proportion of GHGs, including the warming effect (Ward et al., 1996; Goode et al., 2000). Biomass burning is strongly related to CO_2 in terms of loss of vegetation that plays an important role in absorbing CO_2 from the atmosphere (Holtsmark, 2015). In several regions of the world, human activities are known to trigger fires (Mollicone et al., 2006; Field et al., 2009; Jang et al., 2019). Specifically, in Asia, fires are used as a land-clearing tool for conversion of forests to agriculture through slash and burn, planting commercial crops, and clearing agricultural residues and waste (Kant et al., 2000; Prasad et al., 2000, 2001, 2002; Prasad and Badarinath, 2004; Gupta et al., 2001; Vadrevu et al., 2008; Badarinath et al., 2008a, b; Hayasaka et al., 2014; Biswas et al., 2015a, b; Justice et al., 2015; Lasko et al., 2017; 2018a, b; Hayasaka and Sepriando, 2018; Lasko and Vadrevu, 2018; Israr et al., 2018; Saharjo and Yungan, 2018). The start of fires and the duration of fires are crucial factors useful to estimate the amount of CO_2 released (Andreae and Merlet, 2001; Vadrevu and Justice, 2011). For the same, remote sensing data can be effectively used to map and monitor fires, type of vegetation, and thereby CO_2 emissions (Kaufman et al., 1998; Justice et al., 2002; Prasad et al., 2003, 2004, 2005, 2008a, b; Schultz et al., 2008; Vadrevu, 2008; Van der Werf et al., 2010; Wooster et al., 2015). Particularly, polar-orbiting satellites, due to their high resolution and revisit times, can be used for fire mapping and monitoring (Ramachandran, 2018). For example, Moderate Resolution Imaging Spectroradiometer (MODIS) of the National Aeronautics and Space Administration (NASA) is a representative polar-orbiting satellite sensor, and the derived fire product was successfully validated globally (Csiszar et al., 2006, Zhu et al., 2017, Liu et al., 2019, Fusco et al., 2019). The MODIS fire products help in active fire detection due to their visible as well as thermal bands (Kaufman et al., 1998), and several studies in the Asian region demonstrated their utility and other polar-orbiting satellites for fire mapping and monitoring, emission estimation, and air quality studies (Vadrevu et al., 2006, 2012, 2013, 2014a, b, 2015; Vadrevu and Choi, 2011).

The Global Fire Emissions Database (GFED) is a comprehensive fire emission (FEs) product derived from the MODIS fire data (Van der Werf et al., 2010). GFED uses information on the burned area (BA) retrieved from the MODIS data (Giglio et al., 2013, Padilla et al., 2014). The latest GFED product (GFEDv4.1) has a coarse spatial resolution (~25 km), and it was recently updated into two new products: (1) GFED4 without small fires and (2) GFED4s with small fires. These products show increased FE than the previous GFED3.1 (Randerson et al., 2012, Giglio et al., 2013). The GFED uses a revised version of the Carnegie-Ames-Stanford Approach (CASA) biogeochemical model to estimate biomass. The emission factors (EFs) and combustion completeness (CC) values are inferred from the previous studies, including *in situ* measurements (Yokelson et al., 1997; Andreae and Merlet, 2001; Akagi et al., 2011). However, there are still uncertainties on how biomass is estimated (Rossi et al., 2016). Thus, there is a need to explore different approaches for estimating biomass as fuels of a fire. Also, satellite-based combustion ratio such as fire radiative power (FRP) is promising to estimate the dynamics of CO_2 emissions from biomass burning (Wooster et al., 2005; Vadrevu et al., 2012, 2013, 2017, 2018, 2019; Vadrevu and Lasko, 2018). The objectives of this study are (1) to estimate global FE considering dynamic CC by using FRP of MODIS, (2) to characterize seasonal and regional patterns of wildfires, and (3) to discuss uncertainty in GFED4.1s compared to our estimates.

DATASETS AND METHODOLOGY

MODIS ACTIVE FIRES

We used the MODIS (MOD14A1) active fire product as a reference (Giglio et al., 2016). Since its initial release in 2001, the product was revised until collection 6 through adjusting the algorithms (Giglio et al., 2016). Fire events are detected from

MODIS using the brightness temperatures (BTs) of 4 and 11 μm and the difference between BTs (Kaufman et al., 1998). In the product, the daytime detections from Terra and Aqua (MOD/MYD14GD) are captured surface reflectances using thermal thresholds at 10:30 AM and 13:30 local time, respectively. The nighttime detections from two satellites (MOD/MYD14GN) at 10:30 PM and 1:30 AM identify the pixel as a fire. Those data are combined into one daily active fire (MOD14A1, 1 km), which is used in this study.

MODIS MONTHLY BURNED AREA

For burnt areas, we used the MCD45A1 product (Roy et al., 2005). The spectral, temporal, and structural land surface changes from the fire were detected as a fire-affected area in the product, including the date of burning (Figure 7.1) (Roy et al., 2005). Briefly, the Burnt Areas (BA's) and Burn Dates (BD) detection algorithm uses a bidirectional reflectance distribution function (BRDF) to predict daily backward and forward reflectance in each grid. BAs are detected at a 500 m resolution. The burnt date is provided through detecting rapid changes in daily surface reflectance time-series data using the Julian date (Roy et al., 2008).

MODIS MONTHLY NORMALIZED DIFFERENCE VEGETATION INDEX

In this study, we used the MODIS Normalized Difference Vegetation Index (NDVI) product (MOD13A3) (Huete et al., 2002). The NDVI is calculated using the difference between near infrared (NIR) and red (R), representing green vegetation and photosynthesis. Particularly, the monthly product (1 km) was generated by compositing 16-day data (MOD13A2) temporally. In this study, FEs, as well as aboveground

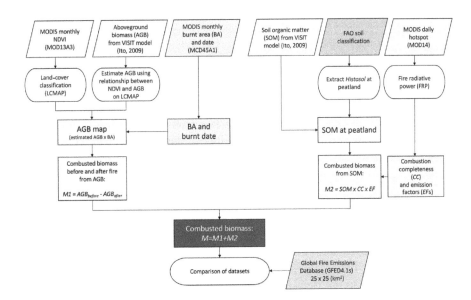

FIGURE 7.1 Flowchart of global carbon dioxide emission estimates from biomass burning using satellite data.

biomass (AGB), was estimated monthly at a 1 km spatial resolution using the monthly 1-km NDVI (MOD13A3).

VISIT Model

The Vegetation Integrative SImulator for Trace gases (VISIT) model is developed for modeling the cycle of GHGs (e.g., CO_2, CH_4, and N_2O). The concepts of carbon (C) assimilation and decomposition rates in terms of CO_2 are almost similar to another simulation model, Sim-CYCLE (Ito and Oikawa, 2002). In Sim-CYCLE, the biomass in terrestrial ecosystems is estimated through an integrated way involving photosynthesis to litter decomposition. The basic data and environmental conditions were obtained from the US National Center for Atmospheric Research (NCEP/NCAR Reanalysis) (Kalnay et al., 1996). Outputs have a daily temporal frequency of 0.5° spatial resolution. Notably, monthly biomass composition was used in this study. For biomass estimation, the dry-matter production theory established by Monsi and Saeki (1953) was used. It recruits the photosynthetic photon flux density at the canopy top (PPFDtop), the biome-specific coefficient (KA), and the cumulative leaf area index (LAI) for estimating gross primary production (GPP). Also, autotrophic plant respiration (AR) and maintenance respiration (ARM) were estimated by a Q10-temperature method (Ryan, 1991). Furthermore, the heterotrophic respiration and the litterfall were estimated using soil moisture data considering microbial activity and using a plant turnover rate, respectively. Hence, this model's biomass denotes total remains after balancing the gain/loss from photosynthesis/respiration (litterfall). More details on the model, including CH_4 and N_2O components, can be found in Ito and Oikawa (2002), Ito (2019), and Inatomi et al. (2010, 2019).

Global Fire Emissions Database

The first version of GFED (GFED2) was released in 2006. In climatic studies, the GFED is used as a representative indicator of comprehensive FE from biomass burning (Shi et al., 2015). The GFED integrates crucial factors involving in situ observations, for example, a BA, fuel loading, CC, and EFs to merge into an FE estimation system (Randerson et al., 2018). There are six kinds of fire types: savanna, boreal forest fires, temperate forest fires, tropical forest fires (deforestation and degradation), peat fires, and agricultural waste burning, and their contribution to the total emission. In GFED, total C emissions are provided; moreover, the dry matter in GFED can be used to partition specific trace gases with EFs in detail (Van der Werf et al., 2017). Global-to-regional scales of EFs make it possible to compare trace gases (Table 7.1). In this study, we considered CO_2 but not the other GHGs (e.g., CO, CH_4, and N_2O). In order to distinguish C emissions from CO_2, total C emissions were used after an adjustment with the proportion of CO_2 (~90% of total C content) among primary GHGs including CO_2, CO, and CH_4 (referred to as GFED4.1s EFs; Table 7.1). Spatial resolution in the first release was 1° (~100 km), which was revised to 0.5° (~50 km) and then to the current 0.25° (~25 km). In this study, we used the latest version of GFED4.1s for comparing C emissions from biomass burning.

TABLE 7.1

Emission Factors (EFs) of GFED4.1s (Randerson et al., 2018) and the Percentage of CO$_2$ Contributed Emissions within Dominant Carbon Contents Including CO$_2$, CO, and CH$_4$ for Adjusting Total C Emissions

	Emission Factor (gC/1000 g of DM) in Each Biome Type					
Name of the Content (Molecular Weight)	Tropical Forest	Temperate Forest	Boreal Forest	Savanna	Agricultural Waste	Peat
CO$_2$ (44.01)	1643	1647	1489	1686	1585	1549
CO (28.01)	93	88	127	63	102	210
CH$_4$ (16.04)	5.07	3.36	5.96	1.94	5.82	13.825
Total carbon (C) content (unit = g/km DM)	491.75	489.42	464.99	488.27	480.35	522.82
C in CO$_2$/total C content (unit = %)	91.12	91.78	87.33	94.17	89.99	80.80
Average of C proportion of CO$_2$ from every biome	89.20 (~90%)					

METHODS

The flowchart of the methodology for FEs generation is given in Figure 7.1. The flow includes (1) estimation of biomass from AGB and soil organic matter (SOM), (2) combination of biomass and FRP, and (3) comparison between FEs and GFED4.1s. Estimating biomass and delineating FRP are the key factors in C emission estimation. Previously, the VISIT model-estimated biomass was validated at 21 different sites on a global scale (Ito and Oikawa, 2002, Ito et al., 2008, Inatomi et al., 2010). The VISIT biomass data (AGB and SOM) were evaluated in terms of r-squares (r^2) reaching almost 1 (0.9999) (Ito and Oikawa, 2002), while there were limitations for a direct comparison of these data since it is a relatively coarse resolution (0.5°) when compared with satellite data. For this reason, we extrapolated large-scale biomass from the VISIT model to correlate with MODIS NDVI (MOD13A3) at a 1 km spatial resolution. There were clear relationships between NDVI from MODIS and AGB from VISIT. For the estimation of AGB, a logistic regression method was used (Equation 7.1):

$$AGB = a * \ln(NDVI) \tag{7.1}$$

where "a" is a constant value in each land-cover (LC) type (Table 7.2). Notably, the "a" value in snow and ice coverage class was explicitly high ($a = 271.2$); however, it might be neglectable because the NDVI in the region would be almost 0. The highest and second "a" appeared in evergreen broadleaf forest and permanent wetlands, respectively. The estimated AGB was applied to the calculation of combusted biomass (M1) by multiplying with BAs from the MCD45A1 product. Post-fire NDVI

TABLE 7.2

Land-Cover Types of the IGBP and Constant Values of Each Type in the Calculation of Aboveground Biomass (AGB) from the Normalized Difference Vegetation Index (NDVI) (Equation 7.1)

Index	Name of the Land-Cover Type (Abbreviations)	Constant Value (a)
0	Water (Water)	0
1	Evergreen needleleaf forests (ENF)	157
2	Evergreen broadleaf forests (EBF)	176.4
3	Deciduous needleleaf forests (DNF)	93.1
4	Deciduous broadleaf forests (DBF)	106.4
5	Mixed forests (MF)	111.7
6	Closed shrublands (C. Shrub)	110.5
7	Open shrublands (O. Shrub)	151.3
8	Woody savannas (W. Savannas)	133.3
9	Savannas (Savannas)	89.7
10	Grasslands (Grass)	78.9
11	Permanent wetlands (P. Wetlands)	167.8
12	Croplands (Crop)	77.9
13	Urban and built-up lands (Urban)	130.7
14	Cropland/natural vegetation mosaics (Crop/Natural)	125.4
15	Permanent snow and ice (Snow and Ice)	271.2
16	Barren	85.6

should contain an already-declined value from intact NDVI before burning. Thus, we used the differences between AGBs before and after fire events from the MODIS BA to calculate the emissions (see the M1 calculation part in Figure 7.1). Three kinds of EFs were tested for AGB (EFAGB) using Agricultural, Boreal, Savanna, Temperate, Tropical categories GFED EFs (min: 0.46; average: 0.475; and max: 0.49; Randerson et al., 2018) using the following equation (Equation 7.2):

$$FE_{AGB} (tC / ha) = AGB * BA * EF_{AGB} \qquad (7.2)$$

As another component of biomass, we also considered SOM. In the VISIT model, SOM was calculated as total C storage from litterfall and mineral soils (Ito and Oikawa, 2002) and validated with *in situ* observations in a rain forest in Malaysia and in a dry evergreen forest in Thailand, which showed high agreement with previous studies (Adachi et al., 2011). In particular, the C emissions from soil respiration were already subtracted in the model, and the SOM biomass was inferred as a balance between storage and loss of C. There are two kinds of layers in SOM: (1) top layer with 0–30 cm below the surface and (2) sub-layer with deeper than 30 cm below the surface. The AGB is conspicuously detectable using satellite data, whereas SOM is not. Thus, a *Histosol* class in the FAO soil classification map was used to reference SOM when we upscaled the SOM in VISIT. The *Histosol* is called "peat soil, bog soil, or organic soil" and consists of *histic* or *folic* horizon, which contains more than 20% (by weight) organic carbon (35% organic matter) within 40-cm soil layers (IUSS Working Group, WRB, 2006). After SOM biomass was estimated with a 1 km spatial resolution, MODIS FRP (MOD14 active fire product) data were used to calculate combusted biomass from SOM (M2). Precise location and date in fire pixels were incorporated in the FEs calculation as it provides information on the land surfaces (Kaufman et al., 1998, Justice et al., 2002, Giglio et al., 2003). Daily (day and night) fire detection result was obtained from MOD14A1. Fire pixels in low, nominal, and high confidence were collected in this study. Particularly, maximum FRP was used to estimate combusted biomass as it is a surrogate measure of combustion rates (Wooster et al., 2005). We used the rate of biomass combustion (kg/sec) as in Wooster et al. (2005) for the CC as (Equations 7.3 and 7.4)

$$CC = \text{Biomass combustion rate (Kg/sec)} = 0.368 * \text{FRP (MW)} \qquad (7.3)$$

$$FE_{SOM}\,(tC/ha) = SOM * CC * EF_{SOM} \qquad (7.4)$$

Similar to EFAGB, three kinds of EFSOM (min: 0.48; mean: 0.525; and max: 0.57) for C contents refer to the *Peat* category in the GFED EF database. These ratios reflect the proportion of biome-specific C contents (CO$_2$, CO, and CH$_4$) equivalent to 48%–57% of the total amounts of biomass (AGB or SOM) when 1 kg of dry matter is burned (Akagi et al., 2011). Additionally, the proportion of total biome-specific C content that CO$_2$ contributed was set as 0.9 (according to Table 7.1, sources from the GFED EF table; Randerson et al., 2018).

$$FE\,(tC/ha) = FE_{AGB} + FE_{SOM} \qquad (7.5)$$

A summation of both FEs calculates total FEs from the AGB and SOM. Notably, EFs for AGB and SOM are tested using nine combinations, as shown in Table 7.3. An optimal combination of EFs was selected by considering R^2, RMSE, and slope of linear fitting function with the monthly GFED and monthly FEs as *x*-axis and *y*-axis, respectively (Figure 7.2).

RESULTS AND DISCUSSION

SELECTED EFS AND ANNUAL EMISSIONS DURING 2001–2017

For investigating the optimal EFs, nine combinations were tested with GFED by considering errors (RMSE) and scatter in distribution of monthly data (R^2 and slope of linear fitting function). In terms of statistical metrics – RMSE, R^2, and slope – 0.46 and 0.57 were selected for AGB and SOM. In particular, RMSE in the case of 1-1 (0.46 and 0.48 for AGB and SOM, respectively; Figure 7.2a) was the smallest, showing 0.106 TgC/month. In contrast, a maximum RMSE (0.1451 TgC/month), i.e., the lowest accuracy between FEs and GFED, was found for EFAGBmax (0.49) and EFSOMmax (0.57) (Table 7.3). These results suggest that RMSE may not be solely used for the

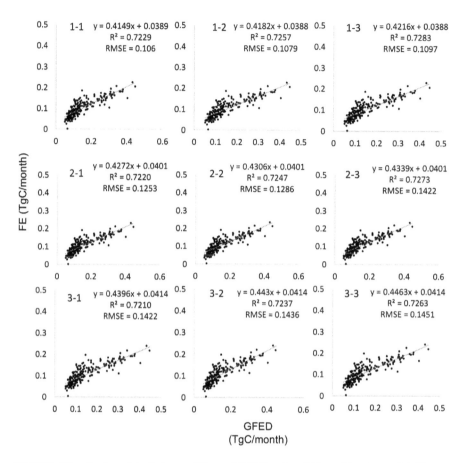

FIGURE 7.2   Scatter plots between FEs and GFED when EF combinations of AGB and SOM were applied to the estimation of C emissions.

TABLE 7.3

RMSEs of FEs with Monthly GFED C Emissions When EFs for AGB and SOM from the GFED EF Table were Applied

	$EF_{AGB(1)} = 0.46$	$EF_{AGB(2)} = 0.475$	$EF_{AGB(3)} = 0.49$
$EF_{SOM(1)} = 0.48$	0.106	0.1253	0.1422
$EF_{SOM(2)} = 0.525$	0.1079	0.1286	0.1422
$EF_{SOM(3)} = 0.57$	0.1097	0.1422	0.1451

optimal EF selection. Thus, we also investigated R^2 and slope and found a maximum for 1–3 (0.46 and 0.57) (Figure 7.2). Hence, the optimal values were set to 0.46 and 0.57 for AGB and SOM, respectively. As a result, the annual average of FEs and GFED was 1.25 ± 0.12 and 1.68 ± 0.15 PgC/year, respectively, when optimal EFs were applied to calculate C of CO_2 emissions (Figure 7.3). In the case of total biome-specific

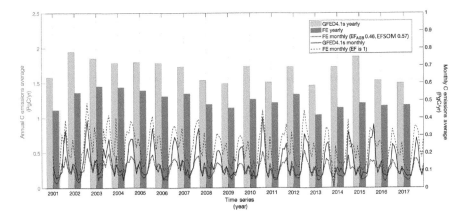

FIGURE 7.3   Time series of carbon emissions from biomass burning in FEs (this study) and GFED4.1s from January 2001 to December 2017. Bars and lines show annual and monthly emissions, respectively. Annual and monthly emissions are referring to the left and right y-axis, respectively (PgC = 10^{15}g of carbon). Annual average of FEs is using optimal EFs (0.46 and 0.57 for AGB and SOM, respectively).

C (including CO$_2$, CO, and CH$_4$) i.e., EF was one or before an EF was applied to), FEs and GFED showed 2.42 ± 0.23 and 1.87 ± 0.52, respectively. The highest C emissions in FEs and GFED were 2003 (2.81 PgC/year) and 2002 (1.94 PgC/year), respectively. Applying various EFs made FEs fluctuate consequently (Figure 7.3). The effect of EFs was higher in large emission seasons, usually during the summer in the Northern Hemisphere. However, in the winter season, the emission strength was small, and the effect of EFs was also small. Particularly, the agreement among FEs (optimal EF), FEs (EF = 1), and GFED was higher in months when smaller emissions occurred (Figure 7.3). We attribute the uncertainties to the MODIS BA product, which might have been influenced due to the cloud cover (Giglio et al., 2010).

SPATIAL DISTRIBUTIONS OF C EMISSIONS

Maps of annual averages of FEs and GFED from 2001 to 2017 are shown in Figure 7.4. The emissions are high in the equatorial and boreal regions. The productivities and fuel densities of forests in these regions are high (Baccini et al., 2018). Potentially, equatorial and boreal regions with large fires will release more emissions because of a relatively large amount of biomass burnt than temperate forestsy. Peat forest fires are a critical source of SOM emissions, which contributes to CO$_2$ emissions (Page et al., 2002). Based on the previous studies, the EFs for peatlands were assigned highest (Andreae and Merlet, 2001, Akagi et al., 2011). Particularly, Indonesia's tropical peatlands experienced large disturbance due to the Mega Rice Project in 1999 (Joosten and Clarke, 2002; Hirano et al., 2012). The drained peatlands made wetlands vulnerable to fires. Deforestation and land clearance by fires are already realized as a huge source of emissions, and the draining of peatlands accelerated it. In that sense, C emissions in Indonesia showed the highest levels in FEs (Figure 7.4). Emissions in FEs seem

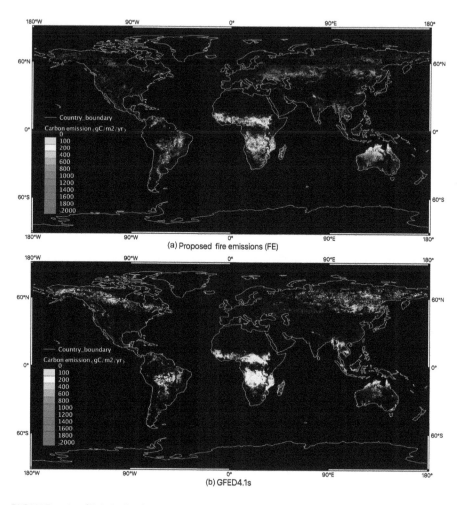

FIGURE 7.4 Global distribution of annual average of CO_2 emissions (gC/m^2/year) for 18 years (2001–2017) based on (a) fire emission (FEs, this study proposed) and (b) the Global Fire Emissions Database (GFED4.1s). Annual averages were 1.25 and 1.68 PgC/year in FEs and GFED, respectively.

sparser than those of GFED; however, the emissions are concentrated. Differences in the spatial resolution might have affected the results. Although GFED4.1s revised the BAs using the MODIS active fire product at 500-m grid, C emissions were regionally regressed to derive the grid products (25×25 km^2) (Giglio et al., 2010). Finally, the unit grid's representativeness between FEs and GFED differs 625 times (1 km^2 vs. 625 km^2). Some trade-offs existed in choosing spatial resolutions. For example, FEs showed the spatial distribution more in detail; however, fires' spatial characteristics would have been overestimated using FRP pixel-based detection. In the relatively low resolution of GFED, the spatial distribution within 625 km^2 is not clearly shown, and regional averages would underestimate the amount. However, in the case of large fires, they would correlate well to mid- to larger-scale products such as GFED.

REGIONAL DISTRIBUTION OF EMISSIONS

As mentioned above, FEs are suitable to estimate the smaller spatial scale fires than GFED. In comparison between GFED4 and GFED4.1s without and with small fires, agricultural waste burning contributes substantially to C emissions and is usually smaller than forest fires (Giglio et al., 2013). In that sense, emissions in Southeast Asia using traditional high-emission areas (e.g., Indochinese Peninsula and Western Indonesia) are shown in Figure 7.5. The BA locations are matched well among FEs and GFED (Figure 7.5a–d). In particular, FEs showed consistency with geomorphological distributions along the high-altitudinal area in Thailand and Cambodia, which would be merged into several grids in GFED because of the product's spatial resolution.

POTENTIAL UNCERTAINTIES

Biomass estimation is a source of uncertainty. In this study, AGB was estimated by polynomial regression between the NDVI and the VISIT-retrieved AGB. AGB of VISIT includes biomass of foliage, stem, and branch. However, NDVI can be an indicator of the greenness of the canopy. If leaves were burnt during a fire event, AGB in our study could capture C emissions correctly; however, stem and branch burning emissions are not easy to capture, thus underestimating total burnt biomass. The uncertainty in SOM is included in the depth of organic soil layers. Satellite imagery is suitable to detect fire spatially. However, the depth of biomass is not detectable,

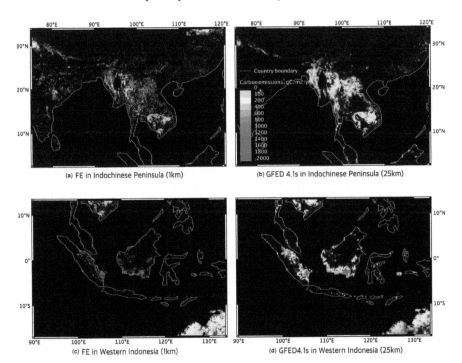

(a) FE in Indochinese Peninsula (1km)

(b) GFED 4.1s in Indochinese Peninsula (25km)

(c) FE in Western Indonesia (1km)

(d) GFED4.1s in Western Indonesia (25km)

FIGURE 7.5 Regional distribution of CO_2 emissions from FEs and GFED. (a and b) Indochinese Peninsula and (c and d) Indonesia. All figures refer to a color legend in (b).

which may also influence the total C emissions. This might be enabled by applying organic soil layer depth data and soil moisture using microwave sensors for describing soil dryness. Not only fuel amount and moisture conditions but also CC would be uncertain. To overcome the limitation related to fixed values of CC in GFED, we used dynamic CC from FRP of MODIS. The agreement with GFED was confirmed; however, FRP itself has uncertainties. FRP would be lost by a process other than radiation (e.g., convection, conduction, and vaporization) (Wooster et al., 2005). Those uncertainties of FRP would affect FEs resulting in lower CC. Finally, EF can provide uncertainties in setting optimal value and the ratio between CO_2 and C regarding combustion type, such as flaming and smoldering. Usually, CO and CO_2 are dominant in flaming and smoldering for early-stage and later fires, respectively (Andreae and Merlet, 2001). It means EFs between CO_2 and the other C contents would vary along with the duration of time. However, FEs and GFED use the same emission ratios without considering different stages of fires, including duration. Thus, the fire duration variable is important to estimate total C emissions, including CO_2. Developing a fire detection system using a geostationary satellite promises to grasp dynamic changes of C emissions with a high frequency of monitoring.

CONCLUSIONS

Satellite-based global fire emissions (FEs) were estimated using the MODIS fire products (e.g., burned area (BA) and radiative firepower (FRP)) at a 1 km spatial resolution. We used polynomial regression between the MODIS NDVI and VISIT model for the above-ground biomass (AGB) estimation. VISIT also estimated biomass in soil organic matter (SOM) in this study, and it was applied to the global peatlands using the FAO Histosol dataset. Dynamic combustion completeness (CC) at a monthly time step was obtained from FRP. In contrast to CC's fixed value in GFED, using dynamic CC in FEs resulted in higher fire emissions (FEs) in Southeast Asia. Emission factors (EFs) for C content from 1 kg of dry matter burning were chosen through comparisons between FEs and GFED for AGB and SOM. The setting of EF is crucial to estimate the amount of C emissions. We infer that capturing CO and CO_2 emission ratios during early and later stages of fires and including the flaming and smoldering fire types can improve C emissions globally.

REFERENCES

Adachi, M., Ito, A., Ishida, A., Kadir, W. R., Ladpala, P., and Yamagata, Y. 2011. Carbon budget of tropical forests in Southeast Asia and the effects of deforestation: an approach using a process-based model and field measurements. *Biogeosciences.* 8(9): 2635–2647.

Akagi, S. K., Yokelson, R. J., Wiedinmyer, C., Alvarado, M. J., Reid, J. S., Karl, T., Crounse, J. D., and Wennberg, P. O. 2011. Emission factors for open and domestic biomass burning for use in atmospheric models. *Atmospheric Chemistry and Physics.* 11(9): 4039–4072.

Andreae, M. O., and Merlet, P. 2001. Emission of trace gases and aerosols from biomass burning. *Global Biogeochemical Cycles.* 15(4): 955–966.

Baccini, M., Walker, W., Farina, M., and Houghton, R. A. 2018. CMS: Estimated Deforested Area Biomass, Tropical America, Africa, and Asia, 2000. ORNL DAAC, Oak Ridge, Tennessee, USA. doi:10.3334/ORNLDAAC/1337.

Badarinath, K. V. S., Kharol, S. K., Krishna Prasad, V., Kaskaoutis, D. G. and Kambezidis, H. D. 2008a. Variation in aerosol properties over Hyderabad, India during intense cyclonic conditions. *International Journal of Remote Sensing*. 29(15): 4575–4597.

Badarinath, K. V. S., Kharol, S. K., Prasad, V. K., Sharma, A. R., Reddi, E. U. B., Kambezidis, H. D. and Kaskaoutis, D. G. 2008b. Influence of natural and anthropogenic activities on UV Index variations–a study over tropical urban region using ground based observations and satellite data. *Journal of Atmospheric Chemistry*. 59(3): 219–236.

Biswas, S., Lasko, K. D. and Vadrevu, K. P. 2015a. Fire disturbance in tropical forests of Myanmar—Analysis using MODIS satellite datasets. *IEEE Journal of Selected Topics in Applied Earth Observations and Remote Sensing*. 8(5): 2273–2281.

Biswas, S., Vadrevu, K. P., Lwin, Z. M., Lasko, K. and Justice, C. O. 2015b. Factors controlling vegetation fires in protected and non-protected areas of Myanmar. *PLoS One*. 10(4), e0124346.

Csiszar, I. A., Morisette, J. T., and Giglio, L. 2006. Validation of active fire detection from moderate-resolution satellite sensors: The MODIS example in northern Eurasia. *IEEE Transactions on Geoscience and Remote Sensing*. 44(7): 1757–1764.

Field, R. D., Van Der Werf, G. R., and Shen, S. S. 2009. Human amplification of drought-induced biomass burning in Indonesia since 1960. *Nature Geoscience*. 2(3): 185–188.

Fusco, E. J., Finn, J. T., Abatzoglou, J. T., Balch, J. K., Dadashi, S., and Bradley, B. A. 2019. Detection rates and biases of fire observations from MODIS and agency reports in the conterminous United States. *Remote Sensing of Environment*. 220: 30–40.

Giglio, L., Descloitres, J., Justice, C. O., and Kaufman, Y. J. 2003. An enhanced contextual fire detection algorithm for MODIS. *Remote Sensing of Environment*. 87(2–3): 273–282.

Giglio, L., Schroeder, W., and Justice, C.O. 2016. The collection 6 MODIS active fire detection algorithm and fire products. Remote Sensing of Environment, 178, pp. 31–41.

Giglio, L., Randerson, J. T., and van der Werf, G. R. 2013. Analysis of daily, monthly, and annual burned area using the fourth-generation global fire emissions database (GFED4). *Journal of Geophysical Research: Biogeosciences*. 118(1): 317–328.

Giglio, L., Randerson, J. T., Van Der Werf, G. R., Kasibhatla, P. S., Collatz, G. J., Morton, D. C., and DeFries, R. S. 2010. Assessing variability and long-term trends in burned area by merging multiple satellite fire products. *Biogeosciences*. 7: 1171–1186

Goode, J. G., Yokelson, R. J., Ward, D. E., Susott, R. A., Babbitt, R. E., Davies, M. A., and Hao, W. M. 2000. Measurements of excess O$_3$, CO$_2$, CO, CH$_4$, C$_2$H$_4$, C$_2$H$_2$, HCN, NO, NH$_3$, HCOOH, CH$_3$COOH, HCHO, and CH$_3$OH in 1997 Alaskan biomass burning plumes by airborne Fourier transform infrared spectroscopy (AFTIR). *Journal of Geophysical Research: Atmospheres*. 105(D17): 22147–22166.

Gupta, P. K., Prasad, V. K., Sharma, C., Sarkar, A. K., Kant, Y., Badarinath, K. V. S. and Mitra, A. P. 2001. CH$_4$ emissions from biomass burning of shifting cultivation areas of tropical deciduous forests–experimental results from ground-based measurements. *Chemosphere-Global Change Science*. 3(2): 133–143.

Hayasaka, H. and Sepriando, A. 2018. Severe air pollution due to peat fires during 2015 super El Niño in Central Kalimantan, Indonesia. In: *Land-Atmospheric Research Applications in South/Southeast Asia*. Vadrevu, K. P., Ohara, T., and Justice, C. (Eds.). Springer, Cham, pp. 129–142.

Hayasaka, H., Noguchi, I., Putra, E. I., Yulianti, N. and Vadrevu, K. 2014. Peat-fire-related air pollution in Central Kalimantan, Indonesia. *Environmental Pollution*. 195: 257–266. doi:10.1016/j.envpol.2014.06.031.

Hirano, T., Segah, H., Kusin, K., Limin, S., Takahashi, H., and Osaki, M. 2012. Effects of disturbances on the carbon balance of tropical peat swamp forests. *Global Change Biology*. 18(11): 3410–3422.

Holtsmark, B. 2015. Quantifying the global warming potential of CO$_2$ emissions from wood fuels. *Global Change Biology Bioenergy*. 7(2): 195–206.

Huete, A., Didan, K., Miura, T., Rodriguez, E. P., Gao, X., and Ferreira, L. G. 2002. Overview of the radiometric and biophysical performance of the MODIS vegetation indices. *Remote Sensing of Environment.* 83(1–2): 195–213.

Inatomi, M., Hajima, T. and Ito, A. 2019. Fraction of nitrous oxide production in nitrification and its effect on total soil emission: A meta-analysis and global-scale sensitivity analysis using a process-based model. *PLOS One.* 14(7): e0219159. doi: 10.1371/journal. pone.0219159

Inatomi, M., Ito, A., Ishijima, K., and Murayama, S. 2010. Greenhouse gas budget of a cool-temperate deciduous broad-leaved forest in Japan estimated using a process-based model. *Ecosystems.* 13(3): 472–483.

Israr, I., Jaya, S. N. I, Saharjo, H. S., Kuncahyo, B., and Vadrevu, K. P. 2018 Spatio-temporal analysis of land and forest fires in Indonesia using MODIS active fire dataset. In: *Land-Atmospheric Research Applications in South/Southeast Asia.* Vadrevu, K. P., Ohara, T., and Justice, C. (Eds.). Springer, Cham, pp.105–128.

Ito, A. 2008. The regional carbon budget of East Asia simulated with a terrestrial ecosystem model and validated using AsiaFlux data. *Agricultural and Forest Meteorology*, 148(5): 738–747.

Ito, A., and Oikawa, T. 2002. A simulation model of the carbon cycle in land ecosystems (Sim-CYCLE): a description based on dry-matter production theory and plot-scale validation. *Ecological Modelling*, 151(2–3): 143–176.

Ito, A., Tohjima, Y., Saito, T., Umezawa, T., Hajima, T., Hirata, R., Saito, M., and Terao, Y. 2019. Methane budget of East Asia, 1990–2015: A bottom-up evaluation. *Science of the Total Environment.* 676: 40–52.

IUSS Working Group, W. R. B. 2006. World reference base for soil resources. World Soil Resources Report, 103.

Jang, E., Kang, Y., Im, J., Lee, D. W., Yoon, J., and Kim, S. K. 2019. Detection and monitoring of forest fires using Himawari-8 geostationary satellite data in South Korea. *Remote Sensing.* 11(3): 271.

Joosten, H., and Clarke, D. 2002. Wise use of mires and peatlands. International Mire Conservation Group and International Peat Society. 304.

Justice, C. O., Giglio, L., Korontzi, S., Owens, J., Morisette, J. T., Roy, D., Descloitres, J., Alleaume, S., Petitcolin, F., and Kaufman, Y. 2002. The MODIS fire products. *Remote Sensing of Environment*, 83 (1–2): 244–262.

Justice, C., Gutman, G. and Vadrevu, K. P. 2015. NASA land cover and land use change (LCLUC): An interdisciplinary research program. *Journal of Environmental Management.* 148(15), 4–9.

Kalnay, E., Kanamitsu, M., Kistler, R., Collins, W., Deaven, D., Gandin, L., Iredell, M., Saha, S., White, G., Woollen, J., Zhu, Y., Chelliah, M., Ebisuzaki, W., Higgins, W., Janowiak, J., Mo, K. C., Ropelewski, C., Wang, J., Leetmaa, A., Reynolds, R., Jenne, R., and Joseph, D. 1996. The NCEP/NCAR 40-year reanalysis project. *Bulletin of the American Meteorological Society*, 77(3): 437–472.

Kant, Y., Ghosh, A. B., Sharma, M. C., Gupta, P. K., Prasad, V. K., Badarinath, K. V. S. and Mitra, A. P. 2000. Studies on aerosol optical depth in biomass burning areas using satellite and ground-based observations. *Infrared Physics & Technology.* 41(1): 21–28.

Kaufman, Y. J., Justice, C. O., Flynn, L. P., Kendall, J. D., Prins, E. M., Giglio, L., Ward, D. E., Menzel, P. W., and Setzer, A. W. 1998. Potential global fire monitoring from EOS-MODIS. *Journal of Geophysical Research: Atmospheres.* 103(D24): 32215–32238.

Lasko, K. and Vadrevu, K. P. 2018. Improved rice residue burning emissions estimates: Accounting for practice-specific emission factors in air pollution assessments of Vietnam. *Environmental Pollution.* 236 (5): 795–806.

Lasko, K., Vadrevu, K. P. and Nguyen, T. T. N. 2018a. Analysis of air pollution over Hanoi, Vietnam using multi-satellite and MERRA reanalysis datasets. *PloS One.* 13(5): e0196629.

Lasko, K., Vadrevu, K. P., Tran, V. T. and Justice, C. 2018b. Mapping double and single crop paddy rice with Sentinel-1A at varying spatial scales and polarizations in Hanoi, Vietnam. *IEEE Journal of Selected Topics in Applied Earth Observations and Remote Sensing.* 11(2): 498–512.

Lasko, K., Vadrevu, K. P., Tran, V. T., Ellicott, E., Nguyen, T. T., Bui, H. Q. and Justice, C. 2017. Satellites may underestimate rice residue and associated burning emissions in Vietnam. *Environmental Research Letters.* 12(8), 085006.

Liu, T., Marlier, M. E., Karambelas, A., Jain, M., Singh, S., Singh, M. K., Gautam, R., and DeFries, R. S. 2019. Missing emissions from post-monsoon agricultural fires in north-western India: regional limitations of MODIS burned area and active fire products. *Environmental Research Communications.* 1(1): 011007.

Mollicone, D., Eva, H. D., and Achard, F. 2006. Ecology: Human role in Russian wild fires. *Nature.* 440(7083): 436–437.

Monsi, M., and Saeki, T. 1953. The light factor in plant communities and its significance for dry matter production. *Japanese Journal of Botany.* 14(1): 22–52.

Padilla, M., Stehman, S. V., and Chuvieco, E. 2014. Validation of the 2008 MODIS-MCD45 global burned area product using stratified random sampling. *Remote Sensing of Environment.* 144: 187–196.

Page, S. E., Siegert, F., Rieley, J. O., Boehm, H. D. V., Jaya, A., and Limin, S. 2002. The amount of carbon released from peat and forest fires in Indonesia during 1997. *Nature,* 420 (6911): 61.

Prasad, V. K. and Badarinth, K. V. S. 2004. Land use changes and trends in human appropriation of above ground net primary production (HANPP) in India (1961–98). *Geographical Journal.* 170 (1): 51–63.

Prasad, V. K., Anuradha, E. and Badarinath, K. V. S. 2005. Climatic controls of vegetation vigor in four contrasting forest types of India—evaluation from National Oceanic and Atmospheric Administration's Advanced Very High Resolution Radiometer datasets (1990–2000). *International Journal of Biometeorology.* 50(1): 6–16.

Prasad, V. K., Badarinath, K. V. S. and Eaturu, A. 2008a. Biophysical and anthropogenic controls of forest fires in the Deccan Plateau, India. *Journal of Environmental Management.* 86(1): 1–13.

Prasad, V. K., Badarinath, K. V. S. and Eaturu, A. 2008b. Effects of precipitation, temperature and topographic parameters on evergreen vegetation greenery in the Western Ghats, India. *International Journal of Climatology: A Journal of the Royal Meteorological Society.* 28(13): 1807–1819.

Prasad, V. K., Badarinath, K. V. S., Yonemura, S. and Tsuruta, H. 2004. Regional inventory of soil surface nitrogen balances in Indian agriculture (2000–2001). *Journal of Environmental Management.* 73(3): 209–218.

Prasad, V. K., Gupta, P. K., Sharma, C., Sarkar, A. K., Kant, Y., Badarinath, K. V. S., Rajagopal, T. and Mitra, A. P. 2000. NO_x emissions from biomass burning of shifting cultivation areas from tropical deciduous forests of India–estimates from ground-based measurements. *Atmospheric Environment.* 34(20): 3271–3280.

Prasad, V. K., Kant, Y. and Badarinath, K. V. S. 2001. CENTURY ecosystem model application for quantifying vegetation dynamics in shifting cultivation areas: A case study from Rampa Forests, Eastern Ghats (India). *Ecological Research.* 16(3), 497–507.

Prasad, V. K., Kant, Y., Gupta, P. K., Elvidge, C. and Badarinath, K. V. S. 2002. Biomass burning and related trace gas emissions from tropical dry deciduous forests of India: A study using DMSP-OLS data and ground-based measurements. *International Journal of Remote Sensing.* 23(14): 2837–2851.

Prasad, V. K., Lata, M. and Badarinath, K. V. S. 2003. Trace gas emissions from biomass burning from northeast region in India—estimates from satellite remote sensing data and GIS. *Environmentalist.* 23(3): 229–236.

Ramachandran, S. 2018. Aerosols and climate change: Present understanding, challenges and future outlook. In: *Land-Atmospheric Research Applications in South/Southeast Asia*. Vadrevu, K. P., Ohara, T., and Justice, C. (Eds.). Springer, Cham. pp. 341–378.

Randerson, J. T., Chen, Y., Van Der Werf, G. R., Rogers, B. M., and Morton, D. C. 2012. Global burned area and biomass burning emissions from small fires. *Journal of Geophysical Research: Biogeosciences*. 117: G4.

Randerson, J. T., G. R. van der Werf, L. Giglio, G. J. Collatz, and P. S. Kasibhatla. 2018. Global Fire Emissions Database, Version 4, (GFEDv4). ORNL DAAC, Oak Ridge, Tennessee, USA. doi:10.3334/ORNLDAAC/1293.

Reddington, C. L., Butt, E. W., Ridley, D. A., Artaxo, P., Morgan, W. T., Coe, H., and Spracklen, D. V. 2015. Air quality and human health improvements from reductions in deforestation-related fire in Brazil. *Nature Geoscience*. 8(10): 768–771.

Rossi, S., Tubiello, F. N., Prosperi, P., Salvatore, M., Jacobs, H., Biancalani, R., Jouse, J. I., and Boschetti, L. 2016. FAOSTAT estimates of greenhouse gas emissions from biomass and peat fires. *Climatic Change*, 135(3–4): 699–711.

Roy, D.P., Boschetti, L., Justice, C.O., and Ju, J. 2008. The collection 5 MODIS burned area product – Global evaluation by comparison with the MODIS active fire product. *Remote sensing of Environment*, 112(9): 3690–3707.

Roy, D. P., Jin, Y., Lewis, P. E., and Justice, C. O. 2005. Prototyping a global algorithm for systematic fire-affected area mapping using MODIS time series data. *Remote Sensing of Environment*, 97(2): 137–162.

Ryan, M. G. 1991. Effects of climate change on plant respiration. *Ecological Applications*, 1(2): 157–167.

Saharjo, B. H., and Yungan, A. 2018. Forest and land fires in Riau province; A Case study in fire prevention, policy implementation with local concession holders. In: *Land-Atmospheric Research Applications in South/Southeast Asia*. Vadrevu, K. P., Ohara, T., and Justice, C. (Eds.). Springer, Cham. pp. 143–170

Schultz, M. G., Heil, A., Hoelzemann, J. J., Spessa, A., Thonicke, K., Goldammer, J. G., Held, A. C., Pereira, J. M., and van Het Bolscher, M. 2008. Global wildland fire emissions from 1960 to 2000. *Global Biogeochemical Cycles*. 22: 2.

Shi, Y., Matsunaga, T., Saito, M., Yamaguchi, Y., and Chen, X. 2015. Comparison of global inventories of CO_2 emissions from biomass burning during 2002–2011 derived from multiple satellite products. *Environmental Pollution*. 206: 479–487.

Vadrevu, K. P., 2008. Analysis of fire events and controlling factors in eastern India using spatial scan and multivariate statistics. *Geografiska Annaler: Series A, Physical Geography*. 90(4): 315–328.

Vadrevu, K. P., and Choi, Y. 2011. Wavelet analysis of airborne CO_2 measurements and related meteorological parameters over heterogeneous landscapes. *Atmospheric Research*. 102(1–2): 77–90.

Vadrevu, K. P. and Justice, C. O. 2011. Vegetation fires in the Asian region: satellite observational needs and priorities. *Global Environmental Change*. 15(1): 65–76.

Vadrevu, K. P., and Lasko, K. 2018b. Intercomparison of MODIS AQUA and VIIRS I-Band fires and emissions in an agricultural landscape—Implications for air pollution research. *Remote Sensing*. 10(7): 978. doi:10.3390/rs10070978.

Vadrevu, K. P., Badarinath, K. V. S. and Anuradha, E. 2008. Spatial patterns in vegetation fires in the Indian region. *Environmental Monitoring and Assessment*, 147(1–3): 1. doi:10.1007/s10661-007-0092-6.

Vadrevu, K. P., Csiszar, I., Ellicott, E., Giglio, L., Badarinath, K. V. S., Vermote, E. and Justice, C. 2012. Hotspot analysis of vegetation fires and intensity in the Indian region. *IEEE Journal of selected topics in applied Earth Observations and Remote Sensing*. 6(1): 224–238.

Vadrevu, K. P., Eaturu, A. and Badarinath, K. V. S. 2006. Spatial distribution of forest fires and controlling factors in Andhra Pradesh, India using spot satellite datasets. *Environmental Monitoring and Assessment*. 123(1–3): 75–96.

Vadrevu, K. P., Giglio, L. and Justice, C. 2013. Satellite based analysis of fire–carbon monoxide relationships from forest and agricultural residue burning (2003–2011). *Atmospheric Environment*. 64: 179–191.

Vadrevu K. P., Ohara T, Justice C. 2014a. Air pollution in Asia. *Environmental Pollution*. 12: 233–235.

Vadrevu, K. P., Lasko, K., Giglio, L. and Justice, C. 2014b. Analysis of Southeast Asian pollution episode during June 2013 using satellite remote sensing datasets. *Environmental Pollution*. 12: 245–256.

Vadrevu, K. P., Lasko, K., Giglio, L. and Justice, C. 2015. Vegetation fires, absorbing aerosols and smoke plume characteristics in diverse biomass burning regions of Asia. *Environmental Research Letters*. 10(10): 105003.

Vadrevu, K. P., Lasko, K., Giglio, L., Schroeder, W., Biswas, S. and Justice, C. 2019. Trends in vegetation fires in south and Southeast Asian countries. *Scientific Reports*. 9(1): 7422. doi:10.1038/s41598-019-43940-x.

Vadrevu, K. P., Ohara, T. and Justice, C. 2017. Land cover, land use changes and air pollution in Asia: A synthesis. *Environmental Research Letters*. 12(12): 120201.

Vadrevu, K. P., Ohara, T. and Justice, C. eds., 2018a. *Land-Atmospheric Research Applications in South and Southeast Asia*. Springer, Cham.

Van der Werf, G. R., Randerson, J. T., Giglio, L., Collatz, G. J., Mu, M., Kasibhatla, P. S., Morton, D. C., DeFries, R. S., Jin, Y., and van Leeuwen, T. T. 2010. Global fire emissions and the contribution of deforestation, savanna, forest, agricultural, and peat fires (1997–2009). *Atmospheric Chemistry and Physics*, 10(23): 11707–11735.

Van Der Werf, G. R., Randerson, J. T., Giglio, L., Van Leeuwen, T. T., Chen, Y., Rogers, B. M., Mu, M., Van Marle, M. J. E., Morton, D. C., Collatz, G. J., Yokelson, R. J. and Kasibhatla, P. S. 2017. Global fire emissions estimates during 1997–2016. *Earth System Science Data*. 9: 697–720.

Ward, D. E., Hao, W. M., Susott, R. A., Babbitt, R. E., Shea, R. W., Kauffman, J. B., and Justice, C. O. 1996. Effect of fuel composition on combustion efficiency and emission factors for African savanna ecosystems. *Journal of Geophysical Research: Atmospheres*. 101(-D19): 23569–23576.

Wooster, M. J., Roberts, G., Freeborn, P. H., Xu, W., Govaerts, Y., Beeby, R., He, J., Lattanzio, A., Fisher, D., and Mullen, R. 2015. LSA SAF Meteosat FRP products–Part 1: Algorithms, product. *Atmospheric Chemistry and Physics* 15: 13217–13239.

Wooster, M. J., Roberts, G., Perry, G. L. W., and Kaufman, Y. J. 2005. Retrieval of biomass combustion rates and totals from fire radiative power observations: FRP derivation and calibration relationships between biomass consumption and fire radiative energy release. *Journal of Geophysical Research: Atmospheres*. 110: D24.

Yokelson, R. J., Susott, R., Ward, D. E., Reardon, J., and Griffith, D. W. 1997. Emissions from smoldering combustion of biomass measured by open-path Fourier transform infrared spectroscopy. *Journal of Geophysical Research: Atmospheres*. 102(D15): 18865–18877.

Zhu, C., Kobayashi, H., Kanaya, Y., and Saito, M. 2017. Size-dependent validation of MODIS MCD64A1 burned area over six vegetation types in boreal Eurasia: Large underestimation in croplands. *Scientific Reports*. 7(1): 4181.

8 Biomass Burning Mercury Emissions over the Maritime Continent

Udaysankar Nair and Aaron Kaulfus
University of Alabama in Huntsville, USA

CONTENTS

INTRODUCTION

Mercury (Hg) has toxic effects on human health, with the exposure being caused by the consumption of its methylated form (MeHg) (Zahir et al. 2005). Inorganic Hg enters aquatic ecosystems primarily through wet deposition and is converted to methylated form by the action of microorganisms. Bioaccumulation of methylated mercury occurs through consumption up the food chain, with human exposure resulting from the consumption of fish. While rice also bioaccumulates methylmercury, concentrations found in rice are much lower compared to fish. However, Hg contamination still poses a risk for populations whose calorific intake is dominated by rice (Li et al. 2010). In this context, it is important to understand the spatial and temporal variability of emissions and dominant processes that control the fate and transport of atmospheric Hg.

Hg occurs in the atmosphere in three forms, namely gaseous elemental (GEM), gaseous oxidized (GOM), and oxidized particulate-bound (HgP). GEM is the dominant form in the atmosphere and contributes to more than 95% of the total atmospheric Hg concentration. The solubility of GEM is low compared to the other two forms of Hg and thus has atmospheric lifetimes on the order of 0.5–2 years (Schroeder and Munthe 1998). The oxidized forms of Hg are highly soluble and have short atmospheric lifetimes, spanning a few days to a week. Because of a long lifetime, GEM is subject to long-range transport and the primary loss mechanism is through oxidation (Holmes, Jacob, and Yang 2006) and subsequent wet removal. Oxidation of GEM in the upper troposphere and stratosphere establishes an upper atmospheric reservoir of GOM and HgP. Stratosphere–troposphere exchange processes play an important role in the transport of oxidized Hg to lower altitudes, which is effectively scavenged by deep convection.

Both natural and anthropogenic emission sources contribute to atmospheric concentrations of Hg. Forest ecosystems are generally net sinks of atmospheric Hg, with it being deposited and stored on the forest canopy and leaf litter on the forest floor (Carpi et al. 2014). Removal of vegetation by burning releases the sequestered Hg, thus making biomass burning an important source, accounting for up to 8% of global atmospheric Hg emissions (Friedli et al. 2009). Cleared surfaces often exhibit enhanced soil Hg emissions, potentially caused by the removal of litter and increased light levels. Thus, land-use and land-cover change (LULCC) also impacts Hg emissions. Further, dry Hg deposition is reduced due to changes in surface characteristics. Note that while natural and prescribed fires alter land cover, LULCC can also lead to increased burning. Land cleared for agriculture often experiences an increase in prescribed burning and has a higher susceptibility to wildfires (Vadrevu et al. 2018). Thus, biomass burning and LULCC impact atmospheric Hg emissions in a coupled manner.

In the maritime continent (Figure 8.1), coupled biomass burning–LULCC forcing of Hg emissions is further subject to modulation by El Niño oscillations. On average, biomass burning emissions in tropical regions increase by 133% during El Niño events, with peak emissions occurring during August–October. The maritime continent has also experienced the highest loss of forest cover in all of the humid tropics over the last several decades (Miettinen, Shi, and Liew 2011), mostly due to the clearing of peat swamp forests. In the last decade, peat swamp forests in this region have been cleared at a rate of ~2.2% per year (Rose et al. 2011). The eastern lowlands of Sumatra and Sarawak's peatlands, Borneo, both lost over half the peatland forest cover between the years 2000 and 2010 (Miettinen, Shi, and Liew 2011). The Center

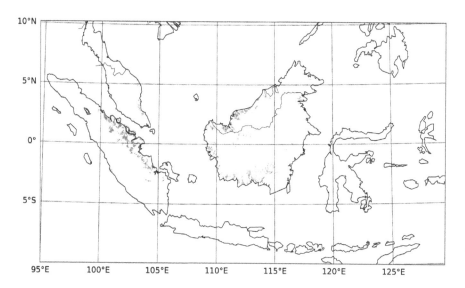

FIGURE 8.1   Land-cover changes in the study area between the years 2000 and 2010. Note that the study area contains the majority of the land regions that are traditionally referred to as the maritime continent. Therefore, from this point on, the manuscript will refer to the study area as the maritime continent. Peat swamp forests, cleared between the years 2000 and 2010, are shown in red color.

for Remote Imaging, Sensing and Processing Land Use and Land Cover (CRISP LULC) maps, derived from 250-m-spatial-resolution bands of NASA Moderate Resolution Imaging Spectroradiometer (MODIS) satellite data, show the drastic nature of land-cover change occurring in these regions (Figure 8.1). Substantial areas of peat swamp forest cover that existed in the year 2000 are found to be replaced by large areas of the patchwork of lowland mosaic, lowland open, and palm plantations by the year 2010 (Figure 8.1). The lowland mosaic and lowland open categories dominate the altered land cover.

Peat swamp forests are forests growing on peat soil, and the accumulated organic matter present on the forest floor functions as an effective biochemical sink for Hg (Kohlenberg et al. 2018). Prior studies show substantial sequestration of Hg in the forest floors of boreal peatlands (Grigal 2003), which is primarily released as GEM during biomass burning (Kohlenberg et al. 2018). Thus, the Hg emissions from boreal peatland fires are expected to have a larger impact on the global cycling of Hg rather than a regional effect (Kohlenberg et al. 2018). While such studies have not been conducted for fires within tropical peatlands, dominant Hg pollution due to biomass burning is also expected to be global rather than local.

DATASETS AND METHODS

To examine the coupled impact of LULCC and biomass burning on Hg pollution in the maritime continent, spatial and temporal variability of total Hg emissions (GEM+GOM+HgP) estimated from the satellite-derived Global Fire Emissions Database, version 4 (GFEDv4) (Van Der Werf et al. 2017; Giglio, Randerson, and Van Der Werf 2013) is utilized. The GFEDv4 includes burned area, emissions, and fractional contribution from different fire types on a $0.25° \times 0.25°$ grid and is available for the time period from 1997 to 2018. Using GFEDV4, we computed the Hg emissions (E) as

$$E = \sum_i DM \times A \times EF_i \times \Delta t$$

where DM, A, EF_i, and Δt are the dry matter emissions (kg/m^2/month), area burned (m^2), emission factor for fire-type source i, and the emission time interval (1 month), respectively. The different types of fire sources are as follows: (1) savanna, grassland, and shrubland fires, (2) boreal forest fires, (3) temperate forest fires, (4) deforestation and degradation, (5) peatland fires, and (6) agricultural waste burning. Note that while emission factors for trace species such as carbon monoxide are well documented for tropical peatlands, the same is not true for Hg emissions. Therefore, a best estimate of Hg emission factor for tropical peatland was derived by scaling up-to-date Hg emission factor for tropical forests (Andreae and Merlet 2019) by the ratio of carbon monoxide emission factors for peatland and tropical forest fires. The basis for this scaling is the high correlation found in laboratory studies between total Hg and carbon monoxide emissions in peatland fires (Kohlenberg et al. 2018).

A salient feature found in the area average, time series of satellite-derived total Hg emissions for the maritime continent, is episodic events of high Hg emissions. These events correlate well with the occurrence of El Niño events (Figure 8.2). End

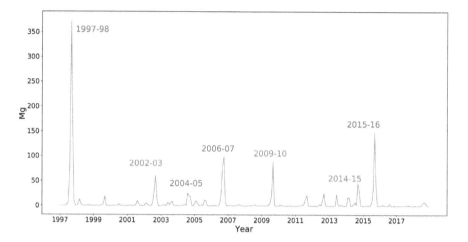

FIGURE 8.2   Times series of total biomass burning mercury emissions for the period 1997–2019. Annotated in the figure are El Niño events that occurred during this time period, with red, green, and blue colors denoting weak, moderate, and very strong events. Intensity classification is based on categories defined by J. Null (https://ggweather.com/enso/oni.htm).

members of events during the El Niño years ordered according to the magnitude of Hg emissions also correlate well with the El Niño events' strength. The highest Hg emissions for the period 1997–2018 occurred during the strong 1997–1998 El Niño event, while the second-highest Hg emissions coincided with another strong 2015–2016 event. The least amount of El Niño-related Hg emissions coincided with the weak event of 2004–2005. For the spectrum of events between the end members, there is a general dependence of emission magnitude to El Niño strength, but the mapping is not strictly one-to-one.

For example, Hg emissions during the moderate 2004–2005 El Niño events are substantially smaller when compared to emissions for all the weak events that occurred during the period 1997–2018. Hg emissions during the La Niña years are generally smaller compared to El Niño years. However, there is substantial variability of Hg emissions between La Niña years. One of the notable features of this variability is the sequentially high emission events that occurred from 2011 to 2013. The spikes in emissions during this time period are anomalously high compared to other La Niña periods and are potentially related to LULCC and associated agricultural/land-clearing fires.

Average biomass burning Hg emissions for the El Niño years during the period 1997–2018 are ~1.5 and 3.7 times higher compared to average emissions for all years and La Niña years, respectively (Table 8.1). Since the biomass burning Hg emission estimates from this study are region-specific, it is difficult to compare against prior studies. The only prior work that analyzed a comparable domain (equatorial Asian region) is Friedli et al. (2009), which considered the time period of 1998–2006. Friedli et al. (2009) report average biomass burning Hg emissions that are ~2.2 times higher (192 ± 216 Mg/year) compared to the present study's estimates.

However, as mentioned, the present study's estimates are based on a 22-year average and thus are more representative.

TABLE 8.1

Mean (μ) and Standard Deviation (σ) of Biomass Burning Mercury Emissions for All El Niño and La Niña years and All years During the Period 1997–2018

Years	M	Σ
All	85.4	142
El Niño	128.6	174
La Niña	23.0	15.5

Dominant source regions of biomass burning-related Hg emissions in the study area (Figure 8.3a) include the eastern regions of Sumatra south of ~2.5° N and Kalimantan's coastal land regions, Sarawak and Sabah. Two local emission maxima occur in the vicinity of Palembang, Sumatra, and Banjarmasin City in South Kalimantan, where the aggregated emissions exceed 25 Mg. Examining the satellite-derived land-cover maps and high-resolution satellite imagery over these areas shows agricultural utilization of land, suggesting the emission sources as agricultural waste burning.

The majority of biomass burning in the maritime continent occurs during the boreal summer (June–August) and fall (September–November) seasons (Figure 8.3b–c). The areal extent of emissions is maximized during the summer months when it is distributed over land areas to both the north and south of the equator (Figure 8.3b). However, during the fall season, the areal extent of emissions reduces and occurs mainly over land areas to the south of the equator (Figure 8.3c).

Since the wildfires are generally suppressed during the La Niña years, the dominant influence on Hg emissions' spatial distribution is expected to be anthropogenic. Spatial patterns of average yearly emissions of Hg for the La Niña years suggest that the significant impact of LULCC on Hg emissions occurs along the coastal regions of Kalimantan and Sarawak (Figure 8.4). Average annual Hg emissions for the periods 1997–2009 and 2010–2018 show major differences over these coastal regions during the fall (Figure 8.4a and b) and summer seasons (Figure 8.4c and d). Further, these differences correlate well with spatial patterns of forest cover changes found between satellite-derived land-cover maps for the years 2000 and 2010 (Figure 8.1). For example, along the central coastal regions of Sarawak, average yearly Hg emissions for 1997–2009 are substantially higher compared to the period 2010–2018. Note the same region also experienced large-scale deforestation during 1997–2009.

The enhanced Hg emissions from this coastal region before the year 2010 are potentially associated with land clearing that occurred over this area. The majority of cleared land has been converted to industrial plantations to cultivate oil palm and pulpwood, and the associated reduction in burning has resulted in diminished emissions during the period 2010–2018 (Miettinen, Shi, and Liew 2011).

On average, burning in the maritime continent contributes ~13% of the global burden of biomass burning Hg emissions, which is estimated to be 612 Mg/year (Kumar et al. 2018). However, note that the burning of peat swamp forests for land

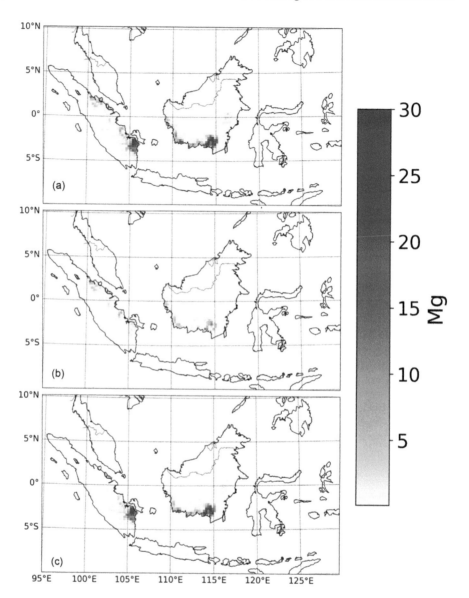

FIGURE 8.3 Spatial distribution of accumulated total gaseous mercury emissions for the period 1997–2018 (top). The middle and bottom panels are the same except for summer (JJA) and fall months (SON).

clearing also indirectly contributes to Hg emissions since they serve as a net sink for atmospheric Hg (Carpi et al. 2014). Also, Hg content in the peat swamp forest soils is higher than in other ecosystems (Grigal 2003). Exposure of this soil due to the forests' burning can also enhance Hg emission for several months following the forest clearing (Carpi et al. 2014). At present, these additional enhancements to Hg emissions caused by biomass burning are not well understood.

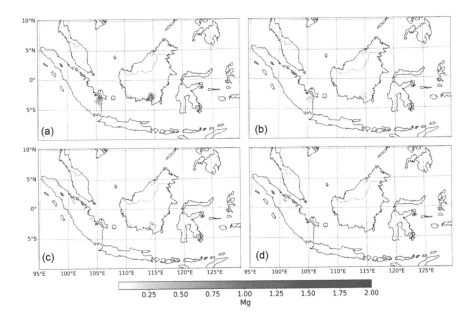

FIGURE 8.4 Total gaseous mercury emissions aggregated for fall-season months of La Niña years for the periods 1997–2009 and 2010–2018 (top left and right). The bottom panels are the same except for summer months.

ACKNOWLEDGMENTS

Udaysankar Nair and Aaron Kaulfus are supported under the NSF CAREER Grant AGS-1352046.

REFERENCES

Andreae, A.O., and P. Merlet. 2019. Emission of Trace Gases and Aerosols from Biomass Burning. Global Biogeochemical. *Atmospheric Chemistry and Physics*. 15 (4): 955–66.

Carpi, A., A.H. Fostier, O.R. Orta, J.C. dos Santos, and M. Gittings. 2014. Gaseous mercury emissions from soil following forest loss and land use changes: Field experiments in the United States and Brazil. *Atmospheric Environment*. 96: 423–29. Elsevier Ltd doi:10.1016/j.atmosenv.2014.08.004.

Friedli, H. R., A. F. Arellano, S. Cinnirella, and N. Pirrone. 2009. Initial estimates of mercury emissions to the atmosphere from global biomass burning. *Environmental Science and Technology*. 43 (10): 3507–13. doi:10.1021/es802703g.

Giglio, L., J.T. Randerson, and G.R. Van Der Werf. 2013. Analysis of daily, monthly, and annual burned area using the fourth-generation global fire emissions database (GFED4). *Journal of Geophysical Research: Biogeosciences*. 118 (1): 317–28. doi:10.1002/jgrg.20042.

Grigal, D F. 2003. Mercury sequestration in forests and peatlands: A review. *Journal of Environmental Quality*. 32 (2): 393–405. doi:10.2134/jeq2003.3930.

Holmes, C.D., D.J. Jacob, and X. Yang. 2006. Global lifetime of elemental mercury against oxidation by atomic bromine in the free troposphere. *Geophysical Research Letters*. 33 (20): 1–5. Elsevier. doi:10.1029/2006GL027176.

Kohlenberg, A.J., M.R. Turetsky, D.K. Thompson, B.A. Branfireun, and C.P.J. Mitchell. 2018. Controls on boreal peat combustion and resulting emissions of carbon and mercury. *Environmental Research Letters.* 13 (3). doi:10.1088/1748-9326/aa9ea8.

Kumar, A., S. Wu, Y. Huang, H. Liao, and J.O Kaplan. 2018. Mercury from wild fi res: Global emission inventories and sensitivity to 2000–2050 global change. *Atmospheric Environment.* 17: 6–15. doi:10.1016/j.atmosenv.2017.10.061.

Li, L., F. Wang, B. Meng, M. Lemes, X. Feng, and G. Jiang. 2010. Speciation of methylmercury in rice grown from a mercury mining area. *Environmental Pollution.* 158 (10): 3103–7. doi:10.1016/j.envpol.2010.06.028.

Miettinen, J., C. Shi, and S.C. Liew. 2011. Deforestation Rates in Insular Southeast Asia between 2000 and 2010. *Global Change Biology.* 17: 2261–70. doi:10.1111/j.1365-2486.2011.02398.x.

Rose, M., C. Posa, L.S. Wijedasa, and R.T. Corlett. 2011. Biodiversity and conservation of tropical peat swamp forests. *BioScience.* 61 (49): 49–57. doi:10.1525/bio.2011.61.1.10.

Schroeder, W.H., and J. Munthe. 1998. Atmospheric mercury—An overview. *Atmospheric Environment.* 32 (5): 809–22. doi:10.1016/S1352-2310(97)00293-8.

Vadrevu, K.P., Ohara, T. and Justice, C. eds. 2018. *Land-Atmospheric Research Applications in South and Southeast Asia.* Springer, Cham.

Van Der Werf, G.R., Randerson, J.T., Giglio, L., Van Leeuwen, T.T., Chen, Y., Rogers, B.M., Mu, M., Van Marle, M.J., Morton, D.C., Collatz, G.J. and Yokelson, R.J., 2017. Global fire emissions estimates during 1997–2016. *Earth System Science Data.* 9(2), 697–720.

Zahir, F., S.J. Rizwi, S.K. Haq, and R.H. Khan. 2005. Low dose mercury toxicity and human health. *Environmental Toxicology and Pharmacology.* 20 (2): 351–60. doi:10.1016/j.etap.2005.03.007.

9 Biomass Burning Emissions in Indonesia and Policy Measures – An Overview

Lailan Syaufina
Bogor Agricultural University (IPB University), Indonesia

Sandhi Imam Maulana
The Ministry of Environment and Forestry, Indonesia

CONTENTS

INTRODUCTION

Biomass burning is an important source of emissions in several countries, including South/Southeast Asia (Gupta et al., 2001; Badarinath et al., 2008a, b; Vadrevu, 2008; Vadrevu and Justice, 2011). Biomass burning can alter the terrestrial ecosystems by disturbing the soil biogeochemistry as well as atmosphere through radiative forcing and energy balance (Kant et al., 2000; Prasad et al., 2000, 2002, 2003; Samset and Myhre, 2011; Yoon et al., 2019). Important causative factors of biomass burning include slash-and-burn agriculture, which is practiced by the indigenous people living in the hilly regions of South/Southeast Asian countries such as in northeast India, northern Myanmar, Laos, and Cambodia (Prasad et al., 2001, 2004, 2008; Prasad and Badarinth, 2004; Vadrevu et al., 2006, 2008, 2012, 2013; 2014a, b; 2015; Vadrevu and Lasko, 2015; Biswas et al., 2015a, b). In addition to slash and burn, in northwestern

India and Thailand, Red River Delta, and Mekong Delta, most of the biomass burning is attributed to agricultural residue burning (Vadrevu et al., 2017;2018; 2019; Vadrevu and Lasko, 2018; Oanh et al., 2018; Lasko et al., 2017, 2018a, b; Lasko and Vadrevu, 2018). Further, forest and peatland burning is most common in Indonesia, Malaysia, and Papua New Guinea (Carlson et al., 2013; Hayasaka et al., 2014; Israr et al., 2018). Indonesia has been experiencing a significant amount of forest loss, degradation, and fires recently. Millions of forest areas are being deforested and drained to make way for oil palm and pulpwood plantations (Hayasaka and Sepriando, 2018). Due to continuous degradation, forest canopies are opened up in Indonesia, making the forest fuels more susceptible to fires. For example, the large forest fire that occurred in 1982/1983 has burned about 3.6 million hectares of tropical forest in East Kalimantan. Burned area statistics obtained from the Ministry of Environment and Forestry, Indonesia, from 2014 to 2019 are shown in Figure 9.1. For monitoring the fire events in Indonesia and elsewhere, satellite remote sensing technology has proved quite efficient for mapping not only the forest cover changes, but also the biomass burning events and pollution (Prasad et al., 2004; Kant et al., 2000; Petropoulous et al., 2013; Justice et al., 2015; Saharjo and Yungan, 2018; Israr et al., 2018).

In Indonesia, support for fire control and research has been coming from different countries. Fire science and fire control and management practices are being implemented, *albeit* with less success. Studies on biomass burning emissions and human dimension aspects are still limited (Syaufina, 2008). During the 1997/1998 fire episode in Indonesia, which has been claimed as the largest fire incident in the 20th century due to the El Niño occurrence, the peat fires caused extensive transboundary haze pollution in Indonesia. As a response, a new era of biomass burning research has opened up that addressed the ecological, environmental, and climate-related consequences of biomass burning in Indonesia and the Southeast Asian region. This study discusses these aspects, specifically the major biomass burning events and emissions since 1997. We also review the future challenges and mitigation aspects for controlling the fires and biomass burning pollution in Indonesia.

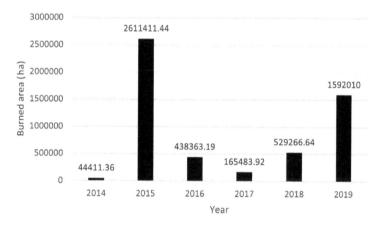

FIGURE 9.1 Burned area in Indonesia during the period 2014–2019. (Ministry of Environment and Forestry.)

BIOMASS BURNING EMISSIONS IN INDONESIA

Using satellite images covering the 2.5-Mha study area in Central Kalimantan, Borneo, during 1997, Page et al. (2002) calculated that 32% (0.79 Mha) of the area had burned, of which peatland accounted for 91.5% (0.73 Mha). The authors have also estimated that 0.19–0.23 gigatons (Gt) of carbon was released to the atmosphere through peat combustion, with a further 0.05 Gt released from the overlying vegetation's burning. Extrapolating these estimates to Indonesia as a whole, Page et al. (2002) estimated ~0.81–2.57 Gt of carbon released to the atmosphere from 1997 from biomass burning of peat and vegetation fires. This is equivalent to 13%–40% of the mean annual global carbon emissions from fossil fuels and contributed significantly to the largest yearly increase in atmospheric CO_2 concentration detected since records began in 1957 (Page et al., 2002).

Kiely et al. (2018) reported that peatland fires emit higher concentrations of particulate matter (PM) than vegetation fires. In Indonesia, vegetation and peatland fires are strongly regulated by climate, with greater fire emissions in drought years. Fire emissions exhibit strong interannual variability, and years with higher emissions can result in severe air pollution events. For example, during 2002, 2004, 2006, 2009, and 2015, Indonesia's fire emissions exceeded the long-term monthly mean by more than one standard deviation, all occurring during the dry season (August–October). The widespread wildfires that raged across Indonesia's island of Sumatra in June 2013 triggered the worst air pollution episode ever recorded in Singapore. The highest 24-hour moving-average $PM_{2.5}$ concentration reported by the authorities reached a whopping 310 g/m^3 on June 20, 2013. Results suggest a maximum likelihood of hourly $PM_{2.5}$ concentrations over 600 g/m^3, twice the maximum 24-h moving average reported by the authorities (Velasco and Rastan, 2015). The 2015–2016 strong El Niño event has dramatically impacted the amount of Indonesian biomass burning, with the El Niño-driven drought further desiccating the already-drier-than-normal landscapes result of decades of peatland draining, widespread deforestation, anthropogenically driven forest degradation, and previous large fire events. Even short, localized fire events in these environments can lead to significant greenhouse gas (GHG) emissions, as demonstrated by Gaveau et al. (2014). They report that a 1-week fire event in Riau Province (Sumatra) was responsible for emitting $172 \pm 59 GtCO_2$ eq., approximately 5%–10% of Indonesia's average annual GHG emissions.

Subsequently, during 2015, Indonesia had experienced an unprecedented number of fires as a result of droughts related to the recent El Niño event and human activities. These events released large amounts of carbon dioxide (CO_2) into the atmosphere. Emission databases such as the Global Fire Assimilation System version 1.2 and the Global Fire Emissions Database version 4s estimated the CO_2 emissions to be approximately 1100 $MtCO_2$ from July to November 2015. This emission was indirectly estimated by using parameters such as burned area, fire radiative power, and emission factors. Further, a unique study by Whitburn et al. (2016) report that among the primary species emitted by fires, ammonia (NH_3) is of particular relevance for air quality when thousands of square kilometers of forest and peatlands in Indonesia went up in flames. In their study on 2015 fires, Whitburn et al. (2016) derived daily and total NH_3 emission fluxes over the affected area using satellite measurements for the years 2008 to 2015. As a result, it is reported that the 2015 fires emitted an estimated 1.4–8.2 Tg of NH_3 (with a maximum of 0.06–0.33 Tg/day). On an annual basis, the 2015 NH_3 emissions are a factor 2–3 larger than in the previous 7 years. In the meantime, Whitburn et al. (2016) also derived NH_3 emission factors for peat soils, which were found to be 2.5–8 times lower than those used in the GFASv1.2 emission inventory but in excellent agreement with those reported in other recent studies. Finally, Whitburn et al. (2016) also estimated that $3.28 \times 109 \, m^3$ peat soil was consumed during these 2015 fires, corresponding to an average burn depth of 39 cm.

Fire episodes in several regions in Indonesia in the middle of August until November 2015 have also left a bitter memory. For example, during this period, Sultan Thaha Saifuddin Airport in Jambi was closed for about 2 months. All the airlines' employees still worked while waiting for the information on the possibility of airplanes to depart from Jakarta and return. The uncertainty lasted for almost 2 months. Schools were closed, and the students were provided with simple masks to face the air pollution from the smoke haze. While the air pollution index at the end of October 2015 was reported about 300–700 Air Pollution Index (API), the actual smoke haze began during August 2015. Almost every year, land fires happen; however, the event of smoke that was most acute after 1997 just happened again in 2015. The reason was almost the same, because of the presence of El Niño in some tropical regions close to the sea in the world plus the mistake of allowing.

OIL PALM EXPANSION AND PEATLAND EMISSIONS

Increasing prices and demand for biofuel and cooking oil from importer countries have caused a remarkable expansion of Indonesia's oil palm plantations in recent years. In particular, oil palm plantations expanded on the peatlands and tropical forests. Using Landsat images from three different periods (the 1990s, the 2000s, and 2012), Ramdani and Hino (2013) have classified the land use and land cover (LULC) of Riau Province, which is the largest oil palm-producing region in Indonesia. In this study, the authors used a hybrid method of integration, generated by combining automatic processing and manual analysis, yielding the best results. They report that tropical forest and peatland were the primary sources of emissions, while the low rain forest cover decreased from 63% in the 1990s to 37% in the 2000s. By 2012, the remaining tropical rain forest cover was only 22%. From the 1990s to the 2000s, total CO_2 emitted to the atmosphere was estimated at 26.6 million tCO_2/year, with 40.62% and 59.38% of the emissions from conversion of peatlands and forests, respectively. Between 2000 and 2012, the total CO_2 emitted to the atmosphere was estimated at 5.2 million tCO_2/year, with 69.94% and 27.62% of the emissions from converted peatlands and forests, respectively. These figures illustrate that there was significant growth in the oil palm industry in Riau Province from 1990 to 2000, with the transformation of tropical forest and peatland as the primary sources of emissions. Meanwhile, the decrease in CO_2 emissions in the period from 2000 to 2012 is possibly due to the enforcement of a moratorium on deforestation (Ramdani and Hino 2013).

From the literature we estimate that total carbon loss from converting peat swamp forests into oil palm is 59.4 ± 10.2 Mg of CO_2 per hectare per year during the first

25 years of LULC change, of which 61.6% arises from the peat. Of the total amount (1.486 ± 183 Mg of CO_2 per hectare over 25 years), 25% is released immediately from land-clearing fire. To maintain high palm oil production, nitrogen inputs through fertilizer are needed, and the magnitude of the resulting increased N_2O emissions compared to CO_2 losses remains unclear (Murdiyarso et al. 2010).

FUTURE CHALLENGES AND POLICY IMPLICATIONS

FIRE PREVENTION LAWS

Forest and land fires in Indonesia may not be easy to control as the burning of biomass has been a traditional practice to clear the land for centuries. However, there is tremendous pressure to carry mitigation measures to curb pollution from the academicians, government, private sector, and the local community (Syaufina 2018). A recent study in regard to current laws and regulations implemented to control forest and land fires in Indonesia conducted by Maulana (2019) shows that although policies and strategies have undergone many developments, several challenges remain on implementation.

To begin with, at a national level, the legal basis for forest and land fire prevention has been stated in Law 41/1999 on Forestry, Law 24/2007 on Disaster Management, and Law 32/2009 on Environmental Protection and Management. In Law 41/1999, fire prevention is included in the section on forest protection and nature conservation. In Law 24/2007, it is regulated as part of disaster management efforts at the pre-disaster stage. Meanwhile, in Law 32/2009, the prevention of forest fires is included in the section on controlling environmental damage. Overall, of those three laws, it seems that there is no dedicated section that regulates forest and land fire control mechanisms specifically. At the central level, this kind of mechanism was elaborated in the form of government regulations (PP), presidential instructions (Inpres), and ministerial regulations (PerMen), and circular (SE). Following up on Law 41/1999, in 2001, the government issued PP 4/2001 concerning Control of Environmental Damage or Pollution Related to Forest and Land Fires. In this government regulation, the prevention is defined as "an effort to maintain the function of forests and land through ways that do not provide opportunities for environmental damage and pollution-related to the forest and land fires." Although, by definition, the government has determined prevention efforts as actions that do not allow the occurrence of damage, in this PP, it is stated that one of the means of prevention of forest fires is an early-detection system to determine the occurrence of forest and land fires. This is contrary to the PP's preventive definition because early detection means that forest and land fires have occurred even on a scale that is not too broad. Three years later, the government issued PP 45/2004 concerning Forest Protection.

In contrast to PP 4/2001, in PP 24/2004 Article 24, early detection is categorized in fire suppression instead of fire prevention. However, in PP 57/2016 concerning amendments to PP 71/2014 on Peat Ecosystems Protection and Management, early detection was again grouped in the field of fire prevention. This different textual content indicates the inconsistency of the prevailing laws and regulations at the central level, as Budiningsih (2017) reported. Other issues related to the unsynchronized content between those PPs related to forest and land fire control were highlighted (Uda et al., 2018; Winarna, 2015; IOPRI, 2017).

Zero-Burning Policy

The government's attention to controlling forest fires on peatlands has begun to increase since 2009. That year, the government started implementing zero-burning policy, which is basically a policy that prohibits open burning but still allows some forms of controlled combustion (Rahmat and Fadli, 2016). Initially, this policy was known as the ASEAN Agreement on Transboundary Haze Pollution (AATHP), wherein Article 9 of this agreement stated that "Each party must take measures to prevent and control activities related to land and/or forest res that may lead to transboundary haze pollution, which includes: Developing and implementing legislative and other regulatory measures to promote zero burning policy to deal with land and/or forest fires resulting in transboundary haze pollution." Although the Indonesian government did not ratify the agreement in 2009, this kind of policy was regulated in Law 41/1999 Article 50, Law 32/2009 Article 69 paragraphs 1 and 2, and PP 4/2001 Article 11. Considering the existing laws and regulations, the Ministry of Agriculture then issued Permentan 14/2009 on Guidelines for the Use of Peatlands for Oil Palm Cultivation. In 2014, the government then ratified the AATHP in the form of Law 26/2014 on AATHP Approval. In line with this step, and as a follow-up to the direction of the President of the Republic of Indonesia at the National Coordination Meeting on October 23, 2015, and November 2015, the Ministry of Environment and Forestry issued two circular letters, namely SE MenLHK S.494/2015 on Prohibition of Peatland Clearing and SE MenLHK S.661/2015 on Peatland Management Instruction.

Subsequently, in anticipation of peatland fires, on June 1, 2016, the Peatland Restoration Agency (BRG), which was just formed through Presidential Regulation 1/2016, issued a circular S.01/BRG-KB/6/2016 concerning Preparedness for the Dry Season. In the same year, considering the need for a general guideline for mitigating forest and land fires, while also considering PP 4/2001 Article 18, the Ministry of Environment and Forestry issued PermenLHK 32/2016 on Forest and Land Fires Control. However, compared with PP 4/2001, the PermenLHK sets more loosely limits on forest fire prevention. Article 1 states that "Prevention of Forest Fire is all efforts, actions or activities carried out to prevent or reduce the possibility of forest fires and land." In contrast, PP 4/2001 explicitly states that prevention is carried out in various ways that do not provide an opportunity for forest fires. Therefore, it is not surprising that in PermenLHK 32/2016, early detection is categorized as one of the means in the field of forest fire prevention.

Fire Prevention in Peatlands

As per the PermenLHK 32/2016, forest fire prevention efforts should implement fire suppression measures as soon as possible when the fire hotspots are visible, as is the working principle of an early-detection system. According to Budiningsih (2017), this kind of prevention approach is less effective if fires occur on peatland because peatland fires are challenging to suppress. For example, even a water-bombing strategy carried out on burned peatlands tends to exacerbate the smoke scale that emerges. Furthermore, this once again indicates that although the direction of the forestry control policy has begun to focus on the prevention efforts, the field's strategy still relies on suppression approaches. In line with this view, several studies related to the implementation of forest and land fire control policies (Meiwanda, 2016; Budiningsih, 2017; Purnomo et al., 2017; Supriyanto et al., 2018) also reveal a similar finding, where the basic steps applied in the field still largely focus on the strategy of monitoring/detecting hotspots and then suppressing the fires as early as possible (detect and suppress). This kind of strategy, in principle, is more of a reactive and non-preventive countermeasure. Moreover, it is also considered ineffective to be applied in controlling peatland fires (Budiningsih, 2017). Based on a solely operational point of view, the application of early detection based on hotspot monitoring does indeed have an advantage in obtaining near-real-time data on fires and the large area that can be monitored. Such hotspot maps are distributed by different sources in Indonesia (Figure 9.2). However, the use of hotspot data also has several disadvantages. Firstly, hotspot monitoring tends to produce low accuracy in indicating the occurrences of forest and land fires, as evidenced in the study by Vetrita and Haryani (2012), who reveal that the accuracy of MODIS hotspot images in Riau Province was only 43%. Secondly, the ground check's need to validate hotspots that emerge is often constrained by fairly heavy terrain conditions. This situation implies a higher budget and longer time needed to conduct such field checks (Vetrita and Haryani, 2012). Thirdly, during ground checking activities in Riau Province, there were often fires, but not detected by the satellite data (i.e., omission errors); thus, much higher resolution remote sensing products are required.

FIRE PREVENTION IN PLANTATION AREAS

Specific to the plantations, the Ministry of Agriculture issued Permentan 5/2018 on the clearing and/or processing of plantation land without burning. In contrast to

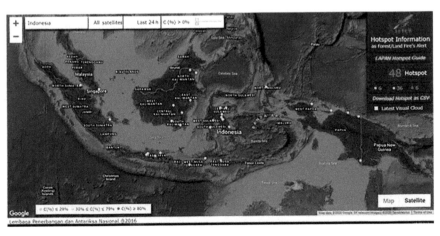

FIGURE 9.2   Hotspot distribution map published by the National Agency for Aviation and Aerospace.

PermenLHK 32/2016, the fire control guidelines at Permentan 5/2018 are devoted to all land/plots that are used for plantation businesses which are indeed under the Ministry of Agriculture. The regulation on plantations states that "The construction of a drainage system consists of a working network consisting of trenches which function to keep the water level always in the range of 50–70 cm." This drainage system also functions as a means for periodically rinsing peatland from excessive acidity. These trenches need to be cleaned periodically so that the circulation of water on the peatland runs smoothly. Determination of peat water level in the range of 50–70 cm is clearly contrary to PP 71/2014 Jo PP 57/2016, which states that peat ecosystems with cultivation functions are declared damaged if the water level on peatland exceeds 40 cm below the peat surface. However, the determination of the water level range of 50–70 cm in Permentan 5/2018 has been in line with the results of scientific research, such as that carried out by Winarna (2015) which shows that the management of groundwater depths ≤ 70 cm below the soil surface is still able to show a decrease in peat soil CO_2 emissions with a cover of 6-year-old oil palm plantations. In addition, the management of groundwater depths ≤ 70 cm below the soil surface can still increase palm oil production by 6 years in the first year of research to 8%–11%. In addition, the peatland with water level of 0.4–0.7 m may not be categorized as damaged peat (Winarna, 2015). For addressing fires in the peatlands, information of water levels is very important. Such information is provided through the Peatland Restoration Agency, Indonesia (Figure 9.3).

Fire Prevention Regulatory Measures and Status

The regulatory analysis results based on the principle of sustainable development conducted by Maulana (2019) in Indonesia indicate that current forest and land fire policies tend to pay more attention to institutional, environmental, and social aspects compared to the economic and technological aspects. This finding is in line with Rosul (2015),

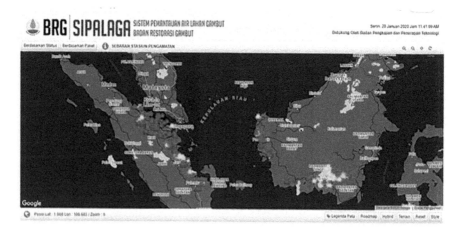

FIGURE 9.3   Peatland groundwater level monitoring system published by the Peatland Restoration Agency.

Purnomo et al. (2017), and Uda et al. (2018), who revealed the lack of government attention to economic and technological aspects in preventing forest and land fires in Indonesia. Specifically, concerning economic aspects, it appears that the most dominant preventive policy motive is the internalization of costs incurred concerning forest and land fires to those responsible for businesses that utilize forest and land resources (Quah, 2002; Maulana, 2019). The embodiment of this motive may be seen from the implementation of three strategic approaches: first, requiring the business authorities to provide forest and land fire control facilities and infrastructure in their concession areas; second, obliging business executives to bear operational costs incurred in relation to forest and land fire control in their concession areas; and third, commanding business responsible parties to provide performance guarantee funds or damage compensation funds in the event of forest and land fires at their concession areas. This motive group's main objective is to put pressure so that concession holders may not use fire anymore and increase their attention so as not to cause forest and land fires (Medrilzam et al., 2017; Uda et al., 2018).

In contrast, providing incentives for zero burning and alternative income sources for people who depend on forest and land resources by the government is still limited. Technically, the government regulates incentives for the implementation of zero-burning land management through the provision of subsidies and assistance with facilities needed for mechanical processing. Meanwhile, the provision of alternative community income is generally implemented by facilitating alternative sources of income/livelihoods for people who are engaged in the forest and land-use sector, as well as providing guarantees for input and output markets for alternative businesses (Herawati and Santoso, 2011; Rosul, 2015; Aminingrum, 2017; Purnomo et al., 2017; Uda et al., 2018). In general, those studies contend that the slash-and-burn method, especially carried out by smallholder plantations and traditional communities, is likely to continue due to the lack of incentives and alternative financial sources provided during zero-burning policy implementation.

Specific to the technology, according to Maulana (2019), policy motives in the field of forest and land fire prevention can be grouped into three, including the development of early-warning systems, the development of early-detection systems, and the development of land management technologies. Technically, early-warning application so far has been applied through two approaches, namely conventional warning facilities and warning facilities based on geographic information systems (GIS). An example of conventional warning facilities' application is by directly installing fire-related information and warning board signs. Meanwhile, GIS-based warning facilities, in general, can be implemented, such as the preparation of fire-prone maps, and advanced prediction systems such as the Fire Danger Rating System (FDRS) that has been built by BMKG with a lead time of 7 days before the incident (Uda et al., 2018). In conjunction with this prediction system, by taking into account the latest developments in modeling the character of seasonal climate and patterns of forest and land burning (McInerney et al., 2013; Field et al., 2016, Maulana et al., 2019), there is still much room for improvements that can be done, in terms of both accuracy and its prediction lead time. For example, a study conducted by Maulana et al. (2019) has successfully established a peatland fire prediction model for the Bengkalis Regency of Riau Province to provide preparation time for control of fires up to 2 months before the occurrence of fire episode.

Like the fire early-warning systems, an early-detection system can also be applied through a direct approach in the field and by utilizing remote sensing technology to monitor the fire hotspots. Technically, early-detection implementation is generally carried out by monitoring efforts using monitoring towers, ultralight trikes, or similar aircraft to drones. As for the use of remote sensing technology, before 2015, the Ministry of Environment and Forestry relied only on one data source; currently, it has increased to three sources of satellite imagery simultaneously, namely MODIS (Terra and Aqua), AVHRR (NOAA), and VIIRS (Suomi-NPP). Although there has been an increase in the data sources, the implementation of early detection based on remote sensing also has problems, such as weak communication and coordination systems in the field, limited budget for ground checks, and low capacity of human resources for data analysis (Purnomo et al., 2017; Supriyanto et al., 2018). Relative to these technological approaches, integrated land management practices involving local people, groups, and communities might effectively control fires.

CONCLUSIONS

Biomass burning in Indonesia is an important topic to address as it releases several GHGs into aerosols. Due to the transboundary nature of the biomass burning pollutants, the smoke emissions from Indonesia also affect other neighboring countries of Singapore, Malaysia, and southern Thailand. The Indonesian government came up with several policies to control fires; the merits and demerits of the same were discussed in the study. Specific to the fire control, we infer that the effectiveness of the government's policies and strategies in combating the recurring fire events is still low in Indonesia. Although there has been a paradigm shift in forest and land fire control policies, where the government is currently more focused on prevention, the strategy applied in the field still relies on reactive efforts through monitoring fire hotspots to then be suppressed as soon as possible (*detect and suppress*), which principally resemble a more non-preventive countermeasure. Such a strategy is not effective if it continues to be applied, particularly to areas dominated by peatlands as most of the fires are smoldering fires, and satellites may not detect the same. In order to improve the effectiveness of fire prevention policies, it is necessary to develop a fire control model that is more grounded in predictive efforts. This kind of control model may not only help the local communities, the government, and the private sector to establish rapid coordination among each other but also enable the authorities to allocate their limited resources appropriately through a linkage between fires prediction and their suitable preventive actions (*predict and prevent*). In addition, mechanisms for providing economic and technological support related to land management are needed, especially at the site level. The role of economic incentives in managing forest fires should also be prioritized since locals cause most of the fires due to economic reasons.

REFERENCES

Aminingrum. 2017. Forest fire contest: the case of forest fire policy design in Indonesia (Theses). The Hague (NL): International Institute of Social Studies.

Badarinath, K.V.S., Kharol, S.K., Krishna Prasad, V., Kaskaoutis, D.G., and Kambezidis, H.D. 2008a. Variation in aerosol properties over Hyderabad, India during intense cyclonic conditions. *International Journal of Remote Sensing*. 29(15): 4575–4597.

Badarinath, K.V.S., Kharol, S.K., Prasad, V.K., Sharma, A.R., Reddi, E.U.B., Kambezidis, H.D., and Kaskaoutis, D.G. 2008b. Influence of natural and anthropogenic activities on UV Index variations–a study over tropical urban region using ground based observations and satellite data. *Journal of Atmospheric Chemistry*. 59(3): 219–236.

Biswas, S., Lasko, K.D. and Vadrevu, K.P. 2015a. Fire disturbance in tropical forests of Myanmar –Analysis using MODIS satellite datasets. *IEEE Journal of Selected Topics in Applied Earth Observations and Remote Sensing*. 8(5): 2273–2281.

Biswas, S., Vadrevu, K.P., Lwin, Z.M., Lasko, K. and Justice, C.O. 2015b. Factors controlling vegetation fires in protected and non-protected areas of Myanmar. *PLoS One*. 10(4): e0124346.

Budiningsih, K. 2017. Implementasi kebijakan pengendalian kebakaran hutan dan lahan di Provinsi Sumatera Selatan. *Jurnal Analisis Kebijakan Kehutanan*. 14(2):165–186.

Carlson, K.M., Heilmayr, R., Gibbs, H.K., Noojipady, P., Burns, D.N., Morton, D.C., Walker, N.F., Paoli, G.D., and Kremen, C. 2018. Effect of oil palm sustainability certification on deforestation and fire in Indonesia. *Proceedings of the National Academy of Sciences*, 115(1): 121–126.

Field, R.D., van der Werf, G.R., Fanin, T., Fetzer, E.J., Fuller, R., Jethva, H., Levy, R., Livesey, N.J., Luo, M., Torres, O., and Worden, H.M. 2016. Indonesian fire activity and smoke pollution in 2015 show persistent nonlinear sensitivity to El Nino-induced drought. *Proceedings of the National Academy of Sciences of the United States of America* (PNAS). 113(33): 9204–9209.

Gaveau, D.L., Salim, M.A., Hergoualc'h, K., Locatelli, B., Sloan, S., Wooster, M., Marlier, M.E., Molidena, E., Yaen, H., DeFries, R., Verchot, L., Murdiyarso, D., Nasi, R., Holmgren, P., and Sheil, D. 2014. Major atmospheric emissions from peat fires in Southeast Asia during non-drought years: Evidence from the 2013 Sumatran fires. *Scientific Reports*. 4(6112). doi:10.1038/srep06112.

Gupta, P.K., Prasad, V.K., Sharma, C., Sarkar, A.K., Kant, Y., Badarinath, K.V.S., and Mitra, A.P. 2001. CH_4 emissions from biomass burning of shifting cultivation areas of tropical deciduous forests–experimental results from ground-based measurements. *Chemosphere-Global Change Science*. 3(2): 133–143.

Hayasaka, H., and A. Sepriando. 2018. Severe air pollution due to peat fires during 2015 super El Niño in Central Kalimantan, Indonesia. In: *Land-Atmospheric Research Applications in South/Southeast Asia*. Vadrevu, K.P., Ohara, T., and Justice, C. (Eds.). Springer, Cham, pp. 129–142.

Hayasaka, H., Noguchi, I., Putra, E.I., Yulianti, N., and Vadrevu, K. 2014. Peat-fire-related air pollution in Central Kalimantan, Indonesia. *Environmental Pollution*. 195: 257–266. doi:10.1016/j.envpol.2014.06.031.

Herawati, H., and Santoso, H. 2011. Tropical forest susceptibility to and risk of fire under changing climate: A review of fire nature, policy and institutions in Indonesia. *Forest Policy and Economics*. 13(4): 227–233.

[IOPRI] Indonesia Oil Palm Research Institute. 2017. *Permasalahan Implementasi PP 57 Tahun 2016 di Perkebunan Kelapa Sawit di Lahan Gambut*. Retrieved 10 October 2017 from https: //www.iopri.org/wp-content/uploads/2017/07/IX-01.-MATERI-PTKS-2017-SOLO_WIN.pdf.

Israr, I., Jaya, S.N.I, Saharjo, H.S., Kuncahyo, B., and Vadrevu, K.P. 2018. Spatio-temporal analysis of land and forest fires in Indonesia using MODIS active fire dataset. In: *Land-Atmospheric Research Applications in South/Southeast Asia*. Vadrevu, K.P., Ohara, T., and Justice, C. (Eds.). Springer, Cham, pp.105–128.

Justice, C., Gutman, G., and Vadrevu, K.P. 2015. NASA land cover and land use change (LCLUC): An interdisciplinary research program. *Journal of Environmental Management.* 148(15): 4–9.

Kant, Y., Prasad, V.K., and Badarinath, K.V.S., 2000. Algorithm for detection of active fire zones using NOAA AVHRR data. *Infrared Physics & Technology,* 41(1): 29–34.

Kiely, L., Spracklen, D., Arnold, S., Marsham, J., Reddington, C., Conibear, L., Knote, C., Kuwata, M., and Budisulistiorini, S.H. 2018. Impact of Indonesian fires on Equatorial Asian air quality between 2002 and 2015. In *Proceedings of the 20th EGU General Assembly 2018 4–13 April,* 2018 in Vienna, Austria, p. 14579.

Lasko, K., Vadrevu, K.P., Tran, V.T., Ellicott, E., Nguyen, T.T., Bui, H.Q., and Justice, C. 2017. Satellites may underestimate rice residue and associated burning emissions in Vietnam. *Environmental Research Letters.* 12(8): 085006.

Lasko, K., and Vadrevu, K.P. 2018. Improved rice residue burning emissions estimates: Accounting for practice-specific emission factors in air pollution assessments of Vietnam. *Environmental Pollution.* 236(5): 795–806.

Lasko, K., Vadrevu, K.P., and Nguyen, T.T.N. 2018a. Analysis of air pollution over Hanoi, Vietnam using multi-satellite and MERRA reanalysis datasets. *PloS One.* 13(5): e0196629.

Lasko, K., Vadrevu, K.P., Tran, V.T., and Justice, C. 2018b. Mapping double and single crop Paddy rice with Sentinel-1A at varying spatial scales and polarizations in Hanoi, Vietnam. *IEEE Journal of Selected Topics in Applied Earth Observations and Remote Sensing.* 11(2): 498–512.

Maulana, S.I., Syaufina, L., Prasetyo, L.B., and Aidi, M.N. 2019. Spatial logistic regression models for predicting peatland fire in Bengkalis Regency, Indonesia. *Journal of Sustainability Science and Management.* 14(3): 55–66.

Maulana. S.I. 2019. Model pengendalian prediktif-preventif kebakaran hutan dan lahan gambut di Kabupaten Bengkalis, Riau (Dissertation). IPB University: Bogor, Indonesia.

McInerney, D., San-Miguel-Ayanz, J., Corti, P., Whitmore, C., Giovando, C., and Camia, A. 2013. Design and Function of the European forest fire information system. *Photogram Engineering and Remote Sensing.* 79: 965–973.

Medrilzam, R.N.H., Widiaryanto, P., Rosylin, L., Firdaus, R., Suprapto, U., Sumantri, Purnomo, H., Wulan, Y.C., Tarigan, M.L.P., and Nugraha, M. 2017. *Grand Design Pencegahan Kebakaran Hutan, Kebun, dan Lahan 2017–2019 Republik Indonesia.* Direktorat Kehutanan dan Sumberdaya Air BAPPENAS: Jakarta, Indonesia.

Meiwanda, G. 2016. Kapabilitas Pemerintah Daerah Provinsi Riau: Hambatan dan Tantangan Pengendalian Kebakaran Hutan dan Lahan. *Jurnal Ilmu Sosial Dan Ilmu Politik.* 19(3): 251–263.

Murdiyarso, D., Hergoualc'h, K., and Verchot, L.V. 2010. Opportunities for reducing greenhouse gas emissions in tropical peatlands. *Proceedings of the National Academy of Sciences of the United States of America.* 107(46): 19655–19660.

Oanh, N.T.K., Permadi, D.A., Dong, N.P., and Nguyet, D.A. 2018. Emission of toxic air pollutants and greenhouse gases from crop residue open burning in Southeast Asia. In: Vadrevu, K.P., Ohara, T., and Justice, C. (Eds). *Land-Atmospheric Research Applications in South/Southeast Asia.* Springer, Cham, pp. 47–68.

Page, S.E., Siegert, F., Rieley, J.O., Boehm, H.D.V., Jaya, A., and Limin, S. 2002. The amount of carbon released from peat and forest fires in Indonesia during 1997. *Nature.* 420:61–65.

Prasad, V.K., Badarinath, K.V.S., and Eaturu, A. 2008. Biophysical and anthropogenic controls of forest fires in the Deccan Plateau, India. *Journal of Environmental Management.* 86(1): 1–13.

Prasad, V.K., Badarinath, K.V.S., Yonemura, S., and Tsuruta, H. 2004. Regional inventory of soil surface nitrogen balances in Indian agriculture (2000–2001). *Journal of Environmental Management.* 73(3): 209–218.

Prasad, V.K., Gupta, P.K., Sharma, C., Sarkar, A.K., Kant, Y., Badarinath, K.V.S., Rajagopal, T., and Mitra, A.P. 2000. NOx emissions from biomass burning of shifting cultivation areas from tropical deciduous forests of India–estimates from ground-based measurements. *Atmospheric Environment*. 34(20): 3271–3280.

Prasad, V.K., Kant, Y., and Badarinath, K.V.S., 2001. CENTURY ecosystem model application for quantifying vegetation dynamics in shifting cultivation areas: A case study from Rampa Forests, Eastern Ghats (India). *Ecological Research*. 16(3): 497–507.

Prasad, V.K., Kant, Y., Gupta, P.K., Elvidge, C., and Badarinath, K.V.S. 2002. Biomass burning and related trace gas emissions from tropical dry deciduous forests of India: A study using DMSP-OLS data and ground-based measurements. *International Journal of Remote Sensing*. 23(14): 2837–2851.

Prasad, V.K., Lata, M., and Badarinath, K.V.S. 2003. Trace gas emissions from biomass burning from northeast region in India—estimates from satellite remote sensing data and GIS. *Environmentalist*. 23(3): 229–236.

Purnomo, H., Shantiko, B., Sitorus, S., Gunawan, H., Achdiawan, R., Kartodihardjo, H., and Dewayani, A.A. 2017. Fire economy and actor network of forest and land fires in Indonesia. *Forest Policy and Economics*. 78: 21–31.

Quah, E. 2002. Transboundary pollution in Southeast Asia: The Indonesian fires. *World Development*. 30(3): 429–441.

Rahmat, F., and Fadli, M. 2016. Reformulasi *zero burning policy* pembukaan lahan di Indonesia. *Jurnal Legislasi Indonesia*. 13(1): 85–96.

Ramdani, F., and Hino, M. 2013. Land use changes and GHG emissions from tropical forest conversion by oil palm plantations in Riau Province, Indonesia. *PLOS ONE*. 8(7): e70323. doi:10.1371/journal.pone.0070323.

Rosul, P. 2015. Analysis of policy effectiveness on forests fires in Riau, Indonesia (Thesis). Duke University: North Carolina, US.

Saharjo, B.H., and Yungan, A. 2018. Forest and land fires in Riau province; A Case study in fire prevention, policy implementation with local concession holders. In: *Land-Atmospheric Research Applications in South/Southeast Asia*. Vadrevu, K.P., Ohara, T., and Justice, C. (Eds.). Springer, Cham, pp. 143–170.

Samset, B.H., and Myhre, G., 2011. Vertical dependence of black carbon, sulphate and biomass burning aerosol radiative forcing. *Geophysical Research Letters*, 38(24): 24802.

Supriyanto, Syarifudin, and Ardi. 2018. Analisis kebijakan pencegahan dan pengendalian kebakaran hutan dan lahan di Provinsi Jambi. *Jurnal Pembangunan Berkelanjutan*. 1(1): 94–104.

Syaufina, L. 2008. *Forest and Land Fires in Indonesia (in Indonesian)*. Bayumedia: Malang, Indonesia. In Bahasa.

Syaufina, L. 2018. Forest and Land Fires in Indonesia: Assessment and Mitigation in: Integrating Disaster Science and Management Global Case Studies in Mitigation and Recovery (P. Samui, D. Kim, C. Ghosh Eds). Elsevier. DOI: 10.1016/B978-0-12-81205 6-9.00008-7

Uda, S.K., Schouten, G., and Hein, L. 2018. The institutional fit of peatland governance in Indonesia. *Land Use Policy*. 1–8. doi:10.1016/j.landusepol.2018.03.031.

Vadrevu, K.P., Ohara, T., and Justice, C. 2014a. Air pollution in Asia. *Environmental Pollution*. 12: 233–235.

Vadrevu, K.P., and Justice, C.O. 2011. Vegetation fires in the Asian region: satellite observational needs and priorities. *Global Environmental Research* 15(1): 65–76.

Vadrevu, K.P., 2008. Analysis of fire events and controlling factors in eastern India using spatial scan and multivariate statistics. *Geografiska Annaler: Series A, Physical Geography*. 90(4): 315–328.

Vadrevu, K.P., and Lasko, K. 2018. Intercomparison of MODIS AQUA and VIIRS I-Band fires and emissions in an agricultural landscape—Implications for air pollution research. *Remote Sensing*. 10(7): 978. doi:10.3390/rs10070978.

Vadrevu, K.P., and Lasko, K.P., 2015. Fire regimes and potential bioenergy loss from agricultural lands in the Indo-Gangetic plains. *Journal of Environmental Management.* 148: 10–20.

Vadrevu, K.P., Badarinath, K.V.S., and Anuradha, E. 2008. Spatial patterns in vegetation fires in the Indian region. *Environmental Monitoring and Assessment,* 147(1–3): 1. doi:10.1007/s10661-007-0092-6.

Vadrevu, K.P., Csiszar, I., Ellicott, E., Giglio, L., Badarinath, K.V.S., Vermote, E., and Justice, C. 2012. Hotspot analysis of vegetation fires and intensity in the Indian region. *IEEE Journal of Selected Topics in Applied Earth Observations and Remote Sensing.* 6(1): 224–238.

Vadrevu, K.P., Eaturu, A., and Badarinath, K.V.S. 2006. Spatial distribution of forest fires and controlling factors in Andhra Pradesh, India using spot satellite datasets. *Environmental Monitoring and Assessment.* 123(1–3): 75–96.

Vadrevu, K.P., Lasko, K., Giglio, L., and Justice, C. 2014b. Analysis of Southeast Asian pollution episode during June 2013 using satellite remote sensing datasets. *Environmental Pollution.* 12: 245–256.

Vadrevu, K.P., Lasko, K., Giglio, L., and Justice, C. 2015a. Vegetation fires, absorbing aerosols and smoke plume characteristics in diverse biomass burning regions of Asia. *Environmental Research Letters.* 10(10): 105003.

Vadrevu, K.P., Lasko, K., Giglio, L., Schroeder, W., Biswas, S., and Justice, C. 2019. Trends in vegetation fires in south and southeast Asian countries. *Scientific Reports.* 9(1): 7422. doi:10.1038/s41598-019-43940-x.

Vadrevu, K.P., Ohara, T., and Justice, C. 2017. Land cover, land use changes and air pollution in Asia: A synthesis. *Environmental Research Letters.* 12(12): 120201.

Vadrevu, K.P., Ohara, T., and Justice, C. eds., 2018. *Land-Atmospheric Research Applications in South and Southeast Asia.* Springer, Cham.

Velasco, E., and Rastan, S. 2015. Air quality in Singapore during the 2013 smoke-haze episode over the Strait of Malacca: Lessons learned. *Sustainable Cities and Society.* 17: 122–131.

Vetrita, Y., and Haryani, N.S. 2012. Validasi *hotspot* MODIS indofire di Provinsi Riau. *Jurnal Ilmiah Geomatika.* 18(1): 17–28.

Whitburn, S., Van Damme, M., Clarisse, L., Turquety, S., Clerbaux, C., and Coheur, P.F. 2016. Doubling of annual ammonia emissions from the peat fires in Indonesia during the 2015 El Niño. *Geophysical Research Letters.* 43: 11007–11014. doi:10.1002/2016GL070620.

Winarna. 2015. Pengaruh kedalaman muka air tanah dan dosis terak baja terhadap hidrofobisitas tanah gambut, emisi karbon, dan produksi kelapa sawit (Dissertation). IPB University: Bogor, Indonesia.

Yoon, J., Chang, D.Y., Lelieveld, J., Pozzer, A., Kim, J., and Yum, S.S. 2019. Empirical evidence of a positive climate forcing of aerosols at elevated albedo. *Atmospheric Research.* 229: 269–279.

10 PM$_{2.5}$ Emissions from Biomass Burning in South/Southeast Asia – Uncertainties and Trade-Offs

Kristofer Lasko
Engineer Research and Development Center, USA

Krishna Prasad Vadrevu
NASA Marshall Space Flight Center, USA

Varaprasad Bandaru
University of Maryland, USA

Thanh Nhat Thi Nguyen and Hung Quang Bui
Vietnam National University, Vietnam

CONTENTS

INTRODUCTION

Nearly 90% of global paddy rice production is concentrated in Asia, with China, India, Indonesia, Bangladesh, Vietnam, and Thailand (FAOSTAT 2018). Rice-planted area in South Asia exceeded 61 million ha in 2017 and 50 million ha in Southeast Asia (FAOSTAT 2018). After the crop is harvested, the mineral-rich crop residues remaining in the field are frequently burned, collected for animal fodder, cookstoves, and bioenergy, reincorporated into the soil, and left remaining to decay the field, or often a combination of the mentioned activities (Duong and Yoshiro 2015). In the South/Southeast Asian region, rice is estimated to account for about 55% of agricultural residues, followed by maize (21%), wheat (18%), and sugarcane (6%) (FAOSTAT 2018).

SATELLITE-BASED DATASETS DO NOT CAPTURE A LARGE PORTION OF AGRICULTURAL BURNING

A robust methodology does not accurately quantify emissions from agricultural burning at a national or regional scale. Crop residue burning, such as that from rice, is unique compared to other biomass burning types such as forest or shrublands. Farmers actively manage rice fields, and the biomass is easy to maneuver. As a result, the residues are sometimes burned in small piles spread throughout the field. Therefore, direct satellite remote sensing of burning is currently limited due to very small fire size; high-temporal-resolution polar-orbiting satellites such as MODIS and VIIRS have difficulty monitoring these agricultural fires (Zhang et al. 2018; Liu et al. 2019), whereas geostationary satellites have limited spatial coverage with coarse spatial resolution. Moreover, cloud coverage can amplify fire monitoring, which is consistent and high throughout many rice-growing areas of South/Southeast Asia (S/SEA), obstructing multispectral satellite observations. Other challenges in monitoring agricultural fires include active field management, which can obfuscate the burn's spectral signature (e.g., tilling), where plowed or wet fields can have similar spectral signatures as burned fields (Hall et al. 2016).

EMISSIONS QUANTIFICATIONS OF AGRICULTURAL BURNING RELY ON TENUOUS ASSUMPTIONS

Challenges remain in burned area-based agricultural fire emissions quantification using globally prominent satellite datasets such as MODIS burned area product

(Hall et al. 2016; Giglio et al. 2018). Within a MODIS burned area pixel, there are several assumptions in emissions assessments: (1) The entire 500 m × 500 m pixel is assumed to be burned; (2) emissions assessments often generalize the land cover as agriculture without a crop, burning practice, or region-specific emission factors (EFs); and (3) imperfect ancillary satellite datasets are used to assume the amount of dry biomass in a given pixel. The MODIS burned area product underestimates agricultural burning due to inherent difficulty in observing these small fires, with difficulty exacerbated by local ground or atmospheric conditions (Hall et al. 2016; Lasko et al., 2017). On the second point, agricultural emissions can vary widely due to spatial and temporal variation in residue amounts (straw and standing stalk variation), amounts burned, EFs for different crops, and burning practice, which impacts combustion efficiency (e.g., piled vs. non-piled burning). For the third point, it is difficult to know exactly how much agricultural biomass is burned because varying proportions of the biomass in the pixel can be burned, especially based on whether the field was hand-harvested or combine-harvested, resulting in different stubble heights and residue amounts (Lasko and Vadrevu 2018; Lasko et al. 2018a). Moreover, our own field experience has shown that residues are occasionally burned in adjacent areas after machine-threshing, away from where they were harvested. In such cases, residues from many fields could be consolidated into one small area and burned, which could reasonably lead to underestimating the amount burned and resulting fire emissions. Further, other types of unrelated burning occur near agricultural fields, such as garbage burning, which could be confused with agricultural burning (Wiedinmyer et al. 2014), and agricultural fires can also get confused with the other fire types in mosaic, smallholder landscapes.

EMISSION FACTORS FOR AGRICULTURAL BURNING SHOULD BE DYNAMIC

Additional challenges remain in agricultural emissions quantification, especially on the use of static, non-dynamic EFs for burning. The amount of pollution emitted varies significantly across seasons and years based on the burning practices and field and weather conditions (Van Leeuwen et al. 2013; Arai et al. 2015; Inomata et al. 2015; Romasanta et al. 2017; Lasko and Vadrevu 2018; Lasko et al. 2018a, b). Piled residues often result in high EFs due to incomplete combustion associated with moist conditions associated with fields harvested by hand, whereas open burning associated with combine-harvested fields where the residues are spread in the field and burned often results in more complete combustion with lower particulate matter EFs as the residues are generally very dry with favorable combustion conditions prior to burning. While studies have provided burning emissions estimates for the different burning practices, research has not yet been able to map the different burning practices and the resulting spatial variation in EFs (piled vs. non-piled), and how much the emissions vary interannually based on changes in meteorology impacting local field moisture conditions. This and other factors inherently result in significant emissions uncertainty in S/SEA, and in the case of Vietnam, fire emissions could vary by more than 30% due to the different burning practices (Lasko 2018).

Interannual Climate Variability Influences Residue Management and Burning

Most of the South/Southeast Asian countries have a monsoon climate. In recent years, many countries have experienced torrential rains and associated flooding, which not only damages crops but, depending on timing, can also lead to changes or reduction in burning due to the difficulty of burning flooded fields with wet residues. While flooding and increased field water levels could lead to a delay or reduction in residue burning and resulting fire emissions, it can, unfortunately, lead to an increase in the emissions of greenhouse gases, such as methane (CH_4), which is one of the most potent greenhouse gases. Also, suppose residues are not burned and reincorporated into the inundated soil; in that case, it can create a trade-off between local/regional air quality degradation from biomass burning emissions ($PM_{2.5}$) and regionally/globally impactful increase in greenhouse gases impacting the broader climate (CH_4, N_2O).

Biomass Burning Emissions from Forest, Peat, and Shrublands

There are other significant sources of biomass burning $PM_{2.5}$ emissions in S/SEA. One of the significant drivers of emissions is attributed to increasing land-cover and land-use change (Prasad et al. 2000, 2001, 2002, 2003, 2004, 2008; Prasad and Badarinath 2004; Kant et al. 2000; Badarinath et al. 2008a, b; Gupta et al. 2001; Biswas et al. 2015a, b; Justice et al. 2015; Vadrevu et al. 2017). Throughout the region, temperate and tropical forests are routinely burned to make way for new plantations or to support agricultural or urban area expansion (Vadrevu et al. 2006; Vadrevu 2008; Vadrevu and Justice 2011; Lasko and Vadrevu 2018; Lasko et al. 2017, 2018a, b). These include peatland fires in Indonesia and other sources such as forest and shrubland fires due to drought and weather conditions. Emissions from forest and peat burning have also been major sources of emissions (Vadrevu et al. 2012, 2013, 2014a, b, 2015, 2017; Vadrevu and Lasko 2015, 2018). Some important studies relating to biomass burning are highlighted in Vadrevu et al. (2018), including the latest trends in fires (Vadrevu et al. 2019). While rice is the predominant agricultural residue burned, there are $PM_{2.5}$ emissions from other crop residues – especially wheat and sugarcane.

Impact of Crop Residue Burning on Soil Organic Carbon and the Resulting Crop Yield

As per the IPCC 2007 report, carbon sequestration in agricultural soils is a low-cost option for CO_2 removal from the atmosphere to mitigate climate change. Soil organic carbon (SOC) is a critical component of soils and directly influences soil quality by affecting water- and nutrient-holding capacity and soil structure, which further impacts crop productivity and yield in the long term (Dormaar et al. 1979; Biederbeck et al. 1980). Crop residue is a primary source of soil carbon sequestration. As such, burning of crop residues leads to loss of 90%–100% of the carbon contained in the residues, contributing to global warming and resulting in the loss of

SOC by increasing soil erosion and affecting carbon mineralization and decomposition. In contrast, retaining residues promotes carbon sequestration and other benefits, including preventing soil erosion and nutrient retention.

Globally, both short- and long-term studies were conducted to understand soil carbon dynamics with the burning of crop residues. These studies found that up-to 50% of SOC is lost with residue burning, irrespective of location, crop, and soil type (Table 10.1). However, the magnitude of the impact varies with soil texture, soil depth, crop, and local biophysical conditions. For instance, clay soils tend to lose less organic carbon with burning practice than sandy soils (Cerri et al. 2011). Similarly, loss of carbon decreases with increasing soil depth (Table 10.1), attributed to less soil vulnerability. Residue burning in perennial crops results in less loss of organic carbon compared to annual crops as the cultivation of perennial crops is less intensive (Bandaru et al. 2013). The halting of residue burning is found to increase the SOC ~0.24 Mg/ha annually (Table 10.1). However, previous studies found that restoration of SOC is a slow process, taking more than 10 years to see significant gains in SOC (Searle and Bitnere 2017).

DATASETS AND PM$_{2.5}$ EMISSIONS CALCULATION FOR RICE, SHRUBLANDS, PEAT, AND FOREST

Rice residue burning emissions are inherently underestimated in satellite-based datasets. To better quantify PM$_{2.5}$ emissions from rice residue burning, we developed a bottom-up approach conducted across the region at the subnational scale, integrating spatially varied estimates of burning practices, EFs, amounts burned, and fuel-loading factors.

$$RR_{PM2.5} = RA \times (FLsw * pbsw) + (FLst * pbst) \times CF \times EFbp$$

where RR$_{PM2.5}$ indicates PM$_{2.5}$ emissions from rice residue, RA is the rice-harvested area, FLsw is the dry biomass fuel loading for straw, pbsw is the percentage of straw burned, FLst is the dry biomass fuel loading for standing stalks, pbst is the percentage of standing stalks burned, CF is the combustion factor, and EFbp is the emission factor in g/kg for the burning practice of pile burning or non-pile burning.

We collected a spatial database of rice-harvested area called Rice Atlas (Laborte et al. 2017) with rice area provided at the subnational scale and based on a combination of agricultural census, FAO data, survey studies, and regional expert knowledge by IRRI.

Fuel-loading factors were obtained and synthesized from regional field studies. These factors estimate the amount of rice residue biomass left in the field. We used factors to represent rice straw as done in previous research but also incorporated a rice standing stubble factor (Figure 10.1), which often remains unaccounted. For Thailand, we used a fuel loading (FL) of 3600 kg/ha for rice straw derived from a local field study (Kanokkanjana and Garivait 2013) and rice stubble FL of 3860 kg/ha Hong Van et al. (2014) conducted in the Mekong River Delta. For the Mekong Delta, Vietnam, we used the rice straw FL of 3470 and 3860 kg/ha for standing stubble. Additional regional FLs were not available; thus, for the remaining regions and countries, we

TABLE 10.1

Change in Soil Organic Carbon Stock (Mg C/ha) under Burning and Non-burning Practices by Soil Depth under Different Cropping Systems Observed in Various Long-Term (Greater than 5 Years) Studies Across the World

Study (S) & Site (Si)	Country	Cropping System	Time Span (Years)	Depth (cm)	Carbon Stock (Mg C/ha)		% Change in C Stock (Mg/ha)	Reference
					Burn/Non-burn	Burn/Non-burn		
S-1 & Si-1	USA	Winter wheat	56	0–30	37.01	39.65	–6.66	Rasmussen and Parton (1994)
S-1 & Si-1	USA	Winter wheat	56	30–60	23_93	25_12	–4.74	
S-1 & Si-2	USA	Winter wheat	56	0–30	38.74	4136	–633	
S-1 & Si-2	USA	Winter wheat	56	30–60	24_50	26_16	–635	
S-1 & Si-3	USA	Winter wheat	56	0–30	39.95	4190	–4.65	
S-1 & Si-3	USA	Winter wheat	56	30–60	24_74	26_26	–5_79	
S-2 & Si-1	Thailand	Sugarcane	>20	0–10	12_51	16_15	–22_54	Sornpoon et al. (2013)
S-2 & Si-1	Thailand	Sugarcane	>20	10–30	2135	23.29	–8.33	
S-2 & Si-1	Thailand	Sugarcane	>20	30–55	9_77	10_00	–230	
S-2 & Si-1	Thailand	Sugarcane	>20	55–72	7.07	6.97	L43	
S-2 & Si-1	Thailand	Sugarcane	>20	72–100	8_50	7_22	17_73	
S-2 & Si-2	Thailand	Sugarcane	>20	0–10	1199	16.74	–2838	
S-2 & Si-2	Thailand	Sugarcane	>20	10–30	20_98	23_07	–9_06	
S-2 & Si-2	Thailand	Sugarcane	>20	30–55	931	10_56	–11_84	
S-2 & Si-2	Thailand	Sugarcane	>20	55–72	6_29	6_65	–5_41	
S-2 & Si-2	Thailand	Sugarcane	>20	72–100	7.22	9.56	–24A8	

(Continued)

TABLE 10.1 (Continued)

Change in Soil Organic Carbon Stock (Mg C/ha) under Burning and Non-burning Practices by Soil Depth under Different Cropping Systems Observed in Various Long-Term (Greater than 5 Years) Studies Across the World

Study (S) & Site (Si)	Country	Cropping System	Time Span (Years)	Depth (cm)	Carbon Stock (Mg C/ha)	% Change in C Stock (Mg/ha)	Reference
S-3 & Si-1	Brazil	Sugarcane	16	0–20	35_80	383	–6_53	Resende et al. (2006)
S-3 & Si-1	Brazil	Sugarcane	16	20–60	9730	98.7	–L42	
S-4 & Si-1	Australia	Rice	>5	0–7.5	11.85	23.21	–48.94	
S-4 & Si-1	Australia	Rice	>5	7_5–15	6_23	8_75	–28_80	
S-4 & Si-2	Australia	Rice	>5	0–7.5	14.04	15A4	–9.07	
S-4 & Si-2	Australia	Rice	>5	7_5–15	1033	11_09	–6_85	Kirby and Fattore (2006)
S-4 & Si-3	Australia	Rice	>5	0–7.5	1033	18.80	–45.05	
S-4 & Si-3	Australia	Rice	>5	7_5–15	9_55	14_74	–35_21	
S-4 & Si-4	Australia	Rice	>5	0–7.5	15.44	16.81	–8.15	
S-4 & Si-4	Australia	Rice	>5	7_5–15	9_55	1109	–13_89	
S-5 & Si-1	Canada	Spring wheat	47	0–20*	34_94	35_14	–0_57	Dormaar et al. (1979)
S-5 & Si-2	Canada	Spring wheat	47	0–20*	43.75	51.87	–15.65	
S-5 & Si-3	Canada	Spring wheat	19	0–20*	31A0	56_63	–44_55	
S-5 & Si-4	Canada	Spring wheat	20	0–20*	36.86	54.10	–3E87	

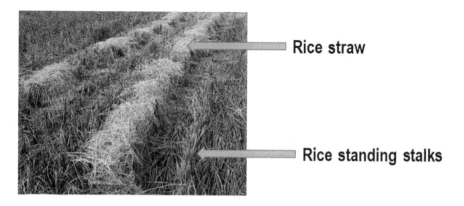

FIGURE 10.1   The figure depicts a combine-harvested field in which the rice stalk is cut by the machine. The cut straw remains on top of the standing stalks which are the portion still intact and in the ground. Both of these residue types are burned; however, the standing stalks are often unaccounted in emissions quantification.

used FL of 2700 kg/ha for rice straw and 6100 kg/ha in Red River Delta, Vietnam, as representative of double-cropped rice most common in S/SEA (Lasko 2019).

The amount of rice straw subjected to burning is the percentage burned (PB). Estimates are based on survey data conducted in geographically limited surveys. We apply factors to different regions based on how they correspond with the literature, and largely follow the recommendations of Oanh et al. (2018), but use country- and region-specific literature when available (Adam 2017; Jain et al. 2014; Quilty et al. 2014; Bunna et al. 2019; Lohan et al. 2018; Oanh et al. 2018; Vongxayya et al. 2019). We inferred from the literature and regional expertise to estimate the amount of rice straw and stubble burned and the prominent burning practice type that influences the amount burned and the combustion factor (CF) and EF (discussed below). Due to the lack of survey data, we inferred about 20% of stubble burned in combine harvest, open burning schemes, whereas for hand harvest often associated with pile burning (or generally, wetter residues with less complete combustion), we assume 5% burned. For rice straw, we used the following PB and burning practice based on the afore-mentioned literature: Bangladesh (51%, pile burning), Bhutan (51%, pile and open burning), Cambodia (17%, open burning), India (51%, open burning – except Punjab 70% PB, Haryana, and Himachal Pradesh 80% PB; Uttar Pradesh 25% PB), Indonesia (62.5% except for Java 37%, pile burning – except Sumatra Island with open burning), Laos (17%, pile burning except for Vientiane with open burning), Malaysia (37%, open burning), Myanmar (65.5%, pile burning except for Mandalay with open burning), Nepal/Pakistan/Sri Lanka (51%, pile and open burning), Thailand (East and Northeast 36.5%, North 65.5%, and South and West 27.5%, open burning), and Vietnam (56.5% Central Highlands, 65.5% Red River Delta, and 62% remaining, open burning in Mekong Delta, 50/50 in Red River Delta, and pile burning elsewhere).

EFs are the several pollution species emitted per unit of combusted material. We used the pile burning and non-pile burning EFs compiled from Lasko and Vadrevu (2018). In this case, pile burning EF is 16.9 g/kg, whereas non-pile burning EF is 8.8 g/kg. Pile burning is most often associated with hand-harvested fields where the rice is harvested, threshed, and then quickly placed in wet piles and burned shortly after, resulting in incomplete combustion with more PM$_{2.5}$, whereas non-pile burning is often associated with combine-harvested fields where the residues are left in neat rows in the field and can become very dry with complete combustion. CFs are related to EFs as they are a measure of combustion completeness. In this case, pile burning is often less complete than non-pile burning, and we used the factors compiled in Lasko (2018) of 0.67 (pile burning) and 0.89 (non-pile burning). We calculate emissions for the most-likely current scenario based on the mentioned factors and also based on 100% machine harvest assuming a future scenario as fields' transition.

Agricultural burning and associated emissions are routinely underestimated in satellite burned area detection datasets. In this regard, we utilize an indirect approach to quantify emissions from rice residue burning based on regional fuel-loading factors, burning-practice-specific EFs and CFs, and the region-specific percentage of residues burned. We incorporate not only rice straw but also rice standing stalks, which are typically unaccounted for (Figure 10.1). We use a spatial database of rice-harvested area, combined with the aforementioned factors applied to the different countries and subregions based on the literature. We calculate current most-likely rice residue emissions and future emissions, assuming fields switch to combine-harvested practices after modernization. Uncertainty is quantified to illustrate that significant uncertainty exists in emissions quantification, and there is a need for studies to address it. Subsequently, we use the Global Fire Emissions Database (GFED) in conjunction with the Rice Atlas harvest calendar to quantify the remaining agricultural emissions (by excluding the GFED agricultural emissions from burning of rice residue months in a given region).

Shrub, Forest, Peat, and Urban-Associated Emissions

For other biomass burning sources, we use the GFED version 4.2s with small fires (van der Werf et al. 2017), which provides dry matter amount burned per major land-cover type. We used region-specific or vegetation-specific EFs from the recent literature, including non-rice agriculture (Li et al. 2007), shrublands (Hosseini et al. 2013), a temperate forest and tropical forest (Aurell and Gullett 2013), and peatlands (Jayarathne et al. 2018). We used MODIS burned area product for burned area uncertainty quantification for 2008–2017 (MCD64A1) collection 6.1 (Giglio et al. 2018). For urban-associated emissions, we used the Regional Emission inventory in ASia (REAS) gridded at 0.25° grid cells for the latest available year 2008 (Kurokawa 2013). We also demonstrate satellite dataset temporal uncertainty in the burned area, which is routinely used for emissions and air quality studies. We discuss trade-offs in biomass burning emissions, a way forward, and impacts on soil carbon and crop yield. An overview flowchart is shown in Figure 10.2.

RESULTS

Total PM$_{2.5}$ Emissions in S/SEA

The total approximate PM$_{2.5}$ emissions for S/SEA are 26.2 million tons, with about 22% attributed to rice residue burning, 1% to non-rice agriculture, 21% to temperate

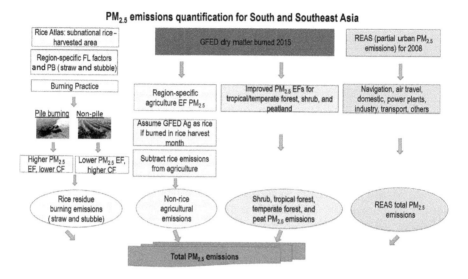

FIGURE 10.2   Overview flowchart of the emissions calculation for South/Southeast Asia.

and tropical forest burning, and 20% to peatlands, 8% to shrublands, and 28% to urban sources accounted for in REAS. Overall, 5.7 million tons of $PM_{2.5}$ was estimated from rice residue burning, 2.0 million tons from shrublands, 0.2 million tons from non-rice agriculture, 0.4 million tons from the temperate forest, 5.1 million tons from tropical forest, 5.2 million from peat, and 7.5 million tons from urban-related emissions. The most biomass burning and urban emissions of $PM_{2.5}$ emissions occur in India, followed by Indonesia, Bangladesh, Myanmar, and Vietnam (Table 10.2). Figure 10.3 shows the spatial distribution at the subnational scale of the $PM_{2.5}$ emissions, including the percentage attributed to rice residue burning. The detailed amounts and percentages are shown at a country scale for each South/Southeast Asian region (Table 10.2) and the top 25 subnational regions (Table 10.3). Overall, the results indicate very high biomass burning emissions across S/SEA with the highest non-agricultural emissions over Indonesia, largely due to peatlands, and the highest percentage of agricultural emissions over parts of Thailand, India, and the Philippines, and other rural delta regions.

We also calculated agricultural emissions in the GFED using the standard $PM_{2.5}$ EF of 6.26 g/kg (Akagi et al. 2011). Overall, the total agricultural $PM_{2.5}$ for S/SEA calculated in this method is only 212,000 tons. Our study found approximately 5.9 million tons of $PM_{2.5}$ from agriculture, indicating a sizeable difference (factor of about 28) in the emissions partially attributed to the EFs, and likely due to missed burned areas from satellite observations.

UNCERTAINTY IN AGRICULTURAL BURNING EMISSIONS ESTIMATIONS

Emissions quantification is fraught with uncertainty. In rice residue burning emissions quantification, significant effort has been placed on rice area and production estimates, which are often used as a basis for emissions. However, there remains

TABLE 10.2

PM$_{2.5}$ Emissions in Tons per Country and by Type of Emission

Country	REAS	Rice Straw and Stubble	Uncertainty (Rice)	Rice Straw/Stubble (All Machine Harvest)	Non-rice Agriculture	Shrub	Temperate Forest	Tropical Forest	Peat	Total Emissions	Rice (%)	Forest (%)	Non-rice Agriculture (%)	Shrub (%)	Peat (%)	REAS (%)
Bangladesh	360,569	223,648	494,418	381,459	14,740	303,604	105,796	328,323	0	1,336,681	16.7	32.5	1.1	22.7	0.0	27.0
Bhutan	7466	514	1137	811	159	607	8734	0	0	17,481	2.9	50.0	0.9	3.5	0.0	42.7
Brunei	14	33	73	46	0	594	0	413	0	1054	3.1	39.2	0.0	56.4	0.0	1.3
Cambodia	633,72	35,559	78,610	36,737	19,440	285,871	0	397,357	0	801,599	4.4	49.6	2.4	35.7	0.0	7.9
East Timor	3508	632	1397	927	19	1678	0	2764	0	8601	7.3	32.1	0.2	19.5	0.0	40.8
India	4,362,173	4,468,201	9,877,852	13,220,454	75,015	150,167	185,780	611,648	75,574	9,928,558	45.0	8.0	0.8	1.5	0.8	43.9
Indonesia	884,773	230,718	510,049	400,925	30,769	390,962	0	3,067,705	5,083,921	9,688,850	2.4	31.7	0.3	4.0	52.5	9.1
Laos	26,924	8476	18,738	9596	4365	174,003	0	192,660	0	406,428	2.1	47.4	1.1	42.8	0.0	6.6
Malaysia	107,084	11,972	26,466	17,566	2505	20,686	0	17,372	70	159,689	7.5	10.9	1.6	13.0	0.0	67.1
Myanmar	213,737	113,029	249,874	198,540	10,130	336,365	112,722	326,844	0	1,112,827	10.2	39.5	0.9	30.2	0.0	19.2
Nepal	117,477	32,233	71,256	50,830	930	543	14,852	0	0	166,035	19.4	8.9	0.6	0.3	0.0	70.8
Pakistan	492,760	54,586	120,674	86,081	7991	603	146	0	0	556,086	9.8	0.0	1.4	0.1	0.0	88.6
Philippines	77,610	104,209	230,374	184,538	9164	34,073	0	3059	0	228,114	45.7	1.3	4.0	14.9	0.0	34.0
Sri Lanka	75,142	22,171	49,014	34716	1219	2485	0	30	0	101,047	21.9	0.0	1.2	2.5	0.0	74.4
Thailand	264,139	216,305	478,185	302,389	28,670	180,414	0	63,241	0	752,769	28.7	8.4	3.8	24.0	0.0	35.1
Vietnam	472,691	178,109	393,745	288,734	33,722	127,130	0	91,530	0	903,180	19.7	10.1	3.7	14.1	0.0	52.3
Total S/SEA	7,529,437	5,700,395	12,601,862	15,214,352	238,839	2,009,785	428,031	5,102,946	5,159,565	26,168,999	21.8	21.1	0.9	7.7	19.7	28.8

Percentage breakdowns are shown at the right and also regional totals at the bottom. We include rice, agriculture, urban (REAS), shrublands, forests, and peat.

TABLE 10.3

Top 30 Subnational Regions Based on Total PM$_{2.5}$ Emissions in Metric Tons

Country	Region	REAS	Rice Straw and Stubble	Uncertainty (Rice)	Rice Straw/ Stubble (100% Machine)	Non-rice Agriculture	Shrub	Temperate Forest	Tropical Forest	Peat	Total Emissions	Percent Rice (%)	Percent Forest (%)	Percent Agriculture (%)
India	Punjab	98,254	3,665,895	8,104,194	11,872,131	46,622	150	0	0	0	3,810,921	96.2	0.0	97.4
Indonesia	Kalimantan Tengah	1,3487	5176	11,442	9132	5827	83,362	0	747,536	2,594,963	3,450,350	0.2	21.7	0.3
Indonesia	Sumatera Selatan	37,046	17,462	38,603	33,032	4896	45,300	0	636,124	1,320,952	2,061,780	0.8	30.9	1.1
India	Andhra Pradesh	428,323	74,175	163,979	128,185	9141	89,592	0	560,489	75,574	1,237,295	6.0	45.3	6.7
India	Uttar Pradesh	671,067	85,681	189,414	99,471	1893	236	570	0	0	759,448	11.3	0.1	11.5
Bangladesh	Dhaka	110,022	53,381	118,010	91,049	9592	221,069	6471	245,854	0	646,390	8.3	39.0	9.7
Indonesia	Kalimantan Selatan	22,464	12,156	26,872	21,448	1173	17,963	0	191,442	368,523	613,720	2.0	31.2	2.2
Indonesia	Kalimantan Barat	19,601	9922	21,935	17,507	1090	25,758	0	220,130	329,863	606,365	1.6	36.3	1.8
India	West Bengal	418,550	111,126	245,666	192,042	449	390	835	0	0	531,349	20.9	0.2	21.0
Indonesia	Papua	25,026	1002	2215	1731	5202	51,430	0	408,874	0	491,533	0.2	83.2	1.3
India	Orissa	306,303	88,974	196,695	153,760	541	4350	0	0	0	400,168	22.2	0.0	22.4
Indonesia	Kalimantan Timur	14,539	3733	8253	6587	1909	38,350	0	269,735	49,571	377,838	1.0	71.4	1.5
Indonesia	Jambi	12,561	3633	8032	6873	1291	26,184	0	118,300	189,577	351,546	1.0	33.7	1.4
India	Bihar	276,370	69,731	15,4154	120,505	440	158	0	0	0	346,699	20.1	0.0	20.2
Pakistan	Punjab	291,606	36,866	81,500	58,137	6322	131	0	0	0	334,925	11.0	0.0	12.9

(Continued)

TABLE 10.3 (Continued)

Top 30 Subnational Regions Based on Total PM$_{2.5}$ Emissions in Metric Tons

Country	Region	REAS	Rice Straw and Stubble	Uncertainty (Rice)	Rice Straw/Stubble (100% Machine)	Non-rice Agriculture	Shrub	Temperate Forest	Tropical Forest	Peat	Total Emissions	Percent Rice (%)	Percent Forest (%)	Percent Agriculture (%)
India	Madhya Pradesh	283,139	35,674	78,864	61,650	3470	710	0	0	0	322,993	11.0	0.0	12.1
India	Maharashtra	258,139	30,131	66,610	52,071	1693	3761	0	0	0	293,724	10.3	0.0	10.8
Indonesia	Riau	41,730	961	2123	1660	1223	25,872	0	139,138	75,574	284,499	0.3	48.9	0.8
India	Karnataka	240,142	28,857	63,794	49,405	1158	3282	0	0	0	273,439	10.6	0.0	11.0
Bangladesh	Chittagong	83,444	32,911	72,755	56,133	2758	35,759	51790	64,365	0	271,027	12.1	42.9	13.2
India	Chhattisgarh	171,711	76,604	169,349	132,383	462	4749	0	0	0	253,526	30.2	0.0%	30.4
India	Tamil Nadu	197,157	37,269	82,391	64,406	659	2413	0	0	0	237,498	15.7	0.0	16.0
Indonesia	Bangka-Belitung	3446	184	406	324	293	7773	0	76,940	147,425	236,061	0.1	32.6	0.2
Thailand	North	71,722	19,290	42,645	30,114	1678	95,848	0	33,124	0	221,662	8.7	14.9	9.5
Thailand	Northeast	51,144	97,734	216,061	127,575	12,808	36,091	0	16,366	0	214,143	45.6	7.6	51.6
Vietnam	Mekong River Delta	94,986	83,852	185,371	130,352	22,090	4606	0	7266	0	212,799	39.4	3.4	49.8
India	Rajasthan	209,049	2723	6021	4706	638	86	0	0	0	212,497	1.3	0.0	1.6
India	Assam	118,256	47,868	105,823	82,724	453	9378	30,568	2349	0	208,873	22.9	15.8	23.1
Bangladesh	Sylhet	76,807	16,425	36,311	28,015	2188	46,752	47,536	18,104	0	207,812	7.9	31.6	9.0
Thailand	Central	90,733	80,889	178,821	122,087	11,025	13,249	0	2322	0	198,218	40.8	1.2	46.4

Several of these high-emission regions are mostly attributed to rice (e.g., Punjab).

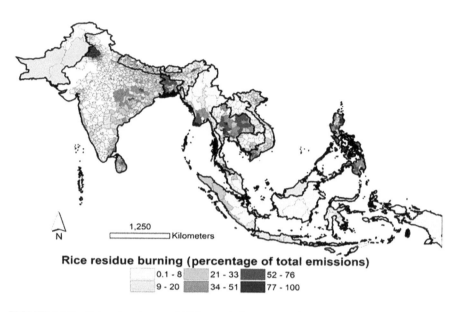

Rice residue burning (percentage of total emissions)

0.1 - 8	21 - 33	52 - 76
9 - 20	34 - 51	77 - 100

FIGURE 10.3   Subnational map of $PM_{2.5}$ emissions across Southeast Asia with percentage of rice residue burning emissions.

significant uncertainty in the other factors – especially when quantifying emissions at a large, regional scale. For example, it is not known how much exact amount of rice residues are burned overall and in a given year due to variation in ban enforcement, alternative uses of residues, variation in harvest practice type (hand harvest or combine harvest) and associated burning practice (pile burning or open burning), and shifts in crop calendar due to climate variation. Moreover, there is uncertainty concerning the biomass fuel-loading factors and how much of the straw and standing stubble are subjected to burning, and the completeness of burning. Accounting for spatial variation in these factors remains incomplete. Currently, we rely on estimates of a general country or subregional estimates. Overall, we estimated uncertainty by assuming 10% uncertainty in rice area estimates, 12% as reported by the fuel-loading factors, 24% and 22% in the machine- and hand-harvested EFs, 80% in the amount subjected to burning, 20% due to climate variation, 9% in CF, and 44% attributed to variation in the burning practices.

In many global emissions studies, satellite data such as the MODIS MCD64A1 product rely on the burned areas. These datasets underdetect agricultural fires. Another type of uncertainty in the satellite data is temporal uncertainty, which becomes important for air quality assessments that rely on weather and dispersion models that change daily. In this regard, we highlight the MCD64A1 burned area temporal uncertainty in the high cloud-covered country of Vietnam and its subregions. Figure 10.4 shows the 90th percentile distribution of burned area uncertainty for 2008–2017 per region, with the average uncertainty below it, and the bottom subfigure containing average burned area. While the median uncertainty is generally under 5 days, the 90th percentile median values exceed 15 days during some months

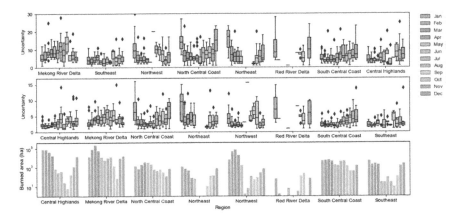

FIGURE 10.4   MODIS MCD64A1 (2008–2017) collection 6 burned area temporal uncertainty (in days) visualized for the regions of the high-uncertainty country of Vietnam. The top is the 90th percentile burned area uncertainty per month, middle is the average burned area uncertainty per month, and bottom is the average monthly burned area on log scale.

in several regions, such as the North Central Coast, highlighting a need for improved temporality (Lasko 2019).

IMPLICATIONS AND DISCUSSION

In this study, we used regionally representative EFs to quantify non-rice biomass burning emissions of PM$_{2.5}$. While we account for varied burn conditions and practices in rice residue burning emissions, we are unable to do so for the other emission sources. Further, we stipulate that the non-rice biomass burning emissions are probably higher than reported due to satellite data underestimation in the amount of dry matter burned and also the use of static EFs which may not reflect regional variation in burn conditions.

While estimating the rice residue burning emissions, we make regional or subregional generalizations about the amount of residues burned, fuel-loading factors, CFs, and EFs, which may have significant spatial and temporal variation. Overall, we approximated the rice residue burning emissions to have about 220% uncertainty and suggest research to continue addressing these uncertainties. There is also likely variation in the amount of residue burned based on the number of crop rotations (e.g., more burning with two or three rice crops per year, less with one as there is usually urgency to clear the field).

We also note that the REAS only provides the PM$_{2.5}$ emissions until 2008, and we used GFED and rice emissions computed from 2015 data to match the year of the Rice Atlas dataset. Therefore, these urban emissions are also very likely to be understated significantly for 2015 due to the high rate of urban growth in S/SEA during 2008–2015. Moreover, we note that REAS does not account for all of the urban-/fossil fuel-related emissions.

These emissions estimates do not convey air quality due to regional and global transport of emissions, lack of temporal component, and local atmospheric conditions that play a significant role in air quality. Recent research also indicates that while neglected, PM_{10} may also be very important for human health and air quality and should also be quantified. Our non-rice agricultural emissions are also likely underestimated due to the difficulty to separate rice from other crop emissions and reliance upon generalized EFs, which are the best currently available.

RESIDUE BURNING TRADE-OFFS AND A WAY FORWARD

The burning of agricultural residues such as those from rice leads to $PM_{2.5}$ emissions. However, depositing the residues in the field can still lead to other non-burning emissions. Residues left to decompose on inundated soils can lead to increased emissions of greenhouse gases, such as CH_4, which is one of the most potent greenhouse gases (Li et al. 1997). Therefore, there is a trade-off between rice residue burning emissions, which lead to air quality degradation and human health ailments, and greenhouse gas emissions from non-burning, which contribute to climate warming and harm on a global scale. Studies have also recommended the use of happy seeders or similar machines that can sow seeds beneath a layer of straw; however, these machines can be costly. We suggest for future research to find an appropriate measure where a combination of burning and residue incorporation or alternative uses are employed in a manner which balances sustainable SOC and crop yield, air quality-related emissions, and greenhouse gas-related emissions, all together with solutions implementable by and economical for smallholder farmers.

CONCLUSION

We quantified total $PM_{2.5}$ emissions estimates for S/SEA at the subnational scale. This is based on estimates of $PM_{2.5}$ emissions from major biomass burning sources in S/SEA including shrublands, temperate forest, tropical forest, peatlands, and non-rice agriculture derived from satellite-based GFED and used region-specific emission factors (EFs) which suggest higher emissions than in Akagi et al. (2011). In addition, rice residue burning emissions are calculated using a bottom-up, indirect approach due to documented underestimation in these emissions. We used subregion- and region-specific biomass fuel-loading factors and burning-practice-specific emission and combustion factors (EFs and CFs), which account for variation in machine-harvested and hand-harvested fields. Urban-, transport-, and industry-related emissions are quantified from the Regional Emissions Inventory in Asia (REAS). We found that approximately 26.2 million tons of $PM_{2.5}$ (18.6 million tons from biomass burning) is emitted in S/SEA with a significant spatial variation across different hotspot regions, with relatively high values over peatland areas of Indonesia. By comparing our results with the GFED agricultural emissions (212,000 tons) using the lower global EFs, we found significantly higher agricultural emissions (5.9 million tons), suggesting there may be a difference of as much as a factor of 28 in missed emissions in the satellite-derived datasets. We discussed the high level of uncertainty in rice residue burning emissions quantification and highlight the need to reduce it; this includes climate variation, which can shift the crop calendar and resulting burn times; burning practice variation; and amount of biomass burned. Moreover, we discussed other aspects of residue burning, such as trade-offs with greenhouse gas emissions and impacts on soil carbon and crop yield. We also provided a way forward to reduce emissions from burning.

DATA DISTRIBUTION

The GIS dataset of subnational PM$_{2.5}$ emissions will be made available for free download on ResearchGate.

REFERENCES

https://www.abc.net.au/news/2017-10-13/vietnam-floods-death-toll-rises-as-swathes-of-land-washed-aw/9048036

Adam, N.S. and El Pebrian, D. 2017. Factors affecting farmers' satisfactions with mechanized rice harvesting in Malaysian paddy fields: A case study of hiring custom operators. *Agricultural Engineering International: CIGR Journal.* 19(2), 120–128.

Akagi, S.K., Yokelson, R.J., Wiedinmyer, C., Alvarado, M.J., Reid, J.S., Karl, T., Crounse, J.D. and Wennberg, P.O. 2011. Emission factors for open and domestic biomass burning for use in atmospheric models. *Atmospheric Chemistry and Physics.* 11(9), 4039–4072.

Arai, H., Hosen, Y., Pham Hong, V.N., Thi, N.T., Huu, C.N. and Inubushi, K. 2015. Greenhouse gas emissions from rice straw burning and straw-mushroom cultivation in a triple rice cropping system in the Mekong Delta. *Soil Science and Plant Nutrition.* 61(4), 719–735.

Aurell, J. and Gullett, B.K. 2013. Emission factors from aerial and ground measurements of field and laboratory forest burns in the southeastern US: PM$_{2.5}$, black and brown carbon, VOC, and PCDD/PCDF. *Environmental Science & Technology.* 47(15), 8443–8452.

Badarinath, K.V.S., Kharol, S.K., Krishna Prasad, V., Kaskaoutis, D.G. and Kambezidis, H.D. 2008a. Variation in aerosol properties over Hyderabad, India during intense cyclonic conditions. *International Journal of Remote Sensing.* 29(15), 4575–4597.

Badarinath, K.V.S., Kharol, S.K., Prasad, V.K., Sharma, A.R., Reddi, E.U.B., Kambezidis, H.D. and Kaskaoutis, D.G. 2008b. Influence of natural and anthropogenic activities on UV Index variations–a study over tropical urban region using ground based observations and satellite data. *Journal of Atmospheric Chemistry.* 59(3), 219–236.

Bandaru, V., Izaurralde, R.C., Manowitz, D., Link, R., Zhang, X. and Post, W.M. 2013. Soil Carbon change and net energy associated with biofuel production on marginal lands: A regional modeling perspective. *Journal of Environment Quality.* 42(6), 1802.

Biederbeck, V.O., Campbell, C.A., Bowren, K.E., Schitzer, M. and McIver, R.N. 1980. Effect of burning straw on soil properties and grain yields in Saskatchewan. *Soil Science Society of America, Journal.* 44:103–111.

Biswas, S., Lasko, K.D. and Vadrevu, K.P. 2015a. Fire disturbance in tropical forests of Myanmar—Analysis using MODIS satellite datasets. *IEEE Journal of Selected Topics in Applied Earth Observations and Remote Sensing.* 8(5), 2273–2281.

Biswas, S., Vadrevu, K.P., Lwin, Z.M., Lasko, K. and Justice, C.O. 2015b. Factors controlling vegetation fires in protected and non-protected areas of Myanmar. *PLoS One* 10(4), e0124346.

Bunna, S., Sereyvuth, H., Somaly, Y., Ngoy, N., Mengsry, L., Chea, S., Ouk, M., Mitchell, J. and Fukai, S. 2019. Head rice yield of crops harvested by combine and hand at different ripening times in Cambodia. *Experimental Agriculture.* 55(1), 132–142.

Cerri, C.C., Galdos, M.V., Maia, S.M.F., Bernoux, M., Feigl, B.J. and Powlson, D. 2011. Effect of sugarcane harvesting systems on soil carbon stocks in Brazil: an examination of existing data. *European Journal of Soil Science.* 62, 23–28.

de Resende, A.S., Xavier, R.P., de Oliveira, O.C., Urquiaga, S., Alves, B.J. and Boddey, R.M. 2006. Long-term effects of pre-harvest burning and nitrogen and vinasse applications on yield of sugar cane and soil carbon and nitrogen stocks on a plantation in Pernambuco, NE Brazil. Plant and *Soil*, 281(1), 339–351.

Dormaar, J.F.; Pittman, U.J. and Spratt, E.D. 1979. Burning crop residues: Effect on selected soil characteristics and long-term wheat yields. *Canadian Journal of Soil Science*. 59: 79–86.

Duong, P.T. and Yoshiro, H. 2015. Current Situation and Possibilities of Rice Straw Management in Vietnam.

FAOSTAT 2018. http://www.fao.org/faostat/en/#data/GB.

Giglio, L., Boschetti, L., Roy, D.P., Humber, M.L. and Justice, C.O. 2018. The Collection 6 MODIS burned area mapping algorithm and product. *Remote Sensing of Environment*. 217, 72–85.

Gupta, P.K., Prasad, V.K., Sharma, C., Sarkar, A.K., Kant, Y., Badarinath, K.V.S. and Mitra, A.P. 2001. CH_4 emissions from biomass burning of shifting cultivation areas of tropical deciduous forests–experimental results from ground-based measurements. *Chemosphere-Global Change Science*. 3(2), 133–143.

Hall, J.V., Loboda, T.V., Giglio, L. and McCarty, G.W. 2016. A MODIS-based burned area assessment for Russian croplands: Mapping requirements and challenges. *Remote Sensing of Environment*. 184, 506–521.

Hong Van, N.P, Nga, T.T., Arai, H., Hosen, Y., Chiem, N.H. and Inubushi, K. 2014. Rice straw management by farmers in a triple rice production system in the Mekong Delta, Viet Nam. *Tropical Agriculture and Development*. 58(4), 155–162.

Hosseini, S., Urbanski, S.P., Dixit, P., Qi, L., Burling, I.R., Yokelson, R.J., Johnson, T.J., Shrivastava, M., Jung, H.S., Weise, D.R. and Miller, J.W., 2013. Laboratory characterization of PM emissions from combustion of wildland biomass fuels. *Journal of Geophysical Research: Atmospheres*. 118(17), 9914–9929.

Inomata, S., Tanimoto, H., Pan, X., Taketani, F., Komazaki, Y., Miyakawa, T., Kanaya, Y. and Wang, Z. 2015. Laboratory measurements of emission factors of nonmethane volatile organic compounds from burning of Chinese crop residues. *Journal of Geophysical Research: Atmospheres*. 120(10), 5237–5252.

Jain, N., Bhatia, A. and Pathak, H. 2014. Emission of air pollutants from crop residue burning in India. *Aerosol and Air Quality Research*. 14(1), 422–430.

Jayarathne, T., Stockwell, C.E., Gilbert, A.A., Daugherty, K., Cochrane, M.A., Ryan, K.C., Putra, E.I., Saharjo, B.H., Nurhayati, A.D., Albar, I. and Yokelson, R.J. 2018. Chemical characterization of fine particulate matter emitted by peat fires in central Kalimantan, Indonesia, during the 2015 El Niño. *Atmospheric Chemistry & Physics*. 18(4).

Justice, C., Gutman, G. and Vadrevu, K.P. 2015. NASA land cover and land use change (LCLUC): An interdisciplinary research program. *Journal of Environmental Management*. 148(15), 4–9.

Kanokkanjana, K. and Garivait, S. 2013. Alternative rice straw management practices to reduce field open burning in Thailand. *International Journal of Environmental Science and Development*. 4(2), 119.

Kant, Y., Ghosh, A.B., Sharma, M.C., Gupta, P.K., Prasad, V.K., Badarinath, K.V.S. and Mitra, A.P. 2000. Studies on aerosol optical depth in biomass burning areas using satellite and ground-based observations. *Infrared Physics & Technology*. 41(1), 21–28.

Kirkby, C.A. and Fattore, A. 2006. Effect of Stubble Burning Versus Retention on Soil Health. Rural Industries Research and Development Corporation (RIRDC) Publication No W05/195. Australia. https://citeseerx.ist.psu.edu/viewdoc/download?doi=10.1.1.494 .8793&rep=rep1&type=pdf

Kurokawa, J., Ohara, T., Morikawa, T., Hanayama, S., Janssens-Maenhout, G., Fukui, T., Kawashima, K. and Akimoto, H. 2013. Emissions of air pollutants and greenhouse gases over Asian regions during 2000–2008: Regional Emission inventory in ASia (REAS) version 2. Atmospheric Chemistry and Physics. 13(21), 11019–11058.

Laborte, A.G., Gutierrez, M.A., Balanza, J.G., Saito, K., Zwart, S.J., Boschetti, M., Murty, M.V.R., Villano, L., Aunario, J.K., Reinke, R. and Koo, J. 2017. Rice Atlas, a spatial database of global rice calendars and production. *Scientific Data.* 4, 170074.

Lasko, K. and Vadrevu, K., 2018. Improved rice residue burning emissions estimates: Accounting for practice-specific emission factors in air pollution assessments of Vietnam. *Environmental Pollution.* 236, 795–806.

Lasko, K., 2019. Incorporating Sentinel-1 SAR imagery with the MODIS MCD64A1 burned area product to improve burn date estimates and reduce burn date uncertainty in wildland fire mapping. *Geocarto International.* doi:10.1080/10106049.2019.1608592.

Lasko, K., Vadrevu, K.P. and Nguyen, T.T.N. 2018a. Analysis of air pollution over Hanoi, Vietnam using multi-satellite and MERRA reanalysis datasets. *PloS One.* 13(5), e0196629.

Lasko, K., Vadrevu, K.P., Tran, V.T. and Justice, C. 2018b. Mapping double and single crop paddy rice with Sentinel-1A at varying spatial scales and polarizations in Hanoi, Vietnam. *IEEE Journal of Selected Topics in Applied Earth Observations and Remote Sensing.* 11(2), 498–512.

Lasko, K., Vadrevu, K.P., Tran, V.T., Ellicott, E., Nguyen, T.T., Bui, H.Q. and Justice, C. 2017. Satellites may underestimate rice residue and associated burning emissions in Vietnam. *Environmental Research Letters.* 12(8), 085006.

Li, C., Frolking, S., Crocker, G.J., Grace, P.R., Klír, J., Körchens, M. and Poulton, P.R. 1997. Simulating trends in soil organic carbon in long-term experiments using the DNDC model. *Geoderma.* 81(1–2), 45–60.

Li, X., Wang, S., Duan, L., Hao, J., Li, C., Chen, Y. and Yang, L. 2007. Particulate and trace gas emissions from open burning of wheat straw and corn Stover in China. *Environmental Science & Technology.* 41(17), 6052–6058.

Liu, T., Marlier, M.E., Karambelas, A., Jain, M., Singh, S., Singh, M.K., Gautam, R. and DeFries, R.S. 2019. Missing emissions from post-monsoon agricultural fires in north-western India: regional limitations of MODIS burned area and active fire products. *Environmental Research Communications.* 1(1), 011007.

Lohan, S.K., Jat, H.S., Yadav, A.K., Sidhu, H.S., Jat, M.L., Choudhary, M., Peter, J.K. and Sharma, P.C., 2018. Burning issues of paddy residue management in north-west states of India. *Renewable and Sustainable Energy Reviews.* 81, 693–706.

Oanh, N.T.K., Permadi, D.A., Hopke, P.K., Smith, K.R., Dong, N.P. and Dang, A.N. 2018. Annual emissions of air toxics emitted from crop residue open burning in Southeast Asia over the period of 2010–2015. *Atmospheric Environment.* 187, 163–173.

Prasad, V.K. and Badarinath, K.V.S. 2004. Land use changes and trends in human appropriation of above ground net primary production (HANPP) in India (1961–98). *Geographical Journal.* 170 (1), 51–63.

Prasad, V.K., Badarinath, K.V.S. and Eaturu, A. 2008. Biophysical and anthropogenic controls of forest fires in the Deccan Plateau, India. *Journal of Environmental Management.* 86(1), 1–13.

Prasad, V.K., Badarinath, K.V.S., Yonemura, S. and Tsuruta, H. 2004. Regional inventory of soil surface nitrogen balances in Indian agriculture (2000–2001). *Journal of Environmental Management.* 73(3), 209–218.

Prasad, V.K., Gupta, P.K., Sharma, C., Sarkar, A.K., Kant, Y., Badarinath, K.V.S., Rajagopal, T. and Mitra, A.P. 2000. NOx emissions from biomass burning of shifting cultivation areas from tropical deciduous forests of India–estimates from ground-based measurements. *Atmospheric Environment.* 34(20), 3271–3280.

Prasad, V.K., Kant, Y. and Badarinath, K.V.S., 2001. CENTURY ecosystem model application for quantifying vegetation dynamics in shifting cultivation areas: A case study from Rampa Forests, Eastern Ghats (India). *Ecological Research.* 16(3), 497–507.

Prasad, V.K., Kant, Y., Gupta, P.K., Elvidge, C. and Badarinath, K.V.S. 2002. Biomass burning and related trace gas emissions from tropical dry deciduous forests of India: A study using DMSP-OLS data and ground-based measurements. *International Journal of Remote Sensing*. 23(14), 2837–2851.

Prasad, V.K., Lata, M. and Badarinath, K.V.S. 2003. Trace gas emissions from biomass burning from northeast region in India—estimates from satellite remote sensing data and GIS. *Environmentalist*. 23(3), 229–236.

Quilty, J.R., McKinley, J., Pede, V.O., Buresh, R.J., Correa Jr., T.Q. and Sandro, J.M., 2014. Energy efficiency of rice production in farmers' fields and intensively cropped research fields in the Philippines. *Field Crops Research*. 168, 8–18.

Rasmussen, P.E. and Parton, W.J., 1994. Long-term effects of residue management in wheat-fallow: I. Inputs, yield, and soil organic matter. Soil Science Society of America Journal, 58(2), 523–530.

Romasanta, R.R., Sander, B.O., Gaihre, Y.K., Alberto, M.C., Gummert, M., Quilty, J., Castalone, A.G., Balingbing, C., Sandro, J., Correa Jr., T. and Wassmann, R. 2017. How does burning of rice straw affect CH_4 and N_2O emissions? A comparative experiment of different on-field straw management practices. *Agriculture, Ecosystems & Environment*. 239, 143–153.

Searle, S. and Bitnere, K., 2017. Review of the impact of crop residue management on soil organic carbon in Europe. Working paper 2017–15, International Council on Clean Transportation.

Sornpoon, W. and Jayasuriya, H.P., 2013. Effect of different tillage and residue management practices on growth and yield of corn cultivation in Thailand. Agricultural Engineering International: CIGR Journal, 15(3), 86–94.

Vadrevu, K.P., Ohara, T. and Justice, C. 2014a. Air pollution in Asia. *Environmental Pollution*. 12, 233–235.

Vadrevu, K., Ohara, T. and Justice, C., 2017. Land cover, land use changes and air pollution in Asia: A synthesis. *Environmental Research Letters*. 12(12), 120201.

Vadrevu, K.P. and Justice, C.O. 2011. Vegetation fires in the Asian region: satellite observational needs and priorities. *Global Environmental Research*. 15(1), 65–76.

Vadrevu, K.P. 2008. Analysis of fire events and controlling factors in eastern India using spatial scan and multivariate statistics. *Geografiska Annaler: Series A, Physical Geography*. 90(4), 315–328.

Vadrevu, K.P. and Lasko, K. 2018. Intercomparison of MODIS AQUA and VIIRS I-Band fires and emissions in an agricultural landscape—Implications for air pollution research. *Remote Sensing*. 10(7), 978. doi:10.3390/rs10070978.

Vadrevu, K.P. and Lasko, K.P. 2015. Fire regimes and potential bioenergy loss from agricultural lands in the Indo-Gangetic Plains. *Journal of Environmental Management*. 148, 10–20.

Vadrevu, K.P., Csiszar, I., Ellicott, E., Giglio, L., Badarinath, K.V.S., Vermote, E. and Justice, C. 2012. Hotspot analysis of vegetation fires and intensity in the Indian region. *IEEE Journal of Selected Topics in Applied Earth Observations and Remote Sensing*. 6(1), 224–238.

Vadrevu, K.P., Eaturu, A. and Badarinath, K.V.S. 2006. Spatial distribution of forest fires and controlling factors in Andhra Pradesh, India using spot satellite datasets. *Environmental Monitoring and Assessment*. 123(1–3), 75–96.

Vadrevu, K.P., Giglio, L. and Justice, C. 2013. Satellite based analysis of fire–carbon monoxide relationships from forest and agricultural residue burning (2003–2011). *Atmospheric Environment*. 64, 179–191.

Vadrevu, K.P., Lasko, K., Giglio, L. and Justice, C. 2014b. Analysis of Southeast Asian pollution episode during June 2013 using satellite remote sensing datasets. *Environmental Pollution*. 12, 245–256.

Vadrevu, K.P., Lasko, K., Giglio, L. and Justice, C. 2015. Vegetation fires, absorbing aerosols and smoke plume characteristics in diverse biomass burning regions of Asia. *Environmental Research Letters.* 10(10), 105003.

Vadrevu, K.P., Lasko, K., Giglio, L., Schroeder, W., Biswas, S. and Justice, C. 2019. Trends in vegetation fires in South and Southeast Asian countries. *Scientific Reports.* 9(1), 7422. doi:10.1038/s41598-019-43940-x.

Vadrevu, K.P., Ohara, T. and Justice, C. 2017. Land cover, land use changes and air pollution in Asia: A synthesis. *Environmental Research Letters.* 12(12), 120201.

Vadrevu, K.P., Ohara, T. and Justice, C. eds., 2018. *Land-Atmospheric Research Applications in South and Southeast Asia.* Springer, Cham.

van der Werf, G.R., Randerson, J.T., Giglio, L., van Leeuwen, T.T., Chen, Y., Rogers, B.M., Mu, M., van Marle, M.J., Morton, D.C., Collatz, G.J. and Yokelson, R.J. 2017. Global fire emissions estimates during 1997–2016. *Earth System Science Data.* 9(2), 697.

Van Leeuwen, T.T., Peters, W., Krol, M.C. and van der Werf, G.R. 2013. Dynamic biomass burning emission factors and their impact on atmospheric CO mixing ratios. *Journal of Geophysical Research: Atmospheres.* 118(12), 6797–6815.

Vongxayya, K., Jothityangkoon, D., Ketthaisong, D., Mitchell, J., Xangsayyasane, P. and Fukai, S., 2019. Effects of introduction of combine harvester and flatbed dryer on milling quality of three glutinous rice varieties in Lao PDR. *Plant Production Science.* 22(1), 77–87.

Wiedinmyer, C., Yokelson, R.J. and Gullett, B.K. 2014. Global emissions of trace gases, particulate matter, and hazardous air pollutants from open burning of domestic waste. *Environmental Science & Technology.* 48(16), 9523–9530.

Zhang, T., Wooster, M., de Jong, M. and Xu, W. 2018. How well does the 'Small Fire Boost' methodology used within the GFED4. 1s fire emissions database represent the timing, location and magnitude of agricultural burning? *Remote Sensing.* 10(6), 823.

11 Greenhouse Gas and Particulate Matter Emissions from Rice Residue Burning in Punjab and Haryana States of India

Niveta Jain and Vinay Kumar Sehgal
Indian Agricultural Research Institute, India

Himanshu Pathak
National Rice Research Institute, India

Om Kumar and Rajkumar Dhakar
Indian Agricultural Research Institute, India

CONTENTS

INTRODUCTION

The Indo-Gangetic Plains (IGP) in India, covering 20% of India's geographical area, contribute nearly 42% to the total food grain production. Rice–wheat cropping system covers ~12 million hectares (Mha) in the IGP and contributes about one-third of the country's total cereal production. Rice and wheat crops contribute 75.5% of the total food grain production and 82.4% of the total cereal production. With the advent of mechanization, harvesting these crops with a machine, "combine harvester," has become very popular with Punjab and Haryana states in north India and is increasingly becoming popular in central India. More than 85% area of rice–wheat cropping system in the states of Punjab and Haryana is harvested by "combine harvester." With a sizeable area and production under rice–wheat, these crops produce a large quantity of residues. India produces approximately 683 Mt of crop residues, 33% from rice, amounting to 225.5 Mt (Jain et al., 2018). These crop residues are used as animal feed, soil mulch, manure, thatching for rural homes, and fuel for domestic and industrial purposes and thus are of tremendous value to the farmers. The residue of wheat crop finds more palatability with the animals than rice residue due to the high silica content in rice residue. However, many of these residues are burned *on-farm* primarily to clear the fields to facilitate the planting of succeeding crops, such as potato and wheat. The problem of *on-farm* burning of residues has further intensified in recent years due to increased mechanization, particularly use of combine harvesters, unavailability of cheap labor, high cost of removing the residues by conventional methods, the declining number of livestock per farm, long period required for *in situ* decomposition, and limited economically viable alternate use of residues.

ISSUES OF CROP RESIDUE BURNING

Crop residues include any biomass left in the field (straw, stubble, and other vegetative parts of crops) after grains and other economic components have been harvested. Processing of crop produce through milling also generates a substantial amount of residues. The disposal of such huge quantities of residues in a short period is a major concern for the government departments of agriculture, environment, and health. The combine harvester cuts the cereal crop at a certain height above the ground, thereby creating two distinct straw components after harvesting in the field: (1) the standing stubble or anchored crop residues; and (2) the windrows of straw or loose crop residues which are big uneven heaped lines of straw which are not easy to dispose-off and create hurdles for sowing and establishment of the subsequent crop. It is estimated that about 3.85–4.25 t/ha of standing stubble is left in the fields after harvesting rice with a combined harvester

(Jain et al., unpublished). To rapidly clear the land for sowing the wheat after rice, farmers burn the standing stubble and heaps of residue in the field. Farmers opt for burning because it is a quick and easy approach to dispose of huge quantities of crop residue without entailing any monetary cost. According to IPCC (2007), about 25% of the crop residue generated is burnt in the agricultural fields. Biomass burning has a strong regional and crop-specific variation with considerable spatial and temporal heterogeneity. The majority of the burning occurs during mid-October to November (coinciding with the harvesting period of rice and other *kharif* crops) and April–May (coinciding with the harvesting period of wheat and other *rabi* crops) periods. Several researchers have quantified the amount of biomass burned. For example, Streets et al. (2003) estimated that around 730 Tg of biomass is burned annually from both anthropogenic and natural sources in Asia. The contribution of India is 18% of the total biomass burned in Asia. Other studies estimated ~72 Mt–132 Mt of crop residue is burnt annually in India (Mehta, 2004; Pathak, 2006; Pathak et al., 2012; Sahai et al., 2011; Jain et al., 2014), of which the major contribution is from rice (43%), wheat (21%), and sugarcane (19%) (Jain et al., 2014).

IMPACTS OF CROP RESIDUE BURNING

Crop residue burning has serious impacts on environmental pollution and the fertility of soils, including the health of humans and animals. Burning of crop residues leads to the release of soot particles, smoke, and air pollutants such as carbon monoxide (CO), NH_3, NO_x, SO_2, nonmethane hydrocarbons (NMHCs), volatile organic compounds (VOCs), semi-volatile organic compounds (SVOCs), and particulates, causing health problems (Badarinath et al., 2006; Vadrevu et al., 2011, 2014, 2018). Emissions of greenhouse gases (GHGs) such as carbon dioxide (CO_2), methane (CH_4), and nitrous oxide (N_2O) cause global warming. These pollutants emitted from biomass burning can be transported over long distances and across political boundaries resulting in elevated oxidant levels, acid deposition, and visibility impairment. Residue burning causes loss of plant nutrients such as N, P, K, and S and adversely impacts soil physical properties and wastage of valuable residues, which could be a source of bioactive compounds for soil biota. Approximately the entire amount of C, 80%–90% of N, 25% of P, 20% of K, and 50% of S available in the crop residues are lost during the burning process as gaseous and particulate matter (Ponnamperuma, 1984; Lefroy et al., 1994). Jain et al. (2014) estimated that in 2010, the burning of sugarcane trash, rice residue, and wheat residue led to the loss of 0.84, 0.45, and 0.14 Mt nutrients per year, respectively, out of which 0.39 Mt was N, 0.30 Mt was P, and 0.014 Mt was K. Other studies by the TERI and the Institute for Social and Economic Change estimated a monetary loss of 76 million rupees as total annual welfare loss, in terms of health damages due to air pollution caused by burning of rice residue in rural Punjab (Kumar et al., 2015).

NEED FOR MONITORING OF RICE AREA UNDER BURNING AND EMISSIONS INVENTORY

Despite an increase in demand for crop residues for alternate uses such as fodder, the residues are still burnt *on-farm* each year. Thus, it is a paradox that the burning of crop residues and scarcity of fodder coexist in this country, and fodder prices have surged significantly in recent years. Burning of rice residue is also a major threat to the environment. In order to devise any effective residue management strategies, it is imperative to understand the spatiotemporal patterns of rice biomass availability, monitoring of residue burning events, and subsequent inventory of emissions of GHGs and particulate matter (PM). Such information shall help policymakers and other stakeholders undertake appropriate mitigation measures and monitor the success of the measures over time. With the above background, we quantified the spatiotemporal rice residue burning patterns in this study, including emissions of GHGs and PM in Punjab and Haryana.

STUDY AREA

The study was conducted for the two northern states of Punjab and Haryana (Figure 11.1). They are predominantly agricultural states with approximately 83.4% and 81.61% of their geographical area under cultivation. About 44.96% of the population in Haryana and 35.6% in Punjab are engaged directly in agriculture. Major crops grown in the two states are rice during the *kharif* (June–October) season and wheat during the *rabi* (November–March) season. About 2.97 Mha of Punjab State and 1.39 Mha of Haryana State are under rice cultivation. These two states alone

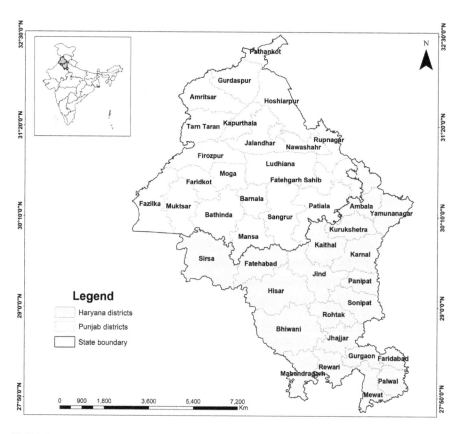

FIGURE 11.1   Map of the study area showing districts of Punjab and Haryana states.

produced 15.97 Mt of rice during 2015–2016, contributing about 15% rice share to the national food stock.

The area under rice in Punjab has increased from 1.2 Mha in 1980–1981 to 2.97 Mha in 2015–2016 with a compound annual growth rate (CAGR) of 2.68%. Production had also increased with a CAGR of 3.78% from 3.2 Mt in 1980–1981 to 11.82 Mt in 2015–2016. The rice productivity improved at a CAGR of 1.07% in the state during this period (Figure 11.2). In Haryana, the area under rice has increased from 0.47 Mha in 1980–1981 to 1.35 Mha in 2015–2016 with a CAGR of 3.06%, and production has increased at a CAGR of 3.48% from 1.25 Mt in 1980–1981 to 4.14 Mt in 2015–2016 and productivity at a CAGR of 0.41%. In both states, two rice types are grown, basmati (long-grained scented) and non-basmati (non-scented). In general, basmati rice is manually harvested, and so its residue is scantly burned, whereas the non-basmati rice is mechanically harvested by combine harvesters, leaving standing stubbles and heaps in rows, and so is mostly burned by the farmers.

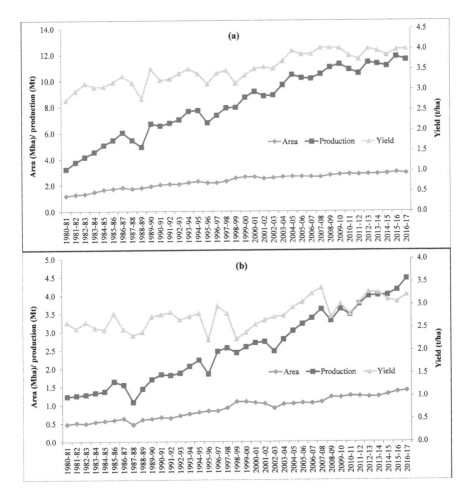

FIGURE 11.2 Trends in area, production, and productivity of rice in (a) Punjab and (b) Haryana.

METHODOLOGY

RICE AREA BURNT

Satellite remote sensing was employed to determine the rice burnt area in this study using the IARI satellite ground station (http://creams.iari.res.in). This study utilized the images from seven different satellites for real-time detection of active fire events between October 18 and November 20, 2015. The details of the satellites with their sensors are presented in Table 11.1. The active fire detection algorithm is allocated into two modules: daytime and nighttime. Both the modules are driven by fire-sensitive short-infrared (SWIR), middle-infrared (MWIR), and long-infrared (LWIR) spectral channels, exploiting the emissivity component of the fires. Fire detection for both day and night is successful for MODIS and VIIRS sensors as they employ a highly sensitive 3.9-μm channel with a strong thermal response to even a smaller portion of the fire in a pixel. The AVHRR on board the NOAA and MetOp satellites was used only for detecting nighttime fires as they lack the 3.9-μm channel. Using the threshold values for both day and night, the anomalous temperatures are detected with background characterization, and thus, the pixel is tagged as fire pixel, and the intensity of the fire is quantified in watts per square meter (W/m^2). For masking clouds, the 12-μm channel was used. A study area bitmap was created at the start of the burning period with a grid size of 1 km × 1 km and bit value = 0. For each active fire location detected from any of the satellites, the corresponding grid in which the fire location lay was flagged by turning the pixel on (bit = 1). If any active fire was detected again for the flagged grid, it was ignored to prevent counting the same grid multiple times for area calculation. At the end of the season, the number of grids flagged "on" in each administrative unit was counted to arrive at the burnt area estimate for that administrative unit.

AMOUNT OF ON-FARM BURNING OF RICE RESIDUE

The amount of residue burned in the field was taken as the amount of straw produced in the rice area burnt estimated from remote sensing. The amount of residue burned was calculated using the following equation:

TABLE 11.1

Details of Satellite Data Received at the IARI Satellite Ground Station and Used in This Study (http://creams.iari.res.in)

S. No.	Satellite Name	Sensor	Resolution (m)	Day/Night Passes
1	Suomi NPP	VIIRS	375/750	Both
2	Terra	MODIS	1000	Both
3	Aqua	MODIS	1000	Both
4	NOAA-18	AVHRR	1000	Night
5	NOAA-19	AVHRR	1000	Night
6	MetOp-1	AVHRR	1000	Night
7	MetOp-2	AVHRR	1000	Night

$$B = S*(A/100) \tag{11.1}$$

where "B" is the amount of residue burnt, "S" is the rice straw produced, and "A" is the percent rice area burnt estimated from remote sensing. The amount of rice straw produced (S) was calculated as the product of paddy production (P), the ratio of straw to grain yield (RCR) estimated using the harvest index (HI), and dry matter fraction (f) as given in Equation 11.2:

$$S = P*\text{RCR}*f \tag{11.2}$$

District-wise data on rice production were obtained from the Ministry of Agriculture and Farmers Welfare (MoAFW; 2015–2016) and from the Department of Agriculture, Govt. of Punjab and Haryana. The value of residue-to-grain ratio (RCR) for the rice grown in the region was taken as 1.5 for non-basmati varieties and 1.86 for basmati varieties. The dry matter fraction of rice residue was taken as 0.86 (Jain et al., 2018). The amount of residue burned infield was also estimated empirically by multiplying the straw produced with the fraction of rice straw burned in the district. Residues of basmati rice are burned to a very limited extent amounting to about 0%–15%, whereas the burning of non-basmati rice residues varied from 25% to 90% in different districts of Punjab and Haryana. These estimates were rationalized based on ground truth data and primary surveys conducted in Punjab and Haryana districts. There was a good correlation between rice residue burning estimates obtained using the empirical method and remote sensing (Figure 11.3).

ESTIMATION OF POLLUTANTS DUE TO RICE RESIDUE BURNING

Pollution load, i.e., emissions of PM and GHGs, is highly dependent on burning conditions, moisture content of straw, and weather conditions, especially the wind speed and wind direction. The pollution load from *on-farm* burning of crop residues was measured in this study using the following methodology. Samples of rice straw were collected from farmers' fields and stored in ziplock bags to estimate moisture content. The samples of gases and PM were collected from *on-farm* burning of rice residues in the fire's downwind direction at 3 m height above ground. Grab samples of air were collected from different locations in the downwind direction in the air sampling bags (5 and 10 L capacity) at a rate of 1 L/min with the help of a diaphragm pump to cover flaming, smoldering, and mixed stages of the burning. A six-channel particulate counter (PC-GW 3016-A Lighthouse, USA) collected data for 30 minutes at the 1-minute frequency for PM 0.1, 0.5, 1.0, 2.5, 5.0, and 10 μm particle size to characterize different particle size distribution during the rice residue burning. The concentrations of CH_4 and CO_2 in the air samples were analyzed using gas chromatography (GC 8A Series, Shimadzu) fitted with a flame ionization detector (FID) and methanizer, respectively, while N_2O concentration was analyzed using gas chromatography (HP 5890) fitted with electron capture detector (ECD). An inventory of PM_{10}, $PM_{2.5}$, CO, and GHG emissions from rice residue burning in Punjab and Haryana states was prepared using Equation 11.3 (IPCC, 2006):

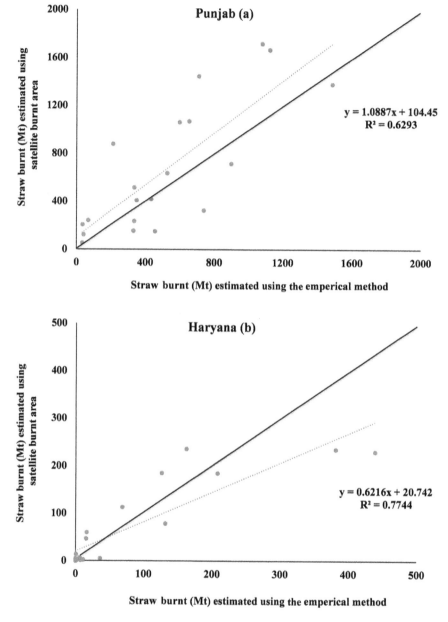

FIGURE 11.3   Relationship between estimates of rice residue burned using empirical and satellite remote sensing methods.

$$E_x = B \times EF_x \times C \qquad (11.3)$$

where "E" is the emission of pollutant x, "B" is the rice residue burnt, "EF_x" is the emission factor for pollutant x, and "C" is the combustion efficiency. The combustion efficiency (Zhang et al., 2008) was estimated by collecting the unburnt residue and ash from the fields randomly and extrapolating it to the total area. To quantify the amount of ash produced in the field, several quadrants of 2×2 m area were marked in such a way that it covered two heaps of straw/stubble left by the combine harvester starting from the midpoint of one heap and ending at the midpoint of second heap. Ash samples were collected from heaps. The mean value of ash/unburnt residue from all the plots was then extrapolated to calculate the combustion efficiency.

RESULTS

AMOUNT OF RICE RESIDUE PRODUCED

The amount of rice residue produced in different districts was estimated to range from 231.9 to 2217.0 thousand tons in Punjab and from 13.5 to 1189.0 thousand tons in Haryana. The total rice residue produced in Punjab and Haryana states during 2015 was estimated to be 22.71 and 8.80 Mt, respectively (Tables 11.2 and 11.3). Maximum rice residue was produced in Firozpur District (2.22 Mt), followed by Sangrur (2.16 Mt), Ludhiana (2.07 Mt), Patiala (1.83 Mt), and Gurdaspur (1.55 Mt) districts, in Punjab. These five districts contributed about 43.2% of the total rice residue produced in Punjab. In Haryana, Karnal (1.19 Mt), Kaithal (1.12 Mt), Kurukshetra (0.85 Mt), Jind (0.82 Mt), and Fatehabad (0.68 Mt) districts together contributed about 52.9% of the total rice residue produced in the state. Basmati rice contributed approximately 27.1% to the total rice residue production and non-basmati rice approximately 72.9% in Punjab. However, in Haryana, basmati rice residues account for 60% of the total. Maximum basmati rice residue production was in Amritsar (0.94 Mt), followed by Tarantaran (0.91 Mt), Firozpur (0.82 Mt), and Gurdaspur (0.69 Mt), whereas non-basmati rice residue production was highest in Sangrur (1.86 Mt), followed by Ludhiana (1.82 Mt), Patiala (1.56 Mt), and Firozpur (1.40 Mt) districts, in Punjab. In Haryana, 51.3% of basmati rice residue was produced from the districts of Jind (625.9 Mt), Karnal (607.7 Mt), Sonipat (537.0 Mt), Kaithal (472 Mt), and Panipat (461.8 Mt), whereas 75.9% of non-basmati rice residue production was from Kaithal (647.2 Mt), Kurukshetra (587.5 Mt), Karnal (581.2 Mt), Ambala (454.7 Mt), and Yamunanagar (403.0 Mt) districts.

AREA UNDER BURNING

Real-time monitoring of residue burning events between October 15 and November 20, 2015, using satellite data available from the IARI satellite ground station is shown in Figure 11.4. In Punjab, the first major burning event happened on October 24, though the majority of the burning occurred between November 7 and 11, 2015. Also, in Haryana, the first major bringing event was observed on October 24, then

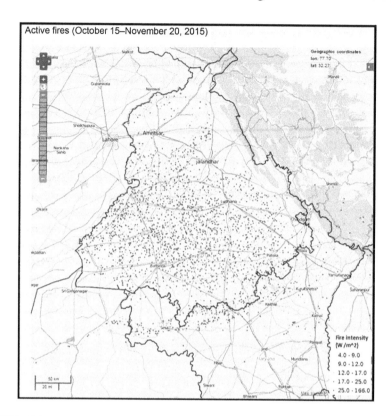

FIGURE 11.4 Map depicting active fire events detected from satellite images between October 15 and November 15, 2015, for Punjab and Haryana states.

around November 2, and later on November 9, 2015. District-wise total rice area burnt is given in Tables 11.2 and 11.3, and at state level, it was 1316.9′ 000 ha (45.5% of rice area) in Punjab and 237.3′ 000 ha (18% of rice area) in Haryana. A study conducted by the National Remote Sensing Agency in Punjab based on analysis of IRS-P6 AWiFS satellite data of India's Punjab State reported 1268.5′ 000 ha of area under rice crop residue burning in 2005 (Badarinath et al., 2006).

Amount of Rice Residue Burned

Using the data-based satellite analysis, the amount of rice residue burnt at the district level varied from 43.4 to 1856.6 thousand tons for Punjab and from 0.0 to 394.8 thousand tons for Haryana (Tables 11.2 and 11.3). The rice residue burning was maximum in Firozpur (1.48 Mt, 60.0%), followed by Ludhiana (1.12 Mt, 55.6%), Sangrur (1.07 Mt, 44.8%), and Bathinda (0.90 Mt, 8.6%), in Punjab State. In Haryana, in the districts of Fatehabad, Sirsa, Kaithal, and Karnal, 0.44 Mt (57.8%), 0.38 (64.2%), 0.21 (16.7%), and 0.16 (12.3%) of rice residues were burned, respectively. The total amount of rice residue burned in Punjab and Haryana was 10.43 Mt and 1.61, respectively.

TABLE 11.2

District-Wise Rice Area and Rice Residue Burnt in Punjab

Districts	Total Area under Rice (000 ha)	Rice Area Burned (000 ha)	Total Rice Residue Generated (000 t)	Total Rice Residue Burned (000 t)
Amritsar	182	45.0	1356.6	335.4
Barnala	108	53	872.1	433.7
Bathinda	119	105.2	944.0	896.8
Faridkot	106	66.3	840.8	525.9
Fatehgarh Sahib	86	41.8	681.1	331.0
Firozpur	286	191.4	2217.2	1483.8
Gurdaspur	202	27.8	1551.7	213.5
Hoshiarpur	71	8.6	562.9	68.2
Jalandhar	167	81.5	1334.7	651.3
Kapurthala	117	43.9	930.2	349.0
Ludhiana	257	140.0	2056.7	1120.4
Mansa	82	57.4	652.3	456.6
Moga	177	75.2	1403.7	596.4
Mohali	29	4.2	231.9	33.5
Muktsar	135	96.9	1027.6	737.6
Nawanshahr	57	4.4	455.7	35.2
Patiala	230	88.6	1833.8	706.4
Ropar (Rupnagar)	36	5.2	286.0	41.4
Sangrur	271	135.7	2162.8	1075.7
Tarantaran	176	45.0	1311.9	335.4
Total	2894.0	1316.9	22,713.7	10,427.4

Of the total rice residue generated, 41.32% and 16.49% were estimated to be burned *on-farm* in Punjab and Haryana. Gadde et al. (2009) also reported that Punjab contributes about 30% to rice residue's total open-field burning after harvest.

EMISSION OF PARTICULATE MATTER

Emissions from field burning of agricultural crop residue in northwest India are a severe environmental hazard. The pollutants released from burning pose a significant health risk for the rural population due to PM emission of different sizes. The concentrations of PM of different sizes (PM 0.3, 0.5, 1.0, 2.5, 5.0, and 10 μm) and GHGs were monitored at multiple sites in Patiala, Ludhiana Sangrur, and Kaithal districts. The district-wise average values are given in Table 11.4. The concentration of PM of different sizes ranged from 57.29 to 594.10 μg/m³. Mittal et al. (2009) also observed poor ambient air quality during rice crop stubble burning episodes in Patiala and reported monthly average concentration of aerosols during the burning of rice stubble ranging from 430 to 442 μg/m³. The general trend of PM emission of different-sized particles from burning events was similar in all the four districts;

TABLE 11.3

District-Wise Rice Area and Rice Residue Burnt in Haryana

Districts	Total Area under Rice (000 ha)	Rice Area Burned (000 ha)	Total Rice Residue Generated (000 t)	Total Rice Residue Burned (000 t)
Ambala	83.00	9.7	591.6	69.1
Bhiwani	21.00	0.2	136.7	1.3
Faridabad	11.00	5.5	72.5	36.2
Fatehabad	100.00	64.4	681.1	438.6
Gurgaon	5.00	0	34.2	0.0
Hisar	48.00	2.4	311.4	15.6
Jhajjar	42.51	0.1	275.2	0.6
Jind	123.00	19.8	818.8	131.8
Kaithal	161.00	29.9	1119.3	207.9
Karnal	173.00	23.6	1189.0	162.2
Kurukshetra	120.00	17.9	846.9	126.3
Mewat	7.00	0	45.3	0.0
Palwal	34.00	0	229.4	0.0
Panchkula	9.00	0	66.2	0.0
Panipat	75.00	1.7	488.3	11.1
Rewari	2.00	0	13.5	0.0
Rohtak	41.00	0.9	268.5	5.9
Sirsa	81.00	57.8	534.5	381.4
Sonipat	88.00	1.1	573.6	7.2
Yamunanagar	70.00	2.3	501.3	16.5
Total	1294.51	237.3	8797.2	1611.8

therefore, their average trend is presented in Figure 11.5. It was observed that during the burning, the concentration of $PM_{1.0}$ was highest, followed by $PM_{0.5}$ and $PM_{2.5}$. In the present study, the respirable particulate matter (RSPM) levels remained higher than those of the National Ambient Air Quality Standards (NAAQS) during measurements. The concentration of fine particulate matter (PM 0.3–1.0 µm) was higher at all the places than RSPM. Fine particulate matter contributed almost 57.5%–85.1% of the RSPM, suggesting that the smaller particles dominated rice crop residue burning. The smaller particles will generally stay in the atmosphere for a longer duration for weeks than large-sized particles that are mostly removed by rainfall. The smaller particles can enter the respiratory system and result in breathing problems and irritation of the lung capillaries, inhibit pulmonary functions, and sometimes result in lung cancer. Besides health impacts, the PM also has serious environmental impacts; the extent of environmental damage depends on the PM's physical and chemical properties. They also can modify the climate by forming smog/clouds; they may absorb solar radiation, resulting in low/reduced visibility. The PM from rice residue burning has silica, elemental carbon, and several anions and cations (Prajapati, 2015) that can contribute to acid deposition.

TABLE 11.4

Average Concentration of Particulate Matter and GHGs during Rice Residue Burning Measured in Different Districts of Punjab and Haryana

Location	Concentration ($\mu g/m^3$) (Particle Size in μm)						Greenhouse Gases			
	0.3	0.5	1	2.5	5	10	CO_2 (ppm)	CH_4 (ppm)	N_2O (ppb)	GWP (g CO_2 eq/m³)
Kaithal	74.94±4.62	202.37±63.39	112.83±70.05	102.21±31.76	185.49±55.19	147.48±60.82	885±24.38	4.9±0.04	403±9.51	1.92±0.01
Sangrur	60.52±16.6	294.3±82.61	219.6±120.6	152.33±77.8	312.88±78.67	223.66±73.97	714±27	3.3±0.1	414±5.7	1.58±0.05
Ludhiana	57.29±6.8	281±70	594±267	193.32±92.4	110.39±31.6	86.57±39.4	536±17.5	2.9±0.1	412±3.8	1.25±0.03
Patiala	62.18±21.4	198.5±96.2	384.68±256	270.71±221.6	81.65±39.6	87.37±53.5	728±155.4	3.9±1.1	411±15.4	1.62±0.31
Average	63.73±12.3	244.04±78.0	327.81±171.8	179.64±107.4	172.6±51.3	136.27±56.9	716±142	3.7±0.9	410±5.0	1.59±0.10

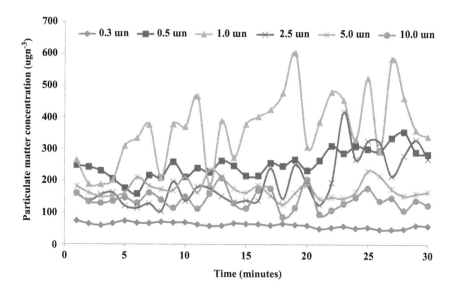

FIGURE 11.5   Average trends in emission of particulate matter of different sizes due to rice residue burning.

Emission of GHGs

Burning of crop residues is a potential source of GHGs such as CH_4, N_2O, and CO_2. On a global scale, crop residue burning contributes ~2% to the total GHG emissions (IPCC, 2006). The concentration of GHGs was monitored in fields of Patiala, Ludhiana, Sangrur, and Kaithal districts (Table 11.4). The concentrations of all three GHGs were higher than the ambient concentration. The measured concentration of CO_2 varied between 536 and 885 ppm, CH_4 between 2.9 and 4.9 ppm, and N_2O between 403 and 412 ppb during the rice residue burning events (Table 11.4). The average increase in CO_2, CH_4, and N_2O was 1.8, 2.1, and 1.3 times the ambient concentration, respectively. The global warming potential (GWP) of these emitted GHGs varied from 1.25 to 1.98 gm CO_2 eq/m³. This may have a direct or indirect effect on the radiation balance of the area affecting the climate.

Emission Estimates of Gaseous and Particulate Matter from Rice Residue Burning

It was estimated that the burning of 12.04 Mt of rice residue in two states led to 15.75 Mt of GHGs having a GWP of 17.71 Mt of CO_2 eq and 0.194 Mt of total particulate matter (TPM) in 2015. The highest emission was of CO_2 (95.7%), and out of the remaining 4.3%, CO accounted for 86.4%, $PM_{2.5}$ for 5.16%, PM_{10} for 3.87%, CH_4 for 3.01%, and N_2O for 0.60% (Figure 11.6). In Punjab, the burning of 10.43 Mt of rice crop residue emitted 13.62 Mt of CO_2, 0.42 Mt of CO, 0.018 Mt of CH_4, 0.004 Mt of N_2O, and 0.17 Mt of PM, of which 0.025 Mt is $PM_{2.5}$ and 0.019 Mt is PM_{10} (Table 11.5).

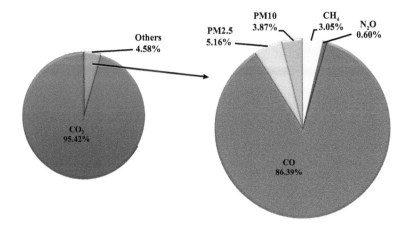

FIGURE 11.6 Contribution of different GHGs and PM from rice residue burning in Punjab and Haryana states.

TABLE 11.5
Emission of GHGs and PM from Rice Residue Burning in Punjab

Districts	CH$_4$ (Gg)	CO$_2$ (Gg)	N$_2$O (Gg)	CO (Gg)	PM$_{2.5}$ (Gg)	PM$_{10}$ (Gg)	TPM (Gg)
Amritsar	0.60	438.13	0.12	13.51	0.81	0.61	5.41
Barnala	0.78	566.56	0.15	17.47	1.04	0.78	7.00
Bathinda	1.60	1171.43	0.31	36.13	2.16	1.62	14.46
Faridkot	0.94	686.93	0.18	21.18	1.27	0.95	8.48
Fatehgarh Sahib	0.59	432.40	0.12	13.34	0.80	0.60	5.34
Firozpur	2.65	1938.19	0.52	59.77	3.57	2.68	23.93
Gurdaspur	0.38	278.94	0.07	8.60	0.51	0.39	3.44
Hoshiarpur	0.12	89.06	0.02	2.75	0.16	0.12	1.10
Jalandhar	1.16	850.80	0.23	26.24	1.57	1.18	10.50
Kapurthala	0.62	455.93	0.12	14.06	0.84	0.63	5.63
Ludhiana	2.00	1463.49	0.39	45.13	2.70	2.02	18.07
Mansa	0.82	596.45	0.16	18.39	1.10	0.82	7.36
Moga	1.07	779.00	0.21	24.02	1.44	1.08	9.62
Mohali	0.06	43.81	0.01	1.35	0.08	0.06	0.54
Muktsar	1.32	963.46	0.26	29.71	1.78	1.33	11.90
Nawanshahr	0.06	45.95	0.01	1.42	0.08	0.06	0.57
Patiala	1.26	922.73	0.25	28.46	1.70	1.28	11.39
Ropar (Rupnagar)	0.07	54.02	0.01	1.67	0.10	0.07	0.67
Sangrur	1.92	1405.11	0.38	43.33	2.59	1.94	17.35
Tarantaran	0.60	438.16	0.12	13.51	0.81	0.61	5.41
Total	18.63	13,620.57	3.65	420.06	25.09	18.82	168.17

The highest emissions of pollutants and GHGs were from Firozpur District (14.23%), followed by Ludhiana (10.74%), Sangrur (10.32%), Bathinda (8.60%), Muktsar (7.07%), and Patiala (6.77%) districts, together accounting for about 57.76% of the total emissions in Punjab.

The burning of 1.61 Mt of rice residue in Haryana emitted 2.11 Mt of CO_2, 0.065 Mt of CO, 0.003 Mt of CH_4, 0.001 Mt of N_2O, 0.0039 Mt of $PM_{2.5}$, and 0.0029 Mt of PM_{10} (Table 11.6). The total PM emission was 0.026 Mt from rice residue burning in Haryana. The districts of Fatehabad, Sirsa, Kaithal, Karnal, Jind, and Kurukshetra contributed 89.58% of pollutant emissions in Haryana State. Gupta and Sahai (2005) estimated that one ton of straw on burning releases 3 kg of PM, 60 kg of CO, 1460 kg of CO_2, 199 kg of ash, and 2 kg of SO_2. Yoshinori and Tadanori (1997) also reported 70% CO_2, 7% CO, 0.7% CH_4, and 2.1% N_2O of gaseous emissions from rice straw burning. A study conducted by the National Remote Sensing Agency in Punjab reported that rice straw/stubble burning contributed 261 Gg of CO, 19.8 Gg of NO_2, 3 Gg of CH_4, 30 Gg of PM_{10}, and 28.3 Gg of $PM_{2.5}$ during October 2005 (Badarinath et al., 2006). Some of the gases such as CO and hydrocarbons which possibly interact with nitrogen dioxide in the presence of sunlight produce photochemical oxidants responsible for formation of photochemical smog. The aerosols released during the burning can affect regional and possibly global radiation budgets by their light-scattering effects

---

TABLE 11.6

Emission of GHGs and PM from Rice Residue Burning in Haryana

Districts	CH_4 (Gg)	CO_2 (Gg)	N_2O (Gg)	CO (Gg)	$PM_{2.5}$ (Gg)	PM_{10} (Gg)	TPM (Gg)
Ambala	0.12	90.31	0.02	2.79	0.17	0.12	1.12
Bhiwani	0.00	1.70	0.00	0.05	0.00	0.00	0.02
Faridabad	0.06	47.34	0.01	1.46	0.09	0.07	0.58
Fatehabad	0.78	572.97	0.15	17.67	1.06	0.79	7.07
Gurgaon	0.00	0.00	0.00	0.00	0.00	0.00	0.00
Hisar	0.03	20.34	0.01	0.63	0.04	0.03	0.25
Jhajjar	0.00	0.85	0.00	0.03	0.00	0.00	0.01
Jind	0.24	172.18	0.05	5.31	0.32	0.24	2.13
Kaithal	0.37	271.52	0.07	8.37	0.50	0.38	3.35
Karnal	0.29	211.86	0.06	6.53	0.39	0.29	2.62
Kurukshetra	0.23	165.01	0.04	5.09	0.30	0.23	2.04
Mewat	0.00	0.00	0.00	0.00	0.00	0.00	0.00
Palwal	0.00	0.00	0.00	0.00	0.00	0.00	0.00
Panchkula	0.00	0.00	0.00	0.00	0.00	0.00	0.00
Panipat	0.02	14.46	0.00	0.45	0.03	0.02	0.18
Rewari	0.00	0.00	0.00	0.00	0.00	0.00	0.00
Rohtak	0.01	7.70	0.00	0.24	0.01	0.01	0.10
Sirsa	0.68	498.24	0.13	15.37	0.92	0.69	6.15
Sonipat	0.01	9.37	0.00	0.29	0.02	0.01	0.12
Yamunanagar	0.03	21.51	0.01	0.66	0.04	0.03	0.27
Total	2.88	2105.34	0.56	64.93	3.88	2.91	25.99

and influence on cloud microphysical processes (Charlson et al., 1991). The situation becomes severe due to atmospheric temperature inversion, which develops from late October to November. It is further accentuated as many of the farmers resort to rice residue burning during evening and night when temperature further declines, thereby causing smog formation near the ground.

Loss of Nutrients and Impacts on Soil Properties

Almost entire amount of C, 80% of N, 25% of P, 50% of S, and 20% of K available in the rice straw are lost due to burning (Dobermann and Fairhurst, 2000). Based on these amounts, it is estimated that burning of 12.04 Mt rice crop residue in two states resulted in the loss of approximately 0.06 Mt of N, 0.003 Mt of P, and 0.03 Mt of K in 2015. It has been reported from the UK that 40%–80% of wheat crop residue N is lost as ammonia when it is burned in the field and emissions of ammonia declined from 20 kt/N/year in 1981 to 3.3 kt/N/year in 1991 as a result of changes in agricultural practices because of an imposed ban on the burning of crop residues (Lee and Atkins, 1994). Kumar et al. (2001) reported that for every ton of wheat residues burned, 2.4 kg of N was lost. Likewise, S losses from the burning of high-S and low-S rice crop residues in Australia were 60% and 40% of S content, respectively (Lefroy et al., 1994). If the rice residue is incorporated or retained in the soil itself, it gets enriched, particularly with organic C and N. Heat from burning residues elevates soil temperature causing death of beneficial soil organisms. Frequent residue burning leads to complete loss of microbial population and reduces N and C in the top 0–15 cm soil profile, which is important for crop root development (Knicker, 2007).

Impacts on Human and Animal Health

Emissions of GHGs and PM have adverse implications not only on the atmosphere but also on human and animal health (Agarwal et al., 2010; Canadian Lung Association, 2007). The health impact of the finest particulate matter ($PM_{2.5}$) is greater because it can penetrate deep into alveolar sections of the lung (Agarwal et al. 2010). Smoke from the burning of residue exacerbates and triggers several respiratory diseases, with infants and the elderly facing the greatest health risks (Goss et al., 2014). According to Singh et al. (2008), more than 60% of the Punjab population live in the rice-growing areas and are exposed to air pollution due to the burning of rice residue. Medical records of Zira town's civil hospital in Firozpur in the rice–wheat belt showed a 10% increase in the number of patients within 20–25 days of the burning period every season. There are about 850,000 asthma patients in Punjab State, and inhalation of fine particulate matter of less than 2.5 µm acts as a trigger and aggravates the disease (Singh et al., 2008). Inhalation of fine particulate matter also adversely affects animal health. Preliminary investigations suggest that air pollution can result in animals' death due to the conversion of hemoglobin to carbaminohemoglobin due to high levels of CO_2 and CO in the blood. It may also cause corneal irritation and temporary blindness, allergic rhinitis, chronic bronchitis, and bronchiolitis leading to asthma-like conditions and a potential decrease in milk yield (Singh et al.,

2008). According to Kumar et al. (2015), medical expenditure and cost of workdays lost in Punjab's rural area due to increased air pollution resulting from rice residue burning were Rs 2.17 and Rs 2.35 per person, respectively. The total cost in rural Punjab was Rs 76.09 million (Rs 36.52 million for medical expenditure and Rs 39.57 million as cost of workdays lost due to health problems). Applying this cost to the total quantity of rice residue burned in Punjab State (10.43 Mt), it is estimated that the health cost associated with the burning of rice residue alone can be ~Rs 8.11 per ton, i.e., about Rs 80 million (US$ 1.2 million) in total.

CONCLUSIONS

On-farm burning of crop residues is becoming a serious environmental and health issue globally. Similarly, the problem of *on-farm* burning of paddy rice after harvest is particularly acute in northern states of Punjab and Haryana, India. Emission of PM from open rice residue burning coupled with meteorological conditions in producing smog/haze poses a severe threat to human and animal health in the region. This study quantified rice residue burnt, GHG emissions, and PM for the year 2015. We integrated satellite remote sensing data and field measurements and ancillary information to estimate the amount of rice residue burnt. Based on the district-wise statistics on the amount of residues burnt and emission of pollutants, we infer that crop residues' burning results in significant loss of nutrients from the soil, including impacts on human and animal health. We also infer the need to conduct similar studies in the other regions to help policymakers design effective crop management strategies, including air pollution mitigation.

ACKNOWLEDGMENT

This study was supported by the National Innovations on Climate Resilient Agriculture (NICRA) project funded by the Indian Council of Agricultural Research (ICAR), New Delhi, India.

REFERENCES

Agarwal, R., Awasthi, A., Singh, N., Gupta, P.K., Mittal, S.K. 2010. Effects of exposure to rice-crop residue burning smoke on pulmonary functions and Oxygen Saturation level of human beings in Patiala (India). *Science of the Total Environment*, 429 (1): 161–166.

Badarinath, K.V.S., Chand, T.K., Prasad, V.K. 2006. Agriculture crop residue burning in the Indo-Gangetic Plains–a study using IRS-P6 AWiFS satellite data. *Current Science*, 91: 1085–1089.

Charlson, R.J., Langner J., Rodhe, H., Leovy C.B. 1991. Perturbation of the Northern Hemisphere radiative balance by backscattering from anthropogenic sulfate aerosols. *Tellus A*, 43:152–163.

Canadian Lung Association. 2007. Pollution & air quality: http://www.lung.ca/protect-protegez/pollution-pollution_e.php.

Dobermann, A., Fairhurst, T.H. 2000. *Nutrient Disorders and Nutrient Management*. Potash and Phosphate Institute, Potash and Phosphate Institute of Canada and International Rice Research Institute, Singapore. http://books.irri.org/9810427425_content.pdf

Gadde, B., Bonnet, S., Menke, C., Garivait, S. 2009. Air pollutant emissions from rice straw open field burning in India, Thailand and the Philippines. *Environmental Pollution*, 157: 1554–1558.

Goss, P.E., Strasser-Weippl, K., Lee-Bychkovsky, B.L., Fan, L., Li, J., Chavarri-Guerra, Y., Liedke, P.E., Pramesh, C.S., Badovinac-Crnjevic, T., Sheikine, Y., Chen, Z. 2014. Challenges to effective cancer control in China, India, and Russia. The lancet oncology, 15(5): 489–538.

Gupta, P.K., Sahai, S. 2005. Residues open burning in rice-wheat cropping system in India: An agenda for conservation of environment and agricultural conservation agriculture. In: *Conservation Agriculture – Status and Prospects*, I.P. Abrol, R.K. Gupta and R.K. Malik (Eds.), Centre for Advancement of Sustainable Agriculture, National Agriculture Science Centre, New Delhi, pp. 50–54.

IPCC (Intergovernmental Panel on Climate Change). 2006. Guidelines for National Greenhouse Gas Inventories (IGES, Japan) (www.ipcc.ch).

IPCC (Intergovernmental Panel on Climate Change). 2007. Fourth Assessment Report. Working Group-I Report – 'The Physical Science Basis'. (www.ipcc.ch).

Jain, N., Bhatia, A., Pathak, H. 2014. Emission of air pollutants from crop residue burning in India. *Aerosol and Air Quality Research*. 14: 422–430. doi:10.4209/aaqr.2013.01.0031.

Jain, N., Sehgal, V.K., Singh, S., Kaushik, N. 2018. Estimation of surplus crop residues in India for biofuel production. TIFAC-IARI Joint Report, TIFAC Publication, New Delhi, p. 434.

Knicker, H. 2007. How does fire affect the nature and stability of soil organic nitrogen and carbon? A review. *Biogeochemistry*. 85: 91–118.

Kumar, K., Goh, K.M., Scott W.R., Frampton, C.M. 2001. Effects of N-labelled crop residues and management practices on subsequent winter wheat yields, nitrogen benefits and recovery under field conditions. *Journal of Agricultural Science*. 136: 35–53.

Kumar, P., Kumar S., and Joshi, L. 2015. Socioeconomic and environmental implications of agricultural residue burning: A case study of Punjab, India. Springer, Open, Springer New Delhi. doi:10.1007/978-81-322-2014-5.

Lee, D.S., Atkins, D.H.F. 1994. Atmospheric ammonia emissions from agricultural waste combustion. *Geophysical Research Letters*. 21: 281–284.

Lefroy, R.D.B., Chaitep, W., Blair, G.J. 1994. Release of sulfur from rice residues under flooded and non-flooded soil conditions. *Australian Journal of Agricultural Research* 45: 657–667.

Mehta, H. 2004. Bioconversion of Different Wastes for Energy Options, Sardar Patel Renewable Energy Research Institute Vallabh Vidyanagar, ppt.

Mittal, S.K., Singh, N., Agarwal, R., Awasthi, A., Gupta, P.K. 2009. Ambient air quality during wheat and rice crop stubble burning episodes in Patiala. Atmospheric Environment, 43(2): 238–244.

MoAFW, Agricultural Statistics at a Glance 2016. 2017. Directorate of Economics and Statistics, Department of Agriculture and Cooperation (DAC), Ministry of Agriculture, Government of India. http://dacnet.nic.in/.

Pathak, B.S. 2006. Crop residue to energy. In: *Environment and Agriculture*, K.L. Chadha and M.S. Swaminathan (Eds.), Malhotra Publishing House, New Delhi, pp. 854–869.

Pathak, H., Jain, N., Bhatia, A. 2012. Crop residues management with conservation agriculture: Potential, constraints and policy needs. IARI, New Delhi, vii+32 p, TB-ICN: 100/2012.

Ponnamperuma, F.N. 1984. Straw as a source of nutrients for wet-land rice. In *Organic Matter and Rice*, S. Banta, and C.V. Mendoza (Eds.), IRRI, Los Banos, pp. 117–136.

Prajapati, M. 2015. Emissions of Air Pollutants from Burning of Rice and Wheat Crop Residues, MSc thesis, India Agricultural Research Institute, New Delhi, India.

Report on 'Haryana Agriculture and Farmers' welfare, ICFA. www.icfa.org.in.

Sahai, S., Sharma, C., Singh, S.K., Gupta, P.K. 2011. Assessment of trace gases, carbon and nitrogen emissions from field burning of agricultural residues in India. *Nutrient Cycling in Agroecosystems*. 89: 143–157.

Singh, R.P., Dhaliwal, H.S., Sidhu, H.S., Manpreet-Singh, Y.S., Blackwell, J. 2008. Economic assessment of the Happy Seeder for rice-wheat systems in Punjab, India. Conference Paper. *AARES 52nd Annual Conference*, Canberra. Australia: ACT.

Streets, D.G., Yarber, K.F., Woo, J.H., Carmichael, G.R. 2003. An inventory of gaseous and primary aerosol emissions in Asia in the year 2000. *Journal of Geophysical Research*. 108: 8809–8823. doi:10.1029/2002JD003093.

Vadrevu, K.P., Ellicott, E., Badarinath, K.V.S. and Vermote, E. 2011. MODIS derived fire characteristics and aerosol optical depth variations during the agricultural residue burning season, north India. *Environmental Pollution*. 159(6): 1560–1569.

Vadrevu, K.P., Ohara, T, Justice, C. 2014. Air pollution in Asia. *Environmental Pollution*. 12: 233–235.

Vadrevu, K.P., Ohara, T. and Justice, C. eds. 2018. *Land-Atmospheric Research Applications in South and Southeast Asia*. Springer, Cham.

Yoshinori, M., Tadanori, K. 1997. Emissions of trace gases (CO_2, CO, CH_4, and N_2O) resulting from rice straw burning. *Soil Science and Plant Nutrition*. 43(4): 849–854. doi:10.108 0/00380768.1997.10414651.

Zhang, H., Ye, X., Cheng, T., Chen, J., Yang, X., Wang, L., Zhang, R. 2008. A laboratory study of agricultural crop residue combustion in China: Emission factors and emission inventory. *Atmospheric Environment*. 36, 8432–8441.

Section III

Aerosol Pollution and Biomass Burning

12 Biomass Burning and Impacts on Aerosols: Optical Properties and Radiative Effects

S. Ramachandran
Physical Research Laboratory, India

Priyadarshini Babu
Bennett University, India

CONTENTS

INTRODUCTION

Emissions from forest fires, burning of crop waste, harvest, and domestic cooking are important sources of biomass burning. Biomass burning emissions include light-absorbing black carbon (BC) and light-scattering organic carbon (OC) aerosols, which can influence the Earth–atmosphere radiation budget and climate (IPCC, 2013). The aerosol particles that make up the smoke from biomass burning emissions are mainly in the fine mode (<1 μm in radius). The radiative forcing due to biomass burning emissions is close to zero because of the positive radiative forcing from BC and negative forcing from OC (IPCC, 2013). Aerosols continue to contribute to the largest uncertainty in total radiative forcing estimates (IPCC, 2013; Vadrevu et al., 2015; Ramachandran, 2018). The net radiative forcing due to aerosols, which is negative (cooling of the Earth–atmosphere system), is highest in the Northern

Hemisphere and near populated and biomass-burning regions (Myhre et al., 2013). The standard deviation in the net aerosol radiative forcing is found to be largest over regions where changes due to vegetation are largest, for example, South Asia (Myhre et al., 2013), which is attributed to the uncertainties in optical properties of biomass burning aerosols. The biomass burning emissions can potentially influence climate on local, regional, and global scales (Vadrevu et al., 2018). The optical properties of aerosols influenced by biomass burning emissions are compared and contrasted with optical properties measured during normal (non-biomass burning emission) conditions. Their radiative effects are analyzed at three different locations in South/ Southeast Asia, which are significantly influenced by biomass burning.

STUDY REGION, LOCATIONS, AND IMPORTANCE

The Indo-Gangetic Plain (IGP) in South Asia and Singapore in Southeast Asia are influenced significantly by biomass burning emissions (Figure 12.1). The IGP is one of the largest river basins globally and one of the densely populated and heavily polluted regions in the world (Kedia et al., 2014). The Himalayas border the IGP in the north, Vindhyan ranges in the south, and the Thar Desert and the Arabian Sea in the west (Figure 12.1). The current study focuses on two environmentally distinct locations in the IGP, namely Kanpur (26.5°N, 80.2°E, 123 m above sea level) and Gandhi College (25.9°N, 84.1°E, 60 m above sea level) (Figure 12.1). Kanpur is an urban, industrial, and densely populated (>4 million) city and is ~500 km east of the megacity Delhi (the capital of India), while Gandhi College is a rural location southeast of Kanpur. Emissions from biofuels (burning of wood, dung cake, and crop waste) are a predominant source of aerosols over rural locations in India, while fossil fuel burning is the primary source over the urban regions (Habib et al., 2006) (Figure 12.1). Singapore is enveloped by the smoke from forest and peat fires almost every year with varying intensities, duration, and impacts (Reid et al., 2013). Singapore, a coastal, urban station, is influenced by emissions from local traffic, industries, sea spray, ships, and long-range transport of biomass burning during the year (Figure 12.1). The primary motivation of this study is not only to delineate the impact of biomass burning emissions on aerosol optical properties and radiative effects but also to unravel the differences in the impact due to the differences in the kinds of biomass burning emissions, as Kanpur and Gandhi College are affected by the burning of agricultural waste, wood, and dung cakes (anthropogenic), while Singapore is affected by the forest and peat fires (natural).

DATA, ANALYSIS, AND RATIONALE

Aerosol Robotic Network (AERONET) Sun/Sky Scanning Radiometer-measured level 2, version 3 quality-assured and cloud-screened daily aerosol optical depths (AODs), single-scattering albedo (SSA), and asymmetry parameter (g) in the wavelength range of 0.44-1.02 μm (namely at 0.44, 0.675, 0.87 and 1.02 μm), and Angstrom wavelength exponent (α) calculated from AODs in the wavelength range of 0.44–0.87 μm (namely at 0.44, 0.675, and 0.87 μm), aerosol radiative forcing and efficiency from January 2012 to December 2016 over Kanpur, Gandhi College, and Singapore

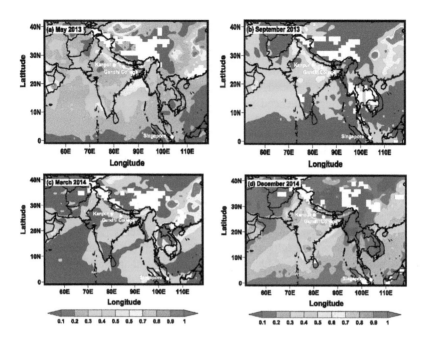

FIGURE 12.1 Regional maps of MODIS-retrieved aerosol optical depths at 0.55 μm corresponding to (a) March 2013 (biomass burning), (b) September 2013 (normal), (c) March 2014 (biomass burning; Table 12.3), and (d) December 2014 (non-biomass burning conditions). The study locations Kanpur, Gandhi College, and Singapore are marked by white circles in each plot.

are utilized. The uncertainty in AOD is less than ±0.02 in the 0.34–1.02 μm wavelength range (Dubovik and King, 2000). The error in SSA is estimated to be 0.03 when AOD at 0.44 μm is >0.2 and becomes higher (0.05–0.07) when AOD is ≤0.2 (Dubovik and King, 2000). Uncertainty in the asymmetry parameter is reported to be in the range of 3%–5% (Andrews et al., 2006). Aerosol properties retrieved in AERONET are used in calculating the solar broadband fluxes in the spectral range from 0.2 to 4.0 μm. The flux simulation relies on the retrieved real and imaginary parts of the refractive index. The spectral integration uses real and imaginary parts of the refractive index that are either extrapolated or interpolated using the refractive index values retrieved at AERONET wavelengths (Dubovik and King, 2000). Similarly, the spectral dependence of surface reflectance is interpolated or extrapolated from the surface albedo values assumed during retrieval at the wavelengths of AERONET. The surface reflectance for land sites of AERONET is approximated by bidirectional reflectance distribution function (BRDF) adapted from the MODIS Ecotype models (Dubovik and King, 2000). The radiative transfer calculation for gaseous absorption is done using the Global Atmospheric Model (Dubuisson et al., 1996). The standard deviation for error in AERONET sky radiance measurements is assumed to be 5% (Dubovik and King, 2000)

Though the aerosol characteristics over the study locations are influenced by biomass burning emissions during the year at varying intensities, the influence was found to be maximum during April–May over Kanpur (Table 12.1) and Gandhi College (Table 12.2) and during June–September over Singapore (Davison et al., 2004), and also during February–March (Table 12.3). Therefore, we have categorized the results obtained during April–May over Kanpur and Gandhi College (12–40 fire counts) as biomass burning-influenced, and those obtained during August–September as non-biomass burning-influenced as the observed fire counts over these locations were the least (0 or 1). During April–May, the higher fire counts are consistent with the springtime agricultural waste/residue burning over the IGP. The fire counts in Gandhi College are higher than those in Kanpur due to the dominance of local biomass burning activities (Kedia et al., 2014).

In Singapore, the results obtained during June–September are classified as influenced by biomass burning, while those obtained during November–December are marked as non-biomass burning influence following Reid et al. (2013) and Davison et al. (2004). We analyze the data obtained during 5 years (2012–2016). Overall, the study locations and results are classified based on the presence or absence of biomass burning emissions, which influenced aerosol characteristics and radiative effects. The fire counts exhibit large month-to-month and year-to-year variations (Tables 12. 1–12.3). We also analyze and present results over the study locations when the fire counts were significantly high and low in the same month during the 5 years. The primary motivation of this analysis is to determine whether there are noticeable changes in aerosol characteristics and forcing when the number of fire counts varies by about an order of magnitude. The aerosol properties are contrasted for May 2013 (fire counts = 10) and May 2016 (1) in Kanpur, April 2016 (103) and April 2015 (7) in Gandhi College, and March 2014 (238) and March 2015 (27) in Singapore, and inferences are discussed.

TABLE 12.1

Monthly Cloud-Corrected Fire Counts for Kanpur during 2012–2016 and Their Mean and Standard Deviation

Month	2012	2013	2014	2015	2016	Mean
January	3	6			1	3 ± 3
February	4		3	1	1	2 ± 2
March	1		1	1	2	1 ± 1
April	4	2	5			4 ± 2
May	9	10	7	5	1	6 ± 4
June	3		4	4		4 ± 1
July						
August			1			1 ± 0
September						
October		1	1		2	1 ± 1
November	3	14	8	2	12	8 ± 5
December	7	3	4	1	2	3 ± 2

TABLE 12.2

Monthly Cloud-Corrected Fire Counts for Gandhi College during 2012–2016 and Their Mean and Standard Deviation

Month	2012	2013	2014	2015	2016	Mean
January	10	4			15	10 ± 6
February	5	1	5	6	13	6 ± 4
March	1	1	6	7	7	4 ± 3
April	41	24	29	7	103	41 ± 37
May	49	33	25	16	6	26 ± 16
June	1	1		4		2 ± 2
July	1	2		2	1	2 ± 1
August	1					1 ± 0
September		1		1		1 ± 1
October				5		5 ± 0
November	12	3	9	9	8	8 ± 3
December	56	35	20	68	5	37 ± 26

TABLE 12.3

Monthly Cloud-Corrected Fire Counts for Singapore during 2012–2016 and Their Mean and Standard Deviation

Month	2012	2013	2014	2015	2016	Mean
January	11	14	23	4	9	12 ± 7
February	6	5	113	13	8	29 ± 47
March	8	26	238	27	22	64 ± 97
April	6	7	8	18	17	11 ± 6
May	5	2	7	6	1	6 ± 2
June	15	13	5	17	8	11 ± 6
July	3	18	5	26	3	12 ± 10
August	6	4	2	11	6	7 ± 4
September	11	4	17	19	11	11 ± 7
October	4	3	18	9	6	7 ± 6
November	2	1	4	6	3	3 ± 2
December		4	1	7	4	3 ± 3

RESULTS

AEROSOL OPTICAL PROPERTIES

The spectral aerosol properties exhibit differing features due to the influence of biomass burning in the study locations (Figure 12.2). AOD, the column-integrated aerosol content, increases due to biomass burning emission-produced aerosols over Kanpur and Gandhi

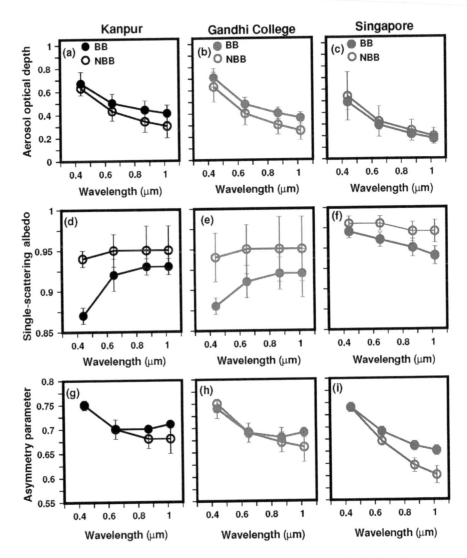

FIGURE 12.2 Aerosol optical depths (a–c), single-scattering albedo (d–f), and asymmetry parameter (g–i) in the spectral wavelength region of 0.44–1.02 μm for Kanpur, Gandhi College, and Singapore, respectively. BB represents biomass burning conditions, and NBB corresponds to aerosol characteristics measured during the non-biomass burning period. Vertical bars indicate ±1σ standard deviation from the mean.

College, but not over Singapore. However, in all the study locations, the mid-visible AOD was >0.3, confirming that all the sites are heavily polluted. SSA, the fraction of energy removed from the incident beam, which reappears as scattered radiation, shows a noticeable decrease as the biomass burning emissions (fire counts) increase. A reduction in SSA confirms the increase in the abundance of light-absorbing aerosols (especially BC). The spectral SSA reveals that the decrease is significant at wavelengths less than 0.5 μm, while above 0.6 µm, the spectra are flat (Figure 12.2d and e). Spectral variation in SSA can provide information on the abundance of absorbing vs. scattering aerosols (e.g., BC, OC, dust, sulfate, sea salt) in different size ranges. SSA spectra will decrease as a function of wavelength when BC is abundant, while SSA will increase with wavelength when the dust is dominant (Russell et al., 2010). Over Kanpur and Gandhi College, SSA shows mixed characters with both BC and dust present in the aerosol size distribution. During pre-monsoon (March–April–May), the winds transport mineral dust from the adjacent arid/semiarid regions to the IGP (Kedia et al., 2014). Dust was found to exhibit a spatial gradient as the abundance of dust was higher over Kanpur than over Gandhi College (Kedia et al., 2014). During the pre-monsoon months of April and May, the fire counts also show an increase. It is interesting to note that SSA spectra remain almost flat in the non-biomass burning scenario with respect to wavelength. In contrast, over Singapore, SSA spectra decrease as wavelength increases, which is characteristic of urban aerosol emissions. Over urban regions where BC aerosols dominate, SSA decreases with wavelength; in Singapore, in both the scenarios, SSA decreases with wavelength; however, SSA is lower due to an increase in biomass burning emissions (Figure 12.2f). SSA is relatively higher in Singapore than in the other two locations during normal and biomass burning-influenced conditions. This is attributed to Singapore's influence more by scattering-type aerosols (sea salt, sulfate, etc.) compared to Kanpur and Gandhi College.

Asymmetry parameter (g), the angular distribution of light scattered by the aerosols, is higher for an aerosol size distribution consisting of larger size particles. g at 0.55 m is low for BC (0.34) and is higher for sulfate aerosols (0.72) at the same wavelength. The value of g ranges from 0.6 to 0.8 for the other aerosol types in the atmosphere. The asymmetry parameter over Kanpur between 0.44 and 0.675 µm decreases and increases thereafter in both the scenarios (Figure 12.2g). In Gandhi College, g exhibits similar features when influenced by biomass burning emissions, while during normal conditions, g decreases with wavelength. In Singapore, the spectral features of g remain the same, confirming that in Singapore during normal and biomass burning-influenced conditions, the aerosol size distribution is dominated by small-sized particles in the fine mode.

Spectral features of aerosols did not exhibit noticeable changes when the fire counts increase significantly (Figure 12.3) over the study locations. The spectral features are compared and contrasted for May 2013 and May 2016 in Kanpur, April 2015 and April 2016 in Gandhi College, and March 2014 and March 2015 in Singapore, respectively, when fire counts showed vast differences (Tables 12.1–12.3). This comparison illustrates that aerosol characteristics and spectral features will exhibit changes when the sources of aerosol emissions vary, resulting in variations in their composition (chemical) and less due to an increase/decrease in the amount of aerosol emissions (physical).

OC, produced from fossil fuel, biofuel, and biomass burning emissions and natural biogenic emissions, is mainly a scattering-type aerosol. OC is the dominant component in biomass burning aerosols (Artaxo et al., 1998); however, OC particles that are freshly emitted can contribute to absorption in the ultraviolet and blue wavelengths (Kirchstetter et al., 2004). In this case, SSA at 0.44 µm is <0.90 (Figures 12.2 and 12.3) in Kanpur and Gandhi College, and SSA in Gandhi College is slightly

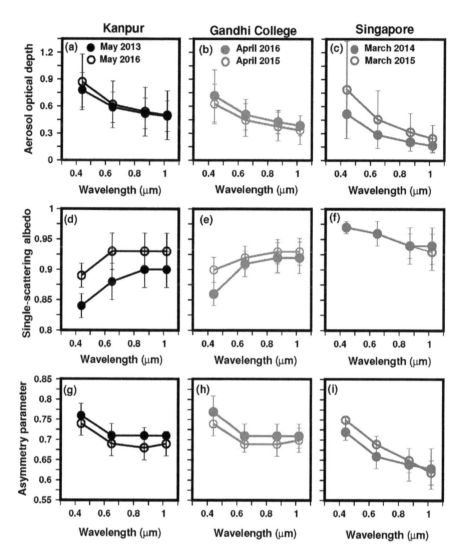

FIGURE 12.3 Aerosol optical depths (a–c), single-scattering albedo (d–f), and asymmetry parameter (g–i) in the spectral wavelength region of 0.44–1.02 μm for Kanpur, Gandhi College, and Singapore, respectively. The aerosol properties are drawn for specific months during the 5-year period when these locations are influenced by more and less fire counts. See Tables 12.1–12.3 for details. Vertical bars indicate ±1σ standard deviation from the mean.

higher than that of Kanpur. Gandhi College, influenced by more amount of biofuel and biomass burning than Kanpur (Kedia et al., 2014), is expected to have more OC and, hence, an enhanced absorption at 0.44 μm. OC was found to enhance aerosol absorption during winter in Kanpur (Arola et al., 2011), leading to a lower SSA. Since SSA is higher over Gandhi College than over Kanpur, we conclude that OC (the main scattering-type aerosol) is not enhancing the aerosol absorption during April–May when it is influenced by biomass burning emissions and its transport.

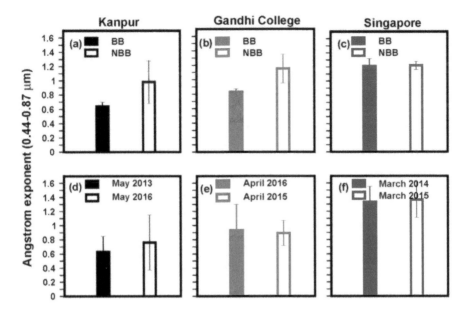

FIGURE 12.4 The Angstrom wavelength exponent obtained for biomass burning in comparison with the non-biomass burning scenarios, and specific months when the fire counts are high and low during the 5-year period for Kanpur (a, d), Gandhi College (b, e), and Singapore (c, f), respectively. Vertical bars represent ±1σ standard deviation from the mean.

The Angstrom wavelength exponent (derived using the spectral AODs in the 0.44–0.87 μm wavelength range) is lower for Kanpur and Gandhi College (Figure 12.4) when the spectral AODs are influenced by biomass burning emissions. The Angstrom exponent can increase due to an increase in fine-mode aerosol concentrations or a decrease in coarse-mode aerosol concentrations in the atmosphere. However, the Angstrom wavelength exponent remains the same for Singapore for biomass burning and non-biomass burning conditions consistent with the near-similar spectral AODs (Figure 12.2). These features are identical to those obtained for specific years when the fire counts are high and low (Figure 12.4 and Table 12.1). The Angstrom wavelength exponent is higher in Singapore than in Kanpur and Gandhi College, indicating that the amount of fine-mode aerosols is always higher in Singapore.

Aerosol Type

The sources of aerosol emissions and aerosol types, namely biomass burning, urban/industrial, and dust, can be deduced by correlating the extinction Angstrom exponent and absorption Angstrom exponent, and extinction Angstrom exponent and SSA (Figures 12.5 and 12.6) (Rupakheti et al., 2019). The aerosol sources are mostly urban/industrial and biomass burning over Kanpur and Gandhi College. The analysis also reveals that there are quite a number of aerosol particles that arise from mixed sources other than the above three (Figures 12.5 and 12.6). As mentioned earlier, during pre-monsoon, the winds transport dust particles in addition to the biomass

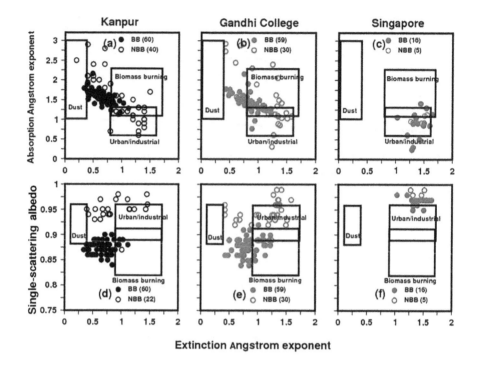

FIGURE 12.5   Classification of aerosol sources and types for Kanpur (a, d), Gandhi College (b, e), and Singapore (c, f) by comparing the absorption Angstrom exponent and extinction Angstrom exponent, and single-scattering albedo and extinction Angstrom exponent, respectively, for the biomass burning and non-biomass burning scenarios.

burning aerosols, as seen here (Figures 12.5 and 12.6). The abundance of dust shows a gradient as we see Kanpur's dust, while dust particles are not present over Gandhi College. In Singapore, all the aerosols can be classified as of urban/industrial and biomass burning origin, and not dust.

Aerosol Radiative Forcing and Efficiency

Aerosol radiative forcing at the surface (SFC), in the atmosphere (ATM), and at the top of the atmosphere (TOA) is higher for the biomass burning emission scenarios for Kanpur and Gandhi College (Figure 12.7). This feature is consistent with the lower SSA values obtained during the months influenced by biomass burning emissions (Figure 12.2). It should be noted that for the same aerosol amount (AOD), radiative forcing will increase when SSA decreases. The AODs for Kanpur and Gandhi College increased or remained the same. The aerosol radiative forcing due to biomass burning aerosols is significantly higher ($25\,W/m^2$) at the surface and in the atmosphere; however, the difference in TOA forcing is negligible. Over Singapore, the radiative forcing in the non-biomass burning scenario is higher (Figure 12.2). Aerosol radiative forcing efficiency (aerosol radiative forcing normalized to the AOD) is a useful measure to quantify the influence of absorbing and scattering aerosols (such as SSA) on radiative forcing. Similar to

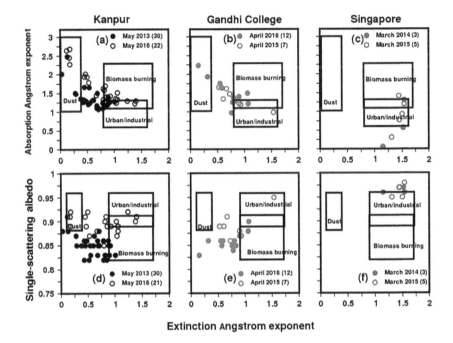

FIGURE 12.6 Aerosol sources and types classified for specific months influenced by biomass burning when the fire counts are high and low during the 5-year period for Kanpur (a, d), Gandhi College (b, e), and Singapore (c, f) by comparing the absorption Angstrom exponent and extinction Angstrom exponent, and single-scattering albedo and extinction Angstrom exponent, respectively.

aerosol radiative forcing, the radiative forcing efficiency of aerosols influenced by biomass burning is higher than that of those in normal conditions over Kanpur and Gandhi College. However, for Singapore, aerosol radiative forcing efficiency increases due to the influence of biomass burning while the radiative forcing is lower. This interesting phenomenon occurs despite the slightly lower AODs observed and accentuates the effect of SSA's lower values observed during biomass burning conditions.

It is clear that even if the AODs are lower or equal, the SSA variations influence the forcing efficiency of aerosols significantly. These features occur during specific years of high- and low-biomass burning periods (Figure 12.8). The aerosol radiative forcing efficiency deduced here is higher than other locations of Asia (Cho et al., 2017). The aerosol radiative forcing efficiency at the surface was about −100 to −120 W/m² per unit AOD over Kathmandu and Beijing. The forcing efficiency in Kanpur and Gandhi College is higher at least by about −25 W/m² per unit AOD when influenced by biomass burning (Figures 12.7 and 12.8). The radiative forcing values of aerosols are higher than those of aerosols estimated during the El Niño event of 1997–1998 over Indonesia (Davison et al. 2004). These significantly higher values of aerosol radiative forcing and efficiency on a regional scale during the months when aerosols are influenced by biomass burning over South and Southeast Asia can be a major contributor to the global mean radiative forcing due to biomass burning smoke.

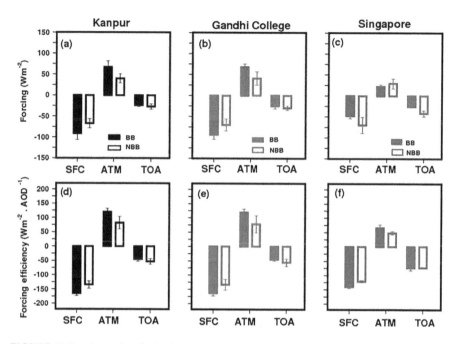

FIGURE 12.7   Aerosol radiative forcing at the surface (SFC), in the atmosphere (ATM), and at the top of the atmosphere (TOA) for (a) Kanpur, (b) Gandhi College, and (c) Singapore during biomass burning (BB) and non-biomass burning (NBB) conditions. Aerosol radiative forcing efficiency (SFC, ATM, and TOA) for (d) Kanpur, (e) Gandhi College, and (f) Singapore. Vertical bars correspond to $\pm 1\sigma$ standard deviation from the mean.

SUMMARY

Biomass burning emissions influence the physical and chemical properties of aerosols and contribute significantly to aerosol radiative forcing. In this study, we examined the influence of biomass burning on aerosol characteristics and radiative forcing over three locations in South (Kanpur and Gandhi College, India) and Southeast Asia (Singapore). Kanpur and Gandhi College are an urban and a rural location in the IGP, respectively, while Singapore is an urban location, and biomass burning sources influence all. Agro-residue burning emissions influence Kanpur and Gandhi College, and emissions from forest and peat fires influence Singapore. Spectral AODs are higher over Kanpur and Gandhi College due to the influence of biomass burning emissions, while the SSA values are lower, confirming that the aerosols are more absorbing. Analysis of aerosol types revealed that aerosols over Kanpur and Gandhi College originate from biomass burning, urban/industrial, and dust emissions. In Singapore, the emissions were dominated more by urban/industrial and biomass burning sources. Dust as an aerosol type was absent over Singapore. The aerosol radiative forcing at the surface (SFC) and in the atmosphere (ATM) is significantly higher ($\sim 25\,W/m^2$) when aerosol characteristics are influenced by biomass burning over Kanpur and Gandhi College. In Singapore, aerosol radiative forcing is higher during non-biomass burning, which could be due to less significant SSA variations,

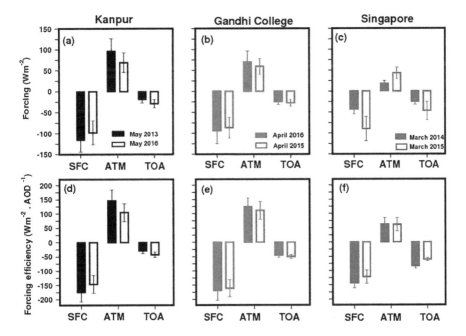

FIGURE 12.8 Aerosol radiative forcing at the surface (SFC), in the atmosphere (ATM), and at the top of the atmosphere (TOA) for (a) Kanpur, (b) Gandhi College, and (c) Singapore for specific months when the fire counts are high and low during the 5-year period. Aerosol radiative forcing efficiency (SFC, ATM, and TOA) for (d) Kanpur, (e) Gandhi College, and (f) Singapore. Vertical bars correspond to $\pm 1\sigma$ standard deviation from the mean.

while the AODs are higher or remain the same during the biomass burning and non-biomass burning periods. The aerosol radiative forcing efficiency (aerosol radiative forcing per unit AOD) at the surface and in the atmosphere is higher during biomass burning events over Kanpur and Gandhi College, highlighting the increase in the abundance of absorbing aerosols due to biomass burning. The radiative forcing and efficiency in the study atmosphere are higher by at least 25 W/m² compared to other locations in Asia, which is significant. These high values indicate that the radiative forcing due to biomass burning aerosols in this region could be a major contributor to the global mean aerosol radiative forcing. The findings remained the same for the low- and high-fire-count scenarios, suggesting that aerosols' properties and their radiative effects are governed more by the variations in the sources of aerosol emissions resulting in variations in their composition (chemical) and less by the amount of aerosol emissions (physical).

ACKNOWLEDGMENTS

Aerosol optical depths are downloaded from https://ladsweb.modaps.eosdis.nasa.gov/archive/allData/61/MOD08_M3/2015/, and fire counts are downloaded from ftp://fuoco.geog.umd.edu/modis/C6/mcd14ml/. We thank the principal investigators of the AERONET sites for their efforts in establishing and maintaining the AERONET sites (https://aeronet.gsfc.nasa.gov), the data of which are used in the study.

REFERENCES

Andrews, E., Sheridan, P.J., Fiebig, M., McComiskey, A., Ogren, J.A., Arnott, P., Covert, D., Elleman, R., Gasparini, R., Collins, D. and Jonsson, H. 2006. Comparison of methods for deriving aerosol asymmetry parameter. *Journal of Geophysical Research: Atmospheres.* 111(D5), D05S04.

Arola, A., Schuster, G., Myhre, G., Kazadzis, S., Dey, S. and Tripathi, S.N., 2011. Inferring absorbing organic carbon content from AERONET data. Atmospheric *Chemistry* and *Physics*, 11(1), 215–225.

Artaxo, P., Fernandas, E.T., Martins, J.V., Yamasoe, M.A., Hobbs, P.V., Maenhaut, W., Longo, K.M. and Castanho, A. 1998. Large-scale aerosol source apportionment in Amazonia. *Journal of Geophysical Research: Atmospheres.* 103(D24), 31837–31847.

Cho, C., Sang-Woo, K., Rupakheti, M., Jin-Soo, P., Panday, A., Soon-Chang, Y., Ji-Hyoung, K., Kim, H., Jeon, H., Sung, M. and Kim, B.M. 2017. Wintertime aerosol optical and radiative properties in the Kathmandu Valley during the SusKat-ABC field campaign. *Atmospheric Chemistry and Physics.* 17(20), 12617.

Davison, P.S., Roberts, D.L., Arnold, R.T. and Colvile, R.N. 2004. Estimating the direct radiative forcing due to haze from the 1997 forest fires in Indonesia. *Journal of Geophysical Research: Atmospheres.* 109(D10). doi:10.1029/2003JD004264.

Dubovik, O. and King, M.D. 2000. A flexible inversion algorithm for retrieval of aerosol optical properties from Sun and sky radiance measurements. *Journal of Geophysical Research: Atmospheres.* 105(D16), 20673–20696.

Dubuisson, P., Buriez, J.C. and Fouquart, Y. 1996. High spectral resolution solar radiative transfer in absorbing and scattering media: Application to the satellite simulation. *Journal of Quantitative Spectroscopy and Radiative Transfer.* 55(1), 103–126.

Giles, D.M., Holben, B.N., Eck, T.F., Sinyuk, A., Smirnov, A., Slutsker, I., Dickerson, R.R., Thompson, A.M. and Schafer, J.S. 2012. An analysis of AERONET aerosol absorption properties and classifications representative of aerosol source regions. *Journal of Geophysical Research: Atmospheres.* 117(D17). doi:10.1029/2012JD018127.

Habib, G., Venkataraman, C., Chiapello, I., Ramachandran, S., Boucher, O. and Reddy, M.S. 2006. Seasonal and interannual variability in absorbing aerosols over India derived from TOMS: Relationship to regional meteorology and emissions. *Atmospheric Environment.* 40(11), 1909–1921.

IPCC 2013. Summary for Policymakers in Climate Change 2013: The Physical Science Basis. Contribution of Working Group I to the Fifth Assessment Report of the Intergovernmental Panel on Climate Change (Stocker TF, Qin D, Plattner G-K, Tignor M, Allen SK, Boschung J, Nauels A, Xia Y, Bex V, Midgley PM (eds.)), 27 pp., Cambridge University Press, Cambridge, United Kingdom and New York, NY, USA.

Kedia, S., Ramachandran, S., Holben, B.N. and Tripathi, S.N. 2014. Quantification of aerosol type, and sources of aerosols over the Indo-Gangetic Plain. *Atmospheric Environment.* 98, 607–619.

Kirchstetter, T.W., Novakov, T. and Hobbs, P.V. 2004. Evidence that the spectral dependence of light absorption by aerosols is affected by organic carbon. *Journal of Geophysical Research: Atmospheres.* 109(D21). doi:10.1029/2004JD004999.

Myhre, G., Samset, B.H., Schulz, M., Balkanski, Y., Bauer, S., Berntsen, T.K., Bian, H., Bellouin, N., Chin, M., Diehl, T. and Easter, R.C. 2013. Radiative forcing of the direct aerosol effect from AeroCom Phase II simulations. *Atmospheric Chemistry and Physics.* 13, 1853–1877.

Ramachandran, S. 2018. Aerosols and climate change: Present understanding, challenges and future outlook. In: *Land-Atmospheric Research Applications in South/Southeast Asia.* Vadrevu, K.P., Ohara, T., and Justice, C. (Eds.). Springer, Cham, pp. 341–378.

Reid, J.S., Hyer, E.J., Johnson, R.S., Holben, B.N., Yokelson, R.J., Zhang, J., Campbell, J.R., Christopher, S.A., Di Girolamo, L., Giglio, L. and Holz, R.E. 2013. Observing and understanding the Southeast Asian aerosol system by remote sensing: An initial review and analysis for the Seven Southeast Asian Studies (7SEAS) program. Atmospheric Research, 122, 403–468.

Reid, J.S., Koppmann, R., Eck, T.F. and Eleuterio, D.P. 2005. A review of biomass burning emissions part II: Intensive physical properties of biomass burning particles. *Atmospheric Chemistry and Physics.* 5(3), 799–825.

Rupakheti, D., Kang, S., Rupakheti, M., Cong, Z., Panday, A.K. and Holben, B.N. 2019. Identification of absorbing aerosol types at a site in the northern edge of Indo-Gangetic Plain and a polluted valley in the foothills of the central Himalayas. Atmospheric Research, 223, 15–23.

Russell, P.B., Bergstrom, R.W., Shinozuka, Y., Clarke, A.D., DeCarlo, P.F., Jimenez, J.L., Livingston, J.M., Redemann, J., Dubovik, O. and Strawa, A. 2010. Absorption Angstrom Exponent in AERONET and related data as an indicator of aerosol composition. Atmospheric Chemistry and Physics, 10(3), 1155–1169.

Vadrevu, K.P., Lasko, K., Giglio, L. and Justice, C. 2015. Vegetation fires, absorbing aerosols and smoke plume characteristics in diverse biomass burning regions of Asia. *Environmental Research Letters.* 10(10), 105003.

Vadrevu, K.P., Ohara, T. and Justice, C. eds. 2018. *Land-Atmospheric Research Applications in South and Southeast Asia.* Springer, Cham.

13 Remotely Sensed Particulate Matter Estimation in Malaysia during the Biomass Burning Season in Southeast Asia

Kasturi Devi Kanniah and
Nurul Amalin Fatihah Kamarul Zaman
Universiti Teknologi Malaysia, Malaysia

CONTENTS

INTRODUCTION

Air pollution has become a serious environmental problem in Southeast Asian countries. The country experiences persistent haze conditions almost every year due to transboundary haze from open burning in Indonesia, especially during the southwest monsoon from May to September (Khan et al., 2015b; Hayasaka et al., 2014; Hayasaka and Sepriando, 2018). Moreover, local sources such as open biomass burning, motor vehicles, and industries (Alias et al., 2014; Khan et al., 2015a) also contribute significantly to the aerosol and pollutant emissions. Biomass burning from the forest and other vegetation in Indonesia releases large amounts of aerosols and their gaseous precursors (Viana et al., 2008). These pollutants are transported to the downwind countries that lie in the main pathway of the Southeast Asian pollution outflow (Reid et al., 2013; Vadrevu et al., 2018). As a primary air pollutant in Malaysia, aerosols with an aerodynamic diameter less than $10\,\mu m$ (PM_{10}) have been proven to cause adverse impacts on climate, air quality, and human health, including respiratory problems, cardiovascular diseases, birth defects, and premature death (Pope et al., 2011; Ramachandran, 2018).

In Malaysia, particulate matter (PM) data is measured, processed, analyzed, and distributed by the Department of Environment (DOE) and the Malaysian Meteorological Department (MET Malaysia). Currently, there are 88 air-quality monitoring stations nationwide, and these limited stations were found to provide inadequate data for describing the spatial variations in PM_{10} throughout the country; thus, it is not easy to address the impacts of PM_{10} on the local climate, including health issues (Zaman et al., 2017). Installing more air quality monitoring stations is limited by financial constraints. Therefore, remote sensing technology has significant potential to monitor PM_{10} over a large spatial extent and in a continuous manner. Remote sensing data are increasingly available for PM studies, and multiple sensors can provide aerosol optical depth (AOD) data. In this study, we analyze various aerosol products and techniques that can be used to estimate PM. Quantifying the PM concentrations and understanding its spatiotemporal patterns, including the transport pathways, are critical for managing the local air quality.

IDENTIFYING SOURCE REGIONS OF PM IN MALAYSIA

Biomass burning is the primary source of pollutants that emit 50% and 67% of global elemental and organic carbon, respectively, in SEA (Gustafsson et al., 2009; Vadrevu et al., 2019). Major regions of biomass burning in SEA are Indonesia, Thailand, Laos, Cambodia, and Vietnam (Israr et al., 2018; Oanh et al., 2018; Inoue, 2018; Lasko and Vadrevu, 2018; Lasko et al., 2017, 2018a, b). Figure 13.1 shows the spatial distribution of hotspot (active fire) in the SEA region for 1 month (1–30 September 2015) as detected by the Moderate Resolution Imaging Spectrometer (MODIS) satellite sensor. Slash and burn agricultural activities occur in the dry season (February–April in the northern part and May–September in the southern part of SEA) on an annual basis. Biomass burning emits a large amount of aerosols, fine particles of $PM_{2.5}$, black carbon, trace gases, volatile organic compounds,

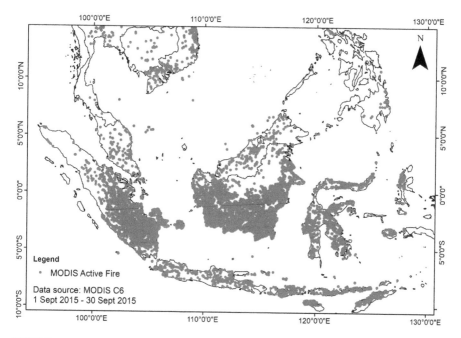

FIGURE 13.1   Active fires (as detected by MODIS sensor) in the Southeast Asian region between 1 and 30 September 2015.

and halogenated carbon into the atmosphere. It is estimated that biomass burning accounts for ~70% of the fine particles ($PM_{2.5}$) in SEA (Fu et al., 2012). In this region, land clearing occurs during the southwest monsoon (May–September), and the low precipitation and wind pattern (southwesterly) carries biomass-burning aerosols to downwind countries, triggering the haze development (Heil et al., 2007; Pavuluri and Kawamura, 2018).

We performed a 5-day back trajectory analysis during the 15th-September dry season to disclose the different air mass sources and transport pathways. It should be noted that 2015 had severe haze from mid-July until October, and the haze was more severe in late September until early October. Based on Figure 13.2, HYSPLIT backward trajectories reveal that air pollution originated mostly from Sumatra. The southwest monsoon transported biomass-burning pollutants from Sumatra within 24 hours. The biomass-burning pollution crossed the western and central regions of Peninsular Malaysia; therefore, both regions are mostly affected by the haze. In the Malaysian part of Borneo (Sabah), the aerosols originated from mid of Borneo Island. Meanwhile, in Sarawak and the southern region of Peninsular Malaysia, the back trajectories showed aerosols' transportation from the northern part of Australia. During the dry season, the southwest monsoon carries mainly biomass-burning aerosols from Sumatra and north Australia to Peninsular Malaysia within 24 hours. In 2015, much more fire hotspots were detected by MODIS (Figure 13.1). At higher altitudes (1500 m), the trajectories mainly originated from the equatorial Indian Ocean, thus being mostly clean at elevated layers. During the wet season, HYSPLIT trajectories

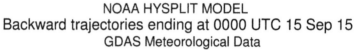

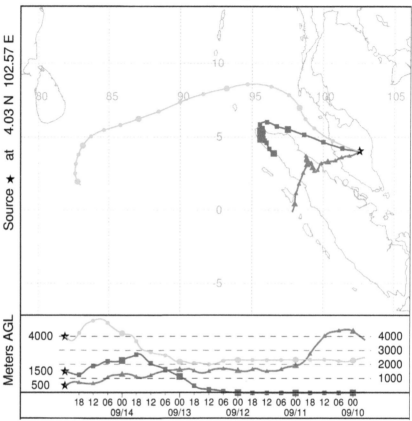

FIGURE 13.2 5-day backward trajectories at Malaysian sites for the dry season (15 September 2015).

(data not shown) revealed that lower air masses get transported from the northeastern direction (South China Sea) due to the northeast monsoon's onset. Higher precipitation from the northeast reduces pollutant transportation over Peninsular Malaysia. The lower aerosol loading during the wet season and the higher precipitation over the area prevent significant aerosol transportation over Peninsular Malaysia, resulting in lower AODs (Kanniah et al., 2014).

PM DATA SOURCES

In Malaysia, PM_{10} monitoring stations were established by DOE and MET Malaysia, and currently, there are 88 air quality monitoring stations nationwide

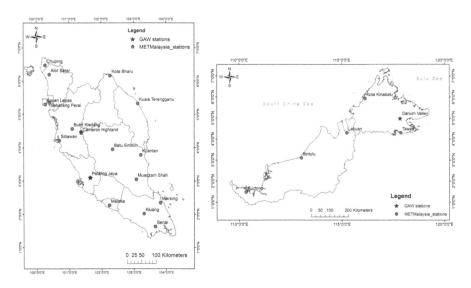

FIGURE 13.3   Distribution of the air pollution stations by MET Malaysia

where 65 continuous air quality monitoring stations are operated by DOE, and MET Malaysia runs 23 stations (Figure 13.3). Since June 2016, DOE awarded a 15-year contract to Pakar Scieno TW Sdn Bhd, a private company to implement the National Environmental Quality Monitoring Program (NEQMP), a new era of air monitoring network in Malaysia. The NEQMP implemented a new method of the Air Pollution Index (API) similar to Singapore's one. Measurements of $PM_{2.5}$ have been included since 2014. Recently, 65 air pollution stations have been strategically distributed including rural, industrial, suburban, and urban areas (Figure 13.4). PM_{10} concentrations are measured using the Beta Attenuation Monitor (BAM 1020) in milligrams or micrograms per cubic meter. Through the application of beta rays, PM_{10} concentration data can be calculated by the BAM 1020 in an hourly manner (Yusof et al., 2010). Even though the ground-based stations are typically capable of providing continuous PM_{10} measurements regularly (hourly) and at high accuracy, quantifying PM_{10} is complex and dynamic and thus, requires a dense network. Therefore, other alternative techniques are required for PM monitoring. The stations run by MET Malaysia measure not only PM_{10} concentrations but also the total suspended particles, tropospheric ozone, reactive gases, and oxidized nitrogen compounds.

The monthly averaged PM_{10} concentrations measured by the DOE for various Malaysian regions for the years 2013 and 2015 are shown in Figure 13.5. Malaysia is divided into central, east, north, and south of Peninsular Malaysia and Sabah and Sarawak for East Malaysia. In Malaysia, severe haze events occurred in 1997, 2005, and 2015 (Latif et al., 2018). As shown in Figure 13.5, all regions in Malaysia show

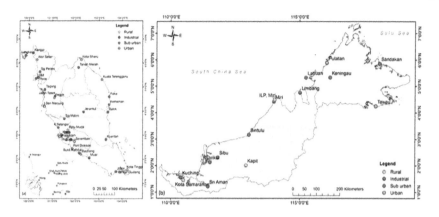

FIGURE 13.4   Locations of the Continuous Air Quality Monitoring (CAQM) stations in Peninsular Malaysia (left panel) and in East Malaysia (right panel).

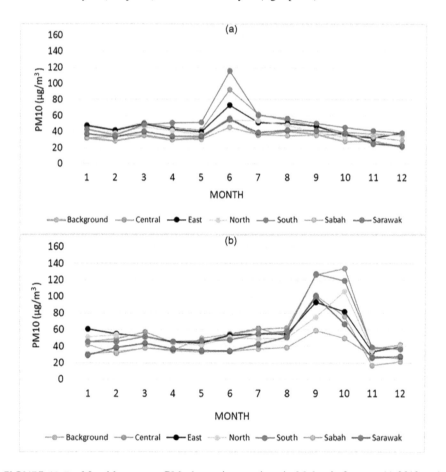

FIGURE 13.5   Monthly average PM_{10} by various regions in Malaysia for year (a) 2013 and (b) 2015.

the highest average PM_{10} in June. The highest PM_{10} was recorded in the southern region in 2013. The rest of the months had average PM_{10}, ranging between 20 and 60 μg/m³. According to DOE (2018), Malaysia experienced a short haze period from 15 to 27 June in 2013 due to transboundary pollution, and south Peninsular Malaysia experienced extremely hazardous (500 μg/m³) PM_{10} levels. The extreme pollution concentration during 2013 is totally not due to El Nino because 2013 is a non-El Nino year (Gaveau et al., 2014; Latif et al., 2018). According to Marlier et al. (2015), the 2013 event was caused by a typhoon and extreme burning in Riau, Indonesia. In 2015, the averaged PM_{10} showed a peak in September and October. The Central region (consisting of Kuala Lumpur and the states of Selangor, Negeri Sembilan, and Melaka) had a higher PM_{10} of 134 μg/m³ in October compared to the southern (120 μg/m³), northern (107 μg/m³), eastern (82 μg/m³), Sarawak (67 μg/m³), and Sabah (50 μg/m³) regions. About 34 areas in Malaysia were severely affected by the 2015 haze episode mainly due to the forest fire in Sumatra and Kalimantan, and the situation became worse due to the El Nino event (Latif et al., 2018).

ESTIMATION OF PM USING AOD MEASUREMENTS

Satellite- and ground-based remote sensing techniques cannot measure PM but are capable of estimating it via AOD observations. AOD is a measure of the extinction of solar beams by aerosols in the atmosphere. A more accurate AOD is provided from the ground-based sun photometers (Snider et al., 2015). The Aerosol Robotic Network (AERONET) is a federation between NASA and PHOTONS to develop ground-based AOD networks worldwide. The network provides global, continuous, and long-term products using a CIMEL (CE318) sun photometer (Holben et al., 1998). Currently, in Malaysia, only two AERONET sites are available, one in Universiti Sains, Malaysia (5.358°N, 100.302°E since Nov 8, 2011) and another in Kuching, Sarawak (1.324°N, 110.349°E, August 2, 2011), making it challenging to estimate PM concentrations (http://aeronet.gsfc.nasa.gov/cgi-bin/site_info/)

AERONET provides temporally resolved cloud-free measurements of spectral AODs (5–7 wavelengths) during daylight hours, which have been extensively used for satellite AOD validation (Munchak et al., 2013; Kanniah et al., 2014). AERONET data from Malaysia have been used to study the aerosol types, their temporal variation, and aerosol optical properties and to predict missing (Lim et al., 2013; Kanniah et al., 2014; Tan et al., 2015). Considering AERONET sites' insufficiency to cover a wide spatial extent, AOD from satellite remote sensing becomes the best approach.

AOD FROM SATELLITE REMOTE SENSING

Various satellites are available to produce AOD with multiple sensors, techniques, retrieval algorithms, characteristics, and capacities. Kanniah et al. (2016) has provided a review of the different aerosol products and their characteristics; hence, this section will only highlight the recent satellite sensors such as Visible Infrared Imager Radiometer Suite (VIIRS) and Advanced Himawari Imager (AHI).

The VIIRS was launched on board the Suomi National Polar-orbiting Partnership satellite in October 2011. It has 22 channels covering the spectrum from 412 to

12,050 nm. Xiao et al. (2016) found that the VIIRS aerosol product was as good as the MODIS AOD product with a higher spatial resolution at about 6 km. Wu et al. (2016) used VIIRS in estimating ground $PM_{2.5}$ concentrations over Beijing–Tianjin–Hebei and obtained a good correlation between AOD and $PM_{2.5}$ with $R^2 = 0.88$ and RMSE $= 13.0574$ μg/m³. Recently, Yao et al. (2017) compared the capability of VIIRS AOD (6 km) and MODIS AOD (3 km) to estimate $PM_{2.5}$ over polluted regions in China. They show that the regression model using VIIRS AOD ($R^2 = 0.71$) was slightly better than that of the MODIS sensor.

The AHI is a new generation of a geostationary meteorological satellite launched on 7 October 2014 by the Japan Meteorological Agency. The AHI is an infrared imager with 16 bands spanning from 0.47 to 13.3 μm (Zhang et al., 2018) and started its operation on 7 July 2015. It has great potential in air quality studies, where it has a relatively moderate spatial resolution of 5 km at 10 minutes intervals (Wei et al., 2019). This capability of the AHI allows high retrieval frequency of AOD data in cloudy regions like the tropics. AOD from the AHI has been validated against AOD from AERONET, and level 3 (L3) data at hourly intervals show better performance than the L2 data (10 minutes). Yang et al. (2020) validated the Himawari-8 AOD L2 product against AERONET over the Asian region. The overestimation of the AHI AOD level 2 product is possibly due to subpixel cloud contamination (Zhang et al., 2019). The cloud contamination issue was minimized by the algorithm developed by Kikuchi et al. (2018), and a new AOD product (L3) was released. The product's validation shows promising results in various regions (Zhang et al., 2018; Jiang et al., 2019). Consequently, the AHI applications in atmospheric studies are increasing (Liu et al., 2019).

MODELLING PM

PM retrieval models from satellite AOD can be divided into scaling factor models, physical analysis models, geostatistical interpolation techniques, and empirical statistical models.

SCALING FACTOR MODELS

The scaling factor model is derived from the chemical transport models (CTMs), which enables the determination of the scale factor between AOD from the satellites and ground. The CTM is used in atmospheric studies because it considers other pollutants' effects and can estimate PM without ground data stations. Specifically, a CTM requires (a) emission data for PM with composition and size resolution for pollutants such as NO_x, SO_2, CO, and NH_3 and (b) meteorological information to simulate transport and removal mechanisms of aerosols and trace gases in the atmosphere. The CTM assimilates meteorological datasets and emission inventories with physics and chemistry algorithms to estimate atmospheric constituents (Van Donkelaar et al., 2010; Engel-Cox et al., 2013). Thus, its uncertainty is dependent on the availability and accuracy of auxiliary datasets and computational cost (Engel-Cox et al., 2013). Crouse et al. (2016) used AOD data from Multi-angle Imaging SpectroRadiometer (MISR), MODIS, and SeaWiFS sensors and $PM_{2.5}$ calculated from the GEOS-CTM model. The scaling factor model yielded a high accuracy when the $PM_{2.5}$ estimated was validated against the $PM_{2.5}$ measured by the North American ground stations ($R = 0.76$). The data were used for long-term health effects study. Nevertheless, CTM is hard to be implemented in developing countries due to the lack of chemical and physical information on pollutants like emission inventories and transformation processes (Van Donkelaar et al., 2010, 2015).

PHYSICAL ANALYSIS MODELS

Physical analysis models estimate PM by using physical parameters. Lin et al. (2015) combined the MODIS 1 km AOD with meteorological parameters such as visibility, relative humidity (RH), hygroscopic growth of aerosols, particle mass extinction efficiency, particle size distribution, fine mode fraction (FMF), and ground-based PM data to estimate $PM_{2.5}$. They reported a good correlation between the satellite-derived $PM_{2.5}$ and ground-based $PM_{2.5}$ at three locations in China with R values ranging between 0.72 and 0.75 (RMSE = 12.86–19.44 μg/m³). In another study, Zhang and Li (2015) used AOD and FMF from the MODIS to obtain $PM_{2.5}$ concentrations over a polluted region in China. They used Fine Mode Fraction (FMF), the columnar volume-to-extinction ratio of fine particulates, planetary boundary layer height, RH, and fine particles' density as inputs for calculating $PM_{2.5}$. Based on this method, they obtained an R value of 0.5 between satellite- and ground-based $PM_{2.5}$. Although this method enables real-time estimation of $PM_{2.5}$ from satellite data, it is not easy to gather the physical parameters; this method may not be suitable for PM estimation in the Southeast Asian region.

GEOSTATISTICAL INTERPOLATION TECHNIQUE

Geostatistical interpolation techniques are the simplest methods to spatially interpolate ground-based PM data over areas located far from the ground monitoring stations. To evaluate satellite data and ground-based parameters on the same scale, ground-based data is generally interpolated using the Kriging technique (Benas et al., 2013). Kloog et al. (2011) incorporated the land use regression (LUR) technique and meteorological variables to estimate $PM_{2.5}$ over the New England region. The authors obtained an R^2 value of 0.81 between AOD and PM_{10} for days without AOD values. It shows that LUR is capable of being used in estimating PM, but this technique depends on in situ data, which are usually limited over large areas (Engel-Cox et al., 2013), especially in the Southeast Asian region.

EMPIRICAL STATISTICAL MODELS

Several researchers developed statistical regression models between the satellite AOD and ground-based PM. The empirical statistical models can produce more accurate estimates of the distribution of $PM_{2.5}$ concentration compared to the physical models and/or scaling factor models (Lin et al., 2015). Empirical statistical models cover simple statistical models such as linear and multiple linear regressions (Zaman et al., 2017) and advanced statistical models. Below, we describe the advanced statistical models:

Artificial Neural Network (ANN)

ANN is an advanced nonlinear model that simulates the human brain process to produce an output from internal calculations (Xiao et al., 2015). ANN requires less formal statistical training, can solve complex dependent and independent parameter relationships, and recognizes interactions between independent parameters (Moustris et al., 2013). Gupta and Christopher (2009a, b) used ANN in estimating $PM_{2.5}$ over Florida. They found that the hourly and daily estimated $PM_{2.5}$ concentrations agreed with measurements with R values 0.74 and 0.78, respectively. ANN was proven to be more robust (Wu et al., 2016). Di et al. (2016) showed that the hybrid of ANN and CTM obtained a better prediction model with $R^2 = 0.84$. Several studies highlighted the usefulness of ANN for predicting particulate pollution, dust storm detection, and AOD retrievals (Gupta and Christopher, 2009b; Wu et al., 2011; El-ossta et al., 2013; Moustris et al., 2013). Recently, Zaman et al. (2017) implemented a multi-layer perceptron (MLP) technique in the ANN model over Malaysia, and the results were more robust compared to multiple linear regression (MLR) with $R^2 = 0.71$ against $R^2 = 0.66$ (Figure 13.6). ANN has intelligent transfer functions that fit dependent and independent parameters and produce satisfactory results for PM estimations (Xiao et al., 2015; Zaman et al., 2017). However, a disadvantage of this method is the "black box" or known as the hidden layer, which complexes the computations and tends to overfitting.

Mixed Effect Model (MEM)

The MEM contains random and fixed effects where the random effect is a variation that is explained by independent variables, while the fixed effect is a variation that is not defined by the independent variables. The model is commonly used to delineate sources of variation that affect the PM concentration. The model assumes PM vertical and diurnal concentration profiles, PM optical properties, and other time-varying parameters that influence the PM and AOD relationship's daily variation. Yap and Hashim (2012) developed a MEM model using the MODIS AOD monthly average values for PM_{10} predictions over Peninsular Malaysia, but the model only generated monthly PM_{10} predictions because of extensive cloud cover and satellite orbital gaps. The MEM may introduce uncertainties due to the assumption of linearity, little spatial variability, and the use of AOD as a single predictor (Xie et al., 2015).

Consequently, some researchers included meteorological parameters and land-use variables to predict $PM_{2.5}$ (Chudnovsky et al., 2012; Kloog et al., 2012). Chudnovsky et al. (2012) used GEOS AOD, surface and atmospheric reflectivity, and aerosol properties to study $PM_{2.5}$ in New England, USA, and the model produced $PM_{2.5}$ that showed high correlation ($R^2 = 0.92$) with the ground data. Zheng et al. (2016) developed a MEM over three major industrial areas in China by using the MODIS AOD at 10 km, meteorological variables, and NO_2 derived from Ozone Monitoring Instrument (OMI) as a proxy for anthropogenic emissions. Based on leave one out cross-validation, the daily estimations yielded R^2 values between 0.77 and 0.8. Since $PM_{2.5}$ and AOD vary spatially (Hu et al., 2013), most researchers

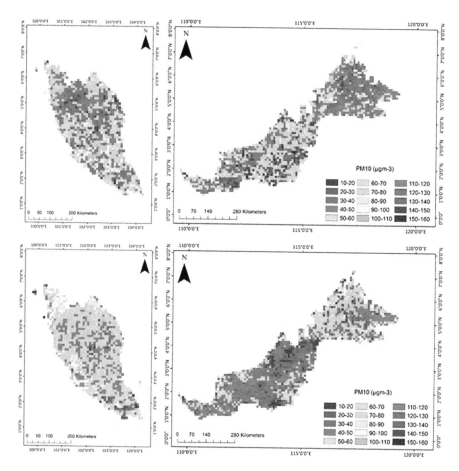

FIGURE 13.6   Spatial distribution of estimated PM$_{10}$ concentration over Malaysia in the dry season (top) and wet season (bottom).

used high-resolution AOD data to estimate PM (Hu et al., 2014; Xie et al., 2015). Beloconi et al. (2016) estimated PM$_{10}$ and PM$_{2.5}$ using synergistic MERIS/AATSR AOD at 1 km, atmospheric stability index, RH, and surface temperature. The MEM obtained accuracies of $R^2 = 0.95$; RMSE $= 2.14$ μg/m^3 (PM$_{10}$), and $R^2 = 0.86$; RMSE $= 2.24$ μg/m^3 (PM$_{2.5}$). In summary, MEM yielded better estimations of PM values over many regions, while the higher AOD resolution tends to improve the model significantly. This enabled many researchers to study PM's critical health effects (Kloog et al., 2012).

Geographic Weighted Regression (GWR)

Unlike MEM, GWR considers spatial variations and nonstationarity into the prediction of PM. Like MEM, GWR is also found to perform better when more parameters that can affect PM's spatial dynamics are considered in the models

(Hu et al., 2013; Ma et al., 2014). You et al. (2016) used a 3-km MODIS AOD product, meteorological variables, and fire emissions from MODIS fire data in the GWR model to estimate ground-level $PM_{2.5}$. The cross-validation of the model produced a low RMSE of 18.6 $\mu g/m^3$. In another study, Zou et al. (2016b) utilized land use, population, meteorological parameters, and AOD at 1-km (derived from simplified high-resolution MODIS aerosol retrieval algorithm) as predictors in their GWR model and ordinary least square (OLS) model to estimate $PM_{2.5}$. They found that the GWR model performed better ($R^2 = 0.75$, RMSE = 10 $\mu g/m^3$) than OLS ($R^2 = 0.53$, RMSE = 16 $\mu g/m^3$). Recently, Bai et al. (2016) pointed out that the spatial resolution is crucial in determining the relationship between $PM_{2.5}$ and anthropogenic sources in China and used the geographically and temporally weighted regression (GTWR) model. In an earlier study, Huang et al. (2010) found that GTWR was better than GWR. GWR was also combined with CTM to estimate PM (van Donkelaar et al., 2015).

GENERALIZED ADDITIVE MODEL (GAM)

GAM is an advanced nonlinear method that allows nonparametric adjustment for the contradicting effects of seasonality, trends, and weather on PM distribution (Dominici et al., 2002). Liu et al. (2009) used a two-stage GAM to study the (1) spatial and (2) temporal variations of $PM_{2.5}$ concentrations by using geostationary operational environmental satellite (GOES) AOD, planetary boundary layer height (PBLH), RH, temperature, wind speed, and land-use parameters (distance to road, road length, traffic volume, land-use type, population, and altitude). The model estimated $PM_{2.5}$ with $R^2 = 0.79$ and a mean prediction error of 0.03 $\mu g/m^3$. Liu et al. (2012) further investigated the capability of the GAM to estimate $PM_{2.5}$ emissions during the Olympic Games in Beijing. They employed MODIS AOD at 10 km, hourly meteorological parameters, and ground $PM_{2.5}$ from the urban and rural sites and concluded that GAM could explain ~70% of the $PM_{2.5}$ variability. Strawa et al. (2013) claimed that GAM could detect high $PM_{2.5}$ values missed by linear models. GAM was also tested in China to estimate $PM_{2.5}$ concentrations (Song et al., 2015). Recently, GAM, in association with LUR, was employed by Zou et al. (2016a) over the Beijing–Tianjin–Hebei (BTH) region in China and GAM was found to perform better than linear models in all seasons ($R^2 = 0.78 - 0.92$) compared to the LUR model ($R^2 = 0.71 - 0.87$). It is also suggested that PM predictions using high-resolution AOD can improve the performance of GAM (Zou et al., 2016a).

TWO STAGE MODELS

The two-stage models are still new, and only a few studies have used them. Hu et al. (2014) used MEM, followed by GWR, to study the spatial and temporal variations in $PM_{2.5}$ over southern USA. They used AOD (1 km) as the main predictor together with meteorological and land-use parameters as ancillary predictors. The model predicted PM well with only a mean prediction error of 1.89 $\mu g/m^3$. Similarly, Ma et al. (2014)

used MEM and GAM to derive $PM_{2.5}$ in BTH over China. This model obtained an overall cross-validation accuracy of $R^2 = 0.79$ and a relative prediction error $= 35.6\%$. Two-stage models can improve PM estimation; however, combining two statistical methods is a complex task. Based on the review, it is recommended to use GWR and GAM for the Southeast Asian region.

$PM_{2.5}$ MONITORING IN MALAYSIA

Most of the air quality and atmospheric studies in Malaysia are conducted using ground PM_{10} measurements (Kanniah et al., 2016). Only a few studies by individual groups have analyzed the spatial distribution of $PM_{2.5}$. Ee-Ling et al. (2015) and Khan et al. (2016) studied $PM_{2.5}$ concentrations and their chemical compositions in residential and urban environments in Peninsular Malaysia and found that $PM_{2.5}$ was largely contributed by motor vehicles and the soil dust. $PM_{2.5}$ distribution in three cities in the northern part of Peninsular Malaysia analyzed by Boon Chun et al. (2012) showed that the $PM_{2.5}$ largely contributed to the cities' turbidity. Previously, the Malaysian Air Pollution Index (API) became an issue when the air pollution reading (specifically in Sothern Johor) was much lower than in Singapore even though haze was bad. Thus, starting mid-August 2018, the Department of Environment, Malaysia improved the calculation of API by using $PM < 2.5\,\mu m$ ($PM_{2.5}$). The latest study by Dahari et al. (2020) examined the seasonal $PM_{2.5}$ concentration in a suburban area in the southern region of Peninsular Malaysia. However, these studies focus on smaller areas; thus, their findings may not represent the real situation at a larger scale. Estimating and monitoring $PM_{2.5}$ at a regional scale using remote sensing is significant for the Southeast Asian countries to address human health concerns.

CONCLUSION

In summary, PM plays a crucial role in the Earth's climate system and global change via radiative forcing and affecting human health and ecosystems. In recent years, air pollution is considered a serious environmental issue in Malaysia since it is influenced by the haze episodes almost every year, impacting air quality and human health. Therefore, estimating fine PM ($PM_{2.5}$) at large spatial scales is important to evaluate Malaysia's pollution levels. The global Environmental Performance Index is currently calculated using $PM_{2.5}$ estimations that need to be refined to characterize the local variations. AOD data from the satellites, coupled with meteorology, land use, and anthropogenic emissions can be effectively used to estimate PM levels in the atmosphere. There are numerous techniques available to estimate PM, and among these models, statistical models are flexible since a variety of datasets can be integrated to provide better estimates of PM. Advanced statistical models such as support vector regression, generalized additive model, and hierarchal models are not tested robustly in this region; thus, we infer the need to explore these models to provide better PM estimates in SEA. The local authorities can use these resulting products for a variety of applications.

ACKNOWLEDGMENT

The authors extend their thanks to WNI WXBUNKA Foundation via research grant (R.J130000.7352.4B406) and the Ministry of Education, Malaysia via the Fundamental Research Grant (R.J130000.7852.5F216) for providing research funding.

REFERENCES

Alias, A., Mat Jafri, M., Lim, H., Saleh, N., Chumiran, S., and Mohamad, A. 2014. Inferring Angstrom exponent and aerosol optical depth from AERONET. *Journal of Environmental Science and Technology.* 7, 166–175.

Bai, Y., Wu, L., Qin, K., Zhang, Y., Shen, Y., and Zhou, Y. 2016. A geographically and temporally weighted regression model for ground-level $PM_{2.5}$ estimation from satellite-derived 500 m resolution AOD. *Remote Sensing.* 8(3), 262.

Beloconi, A., Kamarianakis, Y., and Chrysoulakis, N., 2016. Estimating urban PM10 and PM2. 5 concentrations, based on synergistic MERIS/AATSR aerosol observations, land cover and morphology data. *Remote Sensing of Environment,* 172, 148–s.

Benas, N., Beloconi, A., and Chrysoulakis, N. 2013. Estimation of urban PM_{10} concentration, based on MODIS and MERIS/AATSR synergistic observations. *Atmospheric Environment.* 79, 448–454.

Boon Chun, B., Tan, F., Tan, C., Syahreza, S., Mat Jafri, M. Z., and Lim, H. S. 2012. PM_{10}, $PM_{2.5}$ and PM_1 distribution in Penang Island, Malaysia (vol. 1528).

Chudnovsky, A. A., Lee, H. J., Kostinski, A., Kotlov, T., and Koutrakis, P. 2012. Prediction of daily fine particulate matter concentrations using aerosol optical depth retrievals from the Geostationary Operational Environmental Satellite (GOES). *Journal of the Air & Waste Management Association.* 62(9), 1022–1031.

Crouse, D. L., Philip, S., Van Donkelaar, A., Martin, R. V., Jessiman, B., Peters, P. A., et al. 2016. A new method to jointly estimate the mortality risk of long-term exposure to fine particulate matter and its components. *Scientific Reports.* 6, 18916.

Dahari, N., Latif, M. T., Muda, K., and Hussein, N. 2020. Influence of meteorological variables on suburban atmospheric $PM_{2.5}$ in the Southern Region of Peninsular Malaysia. *Aerosol and Air Quality Research.* 20, 14–25.

Di, Q., Kloog, I., Koutrakis, P., Lyapustin, A., Wang, Y., and Schwartz, J. 2016. Assessing $PM_{2.5}$ exposures with high spatiotemporal resolution across the continental United States. *Environmental Science and Technology.* 50(9), 4712–4721.

DOE. 2018. http://www.doe.gov.my/portalv1/wp-content/uploads/2013/01/Air-Quality-Standard-BI.pdf

Dominici, F., McDermott, A., Zeger, S. L., and Samet, J. M. 2002. On the use of generalized additive models in time-series studies of air pollution and health. *American Journal of Epidemiology.* 156(3), 193–203.

Ee-Ling, O., Mustaffa, N. I. H., Amil, N., Khan, M. F., and Latif, M. T. 2015. Source contribution of PM 2.5 at different locations on the Malaysian Peninsula. *Bulletin of Environmental Contamination and Toxicology.* 94(4), 537–542.

El-ossta, E., Qahwaji, R., and Ipson, S. S. 2013. Detection of dust storms using MODIS reflective and emissive bands. *IEEE Journal of Selected Topics in Applied Earth Observations and Remote Sensing.* 6(6), 2480–2485.

Engel-Cox, J., Oanh, N. T. K., van Donkelaar, A., Martin, R. V., and Zell, E. 2013. Toward the next generation of air quality monitoring: Particulate matter. *Atmospheric Environment.* 80, 584–590.

Environmental Performance Index (EPI). 2018. https://epi.yale.edu/epi-countries Yale University and Columbia University: New Haven, CT, USA.

Fu, J. S., Hsu, N. C., Gao, Y., Huang, K., Li, C., Lin, N. H., et al. 2012. Evaluating the influences of biomass burning during 2006 BASE-ASIA: A regional chemical transport modeling. *Atmospheric Chemistry and Physics.* 12(9), 3837–3855.

Gaveau, D. L. A., Salim, M. A., Hergoualc'h, K., Locatelli, B., Sloan, S., Wooster, M., et al. 2014. Major atmospheric emissions from peat fires in Southeast Asia during non-drought years: Evidence from the 2013 Sumatran fires. *Scientific Reports.* 4, 6112.

Gupta, P., and Christopher, S. A. 2009a. Particulate matter air quality assessment using integrated surface, satellite, and meteorological products: Multiple regression approach. *Journal of Geophysical Research: Atmospheres.* 114(D14). doi:10.1029/2008JD011497.

Gupta, P., and Christopher, S. A. 2009b. Particulate matter air quality assessment using integrated surface, satellite, and meteorological products: 2. A neural network approach. *Journal of Geophysical Research: Atmospheres.* 114(D20). doi:10.1029/2008JD011496.

Gustafsson, O., Kruså, M., Zencak, Z., Sheesley, R., Granat, L., Engström, J., et al. 2009. Brown clouds over South Asia: Biomass or fossil fuel combustion? *Science.* 323, 495–498.

Hayasaka, H., Noguchi, I., Putra, E. I., Yulianti, N., and Vadrevu, K. 2014. Peat-fire-related air pollution in Central Kalimantan, Indonesia. *Environmental Pollution.* 195, 257–266. doi:10.1016/j.envpol.2014.06.031.

Hayasaka, H., and Sepriando, A. 2018. Severe air pollution due to peat fires during 2015 super El Niño in Central Kalimantan, Indonesia. In: *Land-Atmospheric Research Applications in South/Southeast Asia.* Vadrevu, K. P., Ohara, T., and Justice, C. (Eds.). Springer, Cham, pp. 129–142.

Heil, A., Langmann, B., and Aldrian, E. 2007. Indonesian peat and vegetation fire emissions: Study on factors influencing large-scale smoke haze pollution using a regional atmospheric chemistry model. *Mitigation and Adaptation Strategies for Global Change.* 12(1), 113–133.

Holben, B. N., Eck, T. F., Slutsker, I., Tanre, D., Buis, J., Setzer, A., et al. 1998. AERONET—A federated instrument network and data archive for aerosol characterization. *Remote Sensing of Environment.* 66(1), 1–16.

Hu, X., Waller, L. A., Al-Hamdan, M. Z., Crosson, W. L., Estes Jr., M. G., Estes, S. M., et al. 2013. Estimating ground-level $PM_{2.5}$ concentrations in the southeastern US using geographically weighted regression. *Environmental Research.* 121, 1–10.

Hu, X., Waller, L. A., Lyapustin, A., Wang, Y., Al-Hamdan, M. Z., Crosson, W. L., et al. 2014. Estimating ground-level $PM_{2.5}$ concentrations in the Southeastern United States using MAIAC AOD retrievals and a two-stage model. *Remote Sensing of Environment.* 140, 220–232.

Huang, B., Wu, B., and Barry, M. 2010. Geographically and temporally weighted regression for modeling spatio-temporal variation in house prices. *International Journal of Geographical Information Science.* 24(3), 383–401.

Inoue, Y. 2018. Ecosystem carbon stock, atmosphere and food security in slash-and-burn land use: a geospatial study in mountainous region of Laos. In: *Land-Atmospheric Research Applications in South/Southeast Asia.* Vadrevu, K. P., Ohara, T., and Justice, C. (Eds.). Springer, Cham, pp.641–666.

Israr, I., Jaya, S. N. I, Saharjo, H. S., Kuncahyo, B., and Vadrevu, K. P. 2018. Spatio-temporal analysis of land and forest fires in Indonesia using MODIS active fire dataset. In: *Land-Atmospheric Research Applications in South/Southeast Asia.* Vadrevu, K. P., Ohara, T., and Justice, C. (Eds.). Springer, Cham, pp.105–128.

Jiang, T., Chen, B., Chan, K. K. Y., and Xu, B. 2019. Himawari-8/AHI and MODIS aerosol optical depths in China: Evaluation and comparison. *Remote Sensing.* 11(9), 1011.

Kanniah, K. D., Kaskaoutis, D. G., San Lim, H., Latif, M. T., Kamarul Zaman, N. A. F., and Liew, J. 2016. Overview of atmospheric aerosol studies in Malaysia: Known and unknown. *Atmospheric Research.* 182, 302–318.

Kanniah, K. D., Lim, H. Q., Kaskaoutis, D. G., and Cracknell, A. P. 2014. Investigating aerosol properties in Peninsular Malaysia via the synergy of satellite remote sensing and ground-based measurements. *Atmospheric Research.* 138, 223–239.

Khan, F., Latif, M. T., Juneng, L., Amil, N., Mohd Nadzir, M. S., and Syedul Hoque, H. M. 2015a. Physicochemical factors and sources of particulate matter at residential urban environment in Kuala Lumpur. *Journal of the Air & Waste Management Association.* 65(8), 958–969.

Khan, M. F., Latif, M. T., Lim, C. H., Amil, N., Jaafar, S. A., Dominick, D., et al. 2015b. Seasonal effect and source apportionment of polycyclic aromatic hydrocarbons in $PM_{2.5}$. *Atmospheric Environment.* 106, 178–190.

Khan, M. F., Sulong, N. A., Latif, M. T., Nadzir, M. S. M., Amil, N., Hussain, D. F. M., et al. 2016. Comprehensive assessment of $PM_{2.5}$ physicochemical properties during the Southeast Asia dry season (southwest monsoon). *Journal of Geophysical Research: Atmospheres.* 121(14), 589–611.

Kikuchi, M., Murakami, H., Suzuki, K., Nagao, T. M., and Higurashi, A. 2018. Improved hourly estimates of aerosol optical thickness using spatiotemporal variability derived from Himawari-8 geostationary satellite. *IEEE Transactions on Geoscience and Remote Sensing.* 56(6), 3442–3455.

Kloog, I., Koutrakis, P., Coull, B. A., Lee, H. J., and Schwartz, J. 2011. Assessing temporally and spatially resolved $PM_{2.5}$ exposures for epidemiological studies using satellite aerosol optical depth measurements. *Atmospheric Environment.* 45(35), 6267–6275.

Kloog, I., Nordio, F., Coull, B. A., and Schwartz, J. 2012. Incorporating local land use regression and satellite aerosol optical depth in a hybrid model of spatiotemporal $PM_{2.5}$ exposures in the Mid-Atlantic states. *Environmental Science & Technology.* 46(21), 11913–11921.

Lasko, K., Vadrevu, K. P., Tran, V. T., Ellicott, E., Nguyen, T. T., Bui, H. Q., and Justice, C. 2017. Satellites may underestimate rice residue and associated burning emissions in Vietnam. *Environmental Research Letters.* 12(8), 085006.

Lasko, K., and Vadrevu, K. P. 2018. Improved rice residue burning emissions estimates: Accounting for practice-specific emission factors in air pollution assessments of Vietnam. *Environmental Pollution.* 236(5), 795–806.

Lasko, K., Vadrevu, K. P., and Nguyen, T. T. N. 2018a. Analysis of air pollution over Hanoi, Vietnam using multi-satellite and MERRA reanalysis datasets. *PlosOne.* 13(5), e0196629.

Lasko, K., Vadrevu, K. P., Tran, V. T., and Justice, C. 2018b. Mapping double and single crop paddy rice with Sentinel-1A at varying spatial scales and polarizations in Hanoi, Vietnam. *IEEE Journal of Selected Topics in Applied Earth Observations and Remote Sensing.* 11(2), 498–512.

Latif, M. T., Othman, M., Idris, N., Juneng, L., Abdullah, A. M., Hamzah, W. P., et al. 2018. Impact of regional haze towards air quality in Malaysia: A review. *Atmospheric Environment.* 177, 28–44.

Lim, H., Tan, F., Abdullah, K., and Holben, B. 2013. Type of Aerosols Determination Over Malaysia by AERONET Data. Paper presented at the AGU Fall Meeting Abstracts.

Lin, C., Li, Y., Yuan, Z., Lau, A. K., Li, C., and Fung, J. C. 2015. Using satellite remote sensing data to estimate the high-resolution distribution of ground-level $PM_{2.5}$. *Remote Sensing of Environment.* 156, 117–128.

Liu, J., Weng, F., Li, Z., and Cribb, M. C. 2019. Hourly $PM_{2.5}$ estimates from a geostationary satellite based on an ensemble learning algorithm and their spatiotemporal patterns over Central East China. *Remote Sensing.* 11(18), 2120.

Liu, Y., He, K., Li, S., Wang, Z., Christiani, D. C., and Koutrakis, P. 2012. A statistical model to evaluate the effectiveness of $PM_{2.5}$ emissions control during the Beijing 2008 Olympic Games. *Environment International.* 44, 100–105.

Liu, Y., Paciorek, C. J., and Koutrakis, P. 2009. Estimating regional spatial and temporal variability of $PM_{2.5}$ concentrations using satellite data, meteorology, and land use information. *Environmental Health Perspectives.* 117(6), 886.

Ma, Z., Hu, X., Huang, L., Bi, J., and Liu, Y. 2014. Estimating ground-level $PM_{2.5}$ in China using satellite remote sensing. *Environmental Science & Technology.* 48(13), 7436–7444.

Marlier, M. E., DeFries, R. S., Kim, P. S., Koplitz, S. N., Jacob, D. J., Mickley, L. J., et al. 2015. Fire emissions and regional air quality impacts from fires in oil palm, timber, and logging concessions in Indonesia. *Environmental Research Letters.* 10(8), 085005.

Moustris, K., Larissi, I., Nastos, P., Koukouletsos, K., and Paliatsos, A. 2013. Development and application of artificial neural network modeling in forecasting PM_{10} levels in a Mediterranean city. *Water, Air, & Soil Pollution.* 224(8), 1634.

Munchak, L. A., Levy, R. C., Mattoo, S., Remer, L. A., Holben, B. N., Schafer, J. S., Hostetler, C. A., and Ferrare, R. A., 2013. MODIS 3 km aerosol product: Applications over land in an urban/suburban region. *Atmospheric Measurement Techniques.* 6(7), 1747–1759.

Oanh, N. T. K., Permadi, D. A., Dong, N. P., and Nguyet, D. A. 2018. Emission of toxic air pollution and greenhouse gases from crop residue open burning in Southeast Asia. In: *Land-Atmospheric Research Applications in South/Southeast Asia.* Vadrevu, K. P., Ohara, T., and Justice, C. (Eds.). Springer, Cham, pp. 47–68.

Pavuluri, M. C., and Kawamura, K. 2018. Organic aerosols in South and East Asia: Composition and sources. In: *Land-Atmospheric Research Applications in South/Southeast Asia.* Vadrevu, K. P., Ohara, T., and Justice, C. (Eds.). Springer, Cham, pp. 379–408.

Pope, C. A., Brook, R. D., Burnett, R. T., and Dockery, D. W. 2011. How is cardiovascular disease mortality risk affected by duration and intensity of fine particulate matter exposure? An integration of the epidemiologic evidence. *Air Quality, Atmosphere & Health.* 4(1), 5–14.

Ramachandran, S. 2018. Aerosols and climate change: Present understanding, challenges and future outlook. In: *Land-Atmospheric Research Applications in South/Southeast Asia.* Vadrevu, K. P., Ohara, T., and Justice, C. (Eds.). Springer, Cham, pp. 341–378.

Reid, J. S., Hyer, E. J., Johnson, R. S., Holben, B. N., Yokelson, R. J., Zhang, J., et al. 2013. Observing and understanding the Southeast Asian aerosol system by remote sensing: An initial review and analysis for the Seven Southeast Asian Studies (7SEAS) program. *Atmospheric Research.* 122, 403–468.

Snider, G., Weagle, C., Martin, R., Van Donkelaar, A., Conrad, K., Cunningham, D., et al. 2015. SPARTAN: a global network to evaluate and enhance satellite-based estimates of ground-level particulate matter for global health applications.

Song, Y.-Z., Yang, H.-L., Peng, J.-H., Song, Y.-R., Sun, Q., and Li, Y. 2015. Estimating $PM_{2.5}$ concentrations in Xi'an City using a generalized additive model with multi-source monitoring data. *PloS One.* 10(11), e0142149.

Strawa, A., Chatfield, R., Legg, M., Scarnato, B., and Esswein, R. 2013. Improving retrievals of regional fine particulate matter concentrations from Moderate Resolution Imaging Spectroradiometer (MODIS) and Ozone Monitoring Instrument (OMI) multisatellite observations. *Journal of the Air & Waste Management Association.* 63(12), 1434–1446.

Tan, F., San Lim, H., Abdullah, K., Yoon, T. L., and Holben, B. 2015. AERONET data–based determination of aerosol types. *Atmospheric Pollution Research*, 6(4), 682–695.

Vadrevu, K. P., Ohara, T., and Justice, C. eds. 2018. *Land-Atmospheric Research Applications in South and Southeast Asia.* Springer, Cham.

Vadrevu, K. P., Lasko, K., Giglio, L., Schroeder, W., Biswas, S., and Justice, C. 2019. Trends in vegetation fires in South and Southeast Asian countries. *Scientific Reports.* 9(1), 7422. doi:10.1038/s41598-019-43940-x.

Van Donkelaar, A., Martin, R. V., Brauer, M., Kahn, R., Levy, R., Verduzco, C., et al. 2010. Global estimates of ambient fine particulate matter concentrations from satellite-based aerosol optical depth: Development and application. *Environmental Health Perspectives*. 118(6), 847.

Van Donkelaar, A., Martin, R. V., Spurr, R. J., and Burnett, R. T. 2015. High-resolution satellite-derived $PM_{2.5}$ from optimal estimation and geographically weighted regression over North America. *Environmental Science & Technology*. 49(17), 10482–10491.

Viana, M., López, J. M., Querol, X., Alastuey, A., García-Gacio, D., Blanco-Heras, G., et al. 2008. Tracers and impact of open burning of rice straw residues on PM in Eastern Spain. *Atmospheric Environment*. 42(8), 1941–1957.

Wei, J., Li, Z., Sun, L., Peng, Y., Zhang, Z., Li, Z., et al. 2019. Evaluation and uncertainty estimate of next-generation geostationary meteorological Himawari-8/AHI aerosol products. *Science of the Total Environment*. 692, 879–891.

Wu, J., Yao, F., Li, W., and Si, M. 2016. VIIRS-based remote sensing estimation of ground-level $PM_{2.5}$ concentrations in Beijing–Tianjin–Hebei: A spatiotemporal statistical model. *Remote Sensing of Environment*. 184, 316–328.

Wu, Y., Guo, J., Zhang, X., and Li, X. 2011. Correlation between PM concentrations and aerosol optical depth in eastern China based on BP neural networks. *2011 IEEE International Paper presented at the Geoscience and Remote Sensing Symposium (IGARSS)*, pp. 3308–3311.

Xiao, F., Wong, M. S., Lee, K. H., Campbell, J. R., and Shea, Y.-K. 2015. Retrieval of dust storm aerosols using an integrated Neural Network model. *Computers & Geosciences*. 85, 104–114.

Xiao, Q., Zhang, H., Choi, M., Li, S., Kondragunta, S., Kim, J., et al. 2016. Evaluation of VIIRS, GOCI, and MODIS Collection 6 AOD retrievals against ground sunphotometer observations over East Asia. *Atmospheric Chemistry and Physics*. 16, 1255–1269.

Xie, Y., Wang, Y., Zhang, K., Dong, W., Lv, B., and Bai, Y. 2015. Daily estimation of ground-level $PM_{2.5}$ concentrations over Beijing using 3 km resolution MODIS AOD. *Environmental Science & Technology*. 49(20), 12280–12288.

Yang, X., Zhao, C., Luo, N., Zhao, W., Shi, W., and Yan, X. 2020. Evaluation and comparison of Himawari-8 L2 V1.0, V2.1 and MODIS C6.1 aerosol products over Asia and the oceania regions. *Atmospheric Environment*. 220, 117068.

Yao, F., Si, M., Li, W., and Wu, J. 2017. A multidimensional comparison between MODIS and VIIRS AOD in estimating ground-level $PM_{2.5}$ concentrations over a heavily polluted region in China. *The Science of the Total Environment*. 618, 819–828.

Yap, X. Q., and Hashim, M. 2012. A robust calibration approach for PM_{10} prediction from MODIS aerosol optical depth. *Atmospheric Chemistry & Physics Discussions*, 12(12), 31483–31505.

You, W., Zang, Z., Zhang, L., Li, Y., Pan, X., and Wang, W. 2016. National-scale estimates of ground-level $PM_{2.5}$ concentration in China using geographically weighted regression based on 3 km resolution MODIS AOD. *Remote Sensing*. 8(3), 184.

Yusof, N. F. F. M., Ramli, N. A., Yahaya, A. S., Sansuddin, N., Ghazali, N. A., and Al Madhoun, W. 2010. Monsoonal differences and probability distribution of PM 10 concentration. *Environmental Monitoring and Assessment*. 163(1–4), 655–667.

Zaman, N. A. F. K., Kanniah, K. D., and Kaskaoutis, D. G. 2017. Estimating Particulate Matter using satellite based aerosol optical depth and meteorological variables in Malaysia. *Atmospheric Research*. 193, 142–162.

Zhang, W., Xu, H., and Zhang, L. 2019. Assessment of Himawari-8 AHI aerosol optical depth over land. *Remote Sensing*. 11(9), 1108.

Zhang, W., Xu, H., and Zheng, F. 2018. Aerosol optical depth retrieval over East Asia using Himawari-8/AHI Data. *Remote Sensing*. 10(1), 137.

Zhang, Y., and Li, Z. 2015. Remote sensing of atmospheric fine particulate matter ($PM_{2.5}$) mass concentration near the ground from satellite observation. *Remote Sensing of Environment*. 160, 252–262.

Zheng, Y., Zhang, Q., Liu, Y., Geng, G., and He, K. 2016. Estimating ground-level $PM_{2.5}$ concentrations over three megalopolises in China using satellite-derived aerosol optical depth measurements. *Atmospheric Environment*. 124, 232–242.

Zou, B., Chen, J., Zhai, L., Fang, X., and Zheng, Z. 2016a. Satellite based mapping of ground $PM_{2.5}$ concentration using generalized additive modeling. *Remote Sensing*. 9(1), 1.

Zou, B., Pu, Q., Bilal, M., Weng, Q., Zhai, L., and Nichol, J. E. 2016b. High-resolution satellite mapping of fine particulates based on geographically weighted regression. *IEEE Geoscience and Remote Sensing Letters*. 13(4), 495–499.

14 Impacts of Smoke Aerosols over Northern Peninsular Southeast Asia: Results from 7-SEAS Campaigns

Shantanu Kumar Pani and Neng-Huei Lin
National Central University, Taiwan

CONTENTS

INTRODUCTION

Biomass-burning (BB) is a global phenomenon and is mostly regarded as a major source of aerosols and greenhouse and trace gases in the atmosphere (Andreae and Merlet, 2001; Streets et al., 2003). Open BB can significantly influence atmospheric chemistry, air quality, public health, and climate on local, regional, and global scales (Crutzen and Andreae, 1990; Tsay et al., 2016). An extensive range of studies has been conducted worldwide to understand almost all BB aspects, including its sources, emissions, and impact assessments (Chen et al., 2017). BB in tropical and subtropical Africa persists round the year (e.g., Swap et al., 2003; Giglio et al., 2006). Relatively high aerosol loading in Amazonia and nearby South American regions is attributed to BB emissions (*cf.* Martin et al., 2010; Yokelson et al., 2007). Chang and Song (2010) estimated the BB emissions in tropical Asia (extending from India/Nepal southeastward to Indonesia) based on satellite data and reported forest fire as a major source of BB in Southeast Asia (SEA), which is an extensive agricultural region and has witnessed dynamic economic growth and rapid urbanization in recent decades (Badarinath et al.,

2008a, b; Gupta et al., 2001; Kant et al., 2000; Tsay et al., 2016). During boreal spring, BB from natural forest fires and slash-and-burn agricultural practices strongly regulate the regional atmospheric composition over northern peninsular SEA (PSEA; here defined as Myanmar, Laos, Thailand, Vietnam, and Cambodia) and make the climatology of PSEA very different from that of Africa or South America (Biswas et al., 2015a, b; Tsay et al., 2013; Oanh et al., 2018; Inoue, 2018; Lasko et al., 2017, 2018a, b). Intense BB occurs annually during the dry season (February–April), degrades the regional air quality, and modifies the energy balance of the Earth-atmosphere system over PSEA (Lin et al., 2013; Hayasaka et al., 2014; Hayasaka and Sepriando, 2018; Justice et al., 2015; Tsay et al., 2016; Pani et al., 2018, 2021; Ramachandran, 2018; Vadrevu et al., 2018, 2019).

Chemical, physical, and radiative properties of BB emissions has been studied in PSEA mainly during the two grassroot projects, i.e., the Biomass-burning Aerosols in South East Asia (Smoke Impact Assessment; cf. http://smartlabs.gsfc.nasa.gov/; Tsay et al., 2013) and the 7-SEAS (Seven southeast Asian Studies; http://7-seas.gsfc.nasa.gov/; Reid et al., 2013; Lin et al., 2013). Recently, the 7-SEAS/ Biomass-burning Aerosols & Stratocumulus Environment: Lifecycles & Interactions Experiment (BASELInE) campaigns were conducted during spring 2013–2015 to synergize measurements. They included ground-based networks, i.e., Aerosol Robotic Network (AERONET) and Micropulse Lidar Network (MPLNET); sophisticated platforms including Surface-based Mobile Atmospheric Research & Testbed Laboratories (cf. http://smartlabs.gsfc.nasa.gov) mobile facility and regional contributing instruments, along with the satellite retrievals/observations and regional atmospheric chemical transport models to establish a critically needed database and to advance the understanding of BB aerosols and trace gases in northern PSEA (Tsay et al., 2016). The conceptual layout of the 7-SEAS/BASELInE campaigns is presented in Figure 14.1, which highlights a perspective of the region known as the "river of smoke aerosols" from near-source regions in northern PSEA and depicts the confluence of aerosol-cloud-radiation interactions before entering the downwind locations in East Asia including Hong Kong, Dongsha, and Taiwan (Tsay et al., 2016).

In this study, we provide an overview of the observed results from the BASE-ASIA, 7-SEAS/Dongsha Experiment, and 7-SEAS/BASELInE campaigns. Our objective is to characterize the overall impact of BB on air quality and regional climate over northern PSEA. A brief discussion of regional meteorology is given in Section "Regional Meteorology." The analyses of BB impacts on atmospheric composition and aerosol optical properties are discussed in Sections "Impact on Atmospheric Compositions" and "Impact on Aerosol Optical Properties," respectively. Their vertical distributions and implications on regional climate are discussed in Section "Impacts on Regional Radiative Forcing." The BB impact on public health is discussed in Section "Impacts on Public Health," and finally, a summary is given in Section "Concluding Remarks."

REGIONAL METEOROLOGY

The meteorological patterns over PSEA can be characterized by two distinct Asian monsoons, i.e., the winter-northeast and the summer-southwest monsoons (Reid et al., 2013; Yen et al., 2013; Tsay et al., 2016). The regional meteorology normally

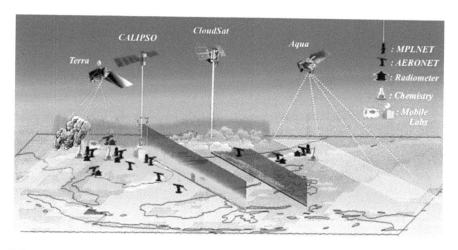

FIGURE 14.1 The conceptual layout of the 7-SEAS/BASELInE spring campaigns. Synergistic deployments of AERONET/MPLNET, SMARTLabs mobile laboratories, and regional contributing instruments (denoted by icons) along the "river of smoke aerosols" under different satellite overpasses to investigate the evolution of atmospheric composition due to springtime BB activities and complex aerosol-cloud-radiation interactions over northern PSEA (adopted from Tsay et al., 2016; Source: "*Aerosol and Air Quality Research*" journal).

followed monsoonal wind fields, driven by the East Asian high and the westward expansion of the northwestern Pacific subtropical high during the spring campaigns of 7-SEAS/BASELInE (2013–2015) over PSEA (Tsay et al., 2016). BB smoke and associated air pollutants were widespread in the planetary boundary layer during the dry-winter monsoon season (Tsay et al., 2016). However, during the peak BB season, the local east–west circulation, enhanced by the well-structured low-level convergent center over northern PSEA, further uplifted the smoke emissions into the lower free-troposphere and carried them downwind to the east through westerly winds aloft (e.g., Lin et al., 2013; Yen et al., 2013; Tsay et al., 2016; Huang et al., 2016). Fire counts obtained from the Moderate Resolution Imaging Spectroradiometer (MODIS) satellite during the dry season of 2014 are shown in Figure 14.2a.

IMPACT ON ATMOSPHERIC COMPOSITIONS

Different kinds of aerosols and trace gases serve as effective markers to distinguish the types and sources of BB emissions (*cf.* Andreae and Merlet, 2001, and references therein). Several papers published recently addressed BB aerosols' physical and chemical properties and trace gases from near-source regions over PSEA to sink–receptor downwind areas. Trace-gas measurements (e.g., CO, CO_2, O_3, SO_2, and nitrogen oxides) obtained from COMMIT (Chemical, Optical, and Microphysical Measurements of In situ Troposphere) mobile laboratory at Doi Ang Khang (DAK), northern Thailand, in 2015 and Sonla, northern Vietnam, during 2012–2013 were analyzed and intercompared to characterize that of regional air masses, classified by meteorological episodes (Pantina et al., 2016; Tsay et al., 2016). Pantina et al.

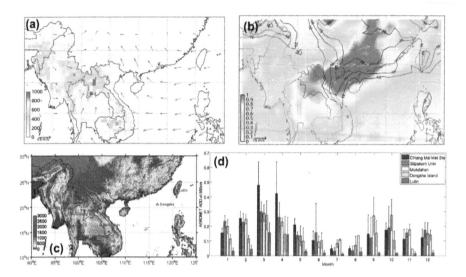

FIGURE 14.2   (a) 0.5° × 0.5° gridded total fire counts merged with 1° × 1° NCEP reanalysis 925 hPa winds (vector) and (b) 1° × 1° AOD (shaded) and COD (contour) at 550 nm obtained from the MODIS instrument aboard the Terra satellite merged with 1° × 1° NCEP reanalysis 700 hPa winds (vector), during March 1 to April 15, 2014. (c) Topographic map with the locations of AERONET sites used in this study. (d) Monthly variation of AOD at five selected AERONET sites (adopted from Wang et al., 2015; Source: *"Aerosol and Air Quality Research"* journal).

(2016) investigated some diagnostic ratios of tracer components to aid in identifying different types of combustion (flaming or smoldering dominant) and sources (BB or urban-pollution). Sayer et al. (2016) investigated the temporal co-variability of trace gases and aerosol data recorded during DAK deployment in 2015. Using radon measurements as an additional trace-gas proxy of atmospheric mixing and transport, Chambers et al. (2016) revealed an upgraded method for representing the transport of transboundary anthropogenic pollution in East Asia. Chen et al. (2016) reported the correlation analyses of total gaseous mercury and trace gas pollutants (including particulate matter, PM) to distinguish the climatic characteristics (e.g., scenario vs. background mercury) of the observations and to verify their discerning factors.

A large concentration of finer $PM_{2.5}$ (PM with aerodynamic diameter $\leq 2.5\,\mu m$), the primary air-quality concern in northern PSEA, is believed to be closely related to large amounts of BB emissions, particularly in the dry season (Pani et al., 2018). Pani et al. (2018) also compared the $PM_{2.5}$ mass during the dry season in 2014 at Chiang Mai with other urban locations under similar monitoring conditions (i.e., either during the dry season or other peak levels of atmospheric pollution) and found Chiang Mai as a highly polluted urban location. Pani et al. (2019b) recently reported that the 24-hour average $PM_{2.5}$ levels at DAK ($118 \pm 36\,\mu g/m^3$) and Chiang Mai ($113 \pm 45\,\mu g/m^3$) were about 4 folds greater than the World Health Organization health-based guideline ($25\,\mu g/m^3$). $PM_{2.5}$ level in the dry season of 2015 at DAK and Chiang Mai was 1.4 and 1.2 times higher than that of 2014, respectively, suggesting the existence of a higher level of BB emissions in 2015 (Pani et al., 2019a). According to the United States Environmental Protection Agency (USEPA), the 24-hour dry $PM_{2.5}$ mass of 40.5–65.4 µg/m³ can pose a moderate health risk, and higher levels are considered to be extremely hazardous to public health (Pani et al., 2018, 2019b).

Several studies investigated the chemical components of BB emissions over the near-source region as well as downwind locations through the analysis of water-soluble inorganic ions, organic carbon (OC), and elemental carbon (EC) contents (e.g., Lee et al., 2011, 2016; Chuang et al., 2013a, b, 2014; Khamkaew et al., 2016; Pani et al., 2018, 2019). Carbonaceous matter (OC and EC) comprised the major portion of $PM_{2.5}$ mass burden in the BB regions, in contrast to typical observations in urban areas, primarily composed of SO_4^{2-} (Tsay et al., 2016). Table 14.1 summarizes the aerosol chemical compositions (in µg/m³) measured at different sites during BASE-ASIA, 7-SEAS/Dongsha Experiment, 7-SEAS/BASELInE, and related studies. Levoglucosan (1, 6-anhydro-b-D-glucopyranose; written as LG hereafter) is generally considered as a highly specific tracer for BB emissions (Tsai et al., 2013; Lee et al., 2016; Pani et al., 2018) and was found in significant levels over PSEA (Table 14.1). LG concentration obtained at Chiang Mai city during the dry season of 2014 (1.13 µg/m³) was higher than those reported at other worldwide urban cities by Pani et al. (2018). Furthermore, Pani et al. (2018) estimated BB's contribution to ambient $PM_{2.5}$ to be 25%–79% by using LG-to-$PM_{2.5}$ emission ratio in the spring of 2014. In addition to LG, non-sea-salt potassium ion (nss-K^+), NO_3^-, OC_3 (evolved at 280°C–480°C), and EC1-OP (EC evolved at 580°C minus the pyrolyzed OC) were found as a marker for PSEA BB (e.g., Lee et al., 2011, 2016; Chuang et al., 2013a, b, 2014; Pani et al., 2018, 2019). OC was the most abundant component in $PM_{2.5}$, and the contribution of BB to ambient OC level was found to be ~90% during the dry season of 2015 over northern PSEA (Pani et al., 2019). The OC/EC ratio ranged between 4.9 and 6.8 over PSEA (Chuang et al., 2013a; Lee et al., 2016; Pani et al., 2018, 2019). Chuang et al. (2013a) inferred that the smoldering of softwood burning mainly contributes the BB aerosols over PSEA based on K^+/LG and LG/Mannosan ratios. Pani et al. (2019) recently estimated, for the first time, the conversion factor of OC to organic matter based on the mass closure approach to be about 1.7 ± 0.3 and 1.6 ± 0.3 at DAK and Chiang Mai, respectively.

Furthermore, Nguyen et al. (2016) investigated the $PM_{2.5}$ organic molecular composition and stable carbon isotope ratios of BB smoke at the Sonla site in northern Vietnam. $PM_{2.5}$ samples were collected in the spring of 2013 and analyzed for more than 50 organic compounds, i.e., primary (i.e., anhydrosugars, lignin and resin products, sugars, sugar alcohols, fatty acids, and phthalate esters) and secondary (i.e., aromatic acids, polyacids, and biogenic oxidation products, such as 2-methyltetrols, alkenes triols, and 3-hydroxyglutaric acid) organic aerosol components. The selected $PM_{2.5}$ samples were composed of mixtures of burned grass, hardwood, softwood, and nonwoody vegetation. By using a variety of approaches (e.g., Fourier transform infrared spectroscopy, scanning electron microscopy-energy dispersive spectroscopy, and hierarchical clusteranalysis), analyses of carbon fractions, organic/inorganic functionalities, and individual particle grouping, Popovicheva et al. (2016) concluded that

TABLE 14.1

Comparison of Aerosol Chemical Compositions (in μg/m³) Measured at Sites of BASE-ASIA, 7-SEAS/Dongsha Experiment, 7-SEAS/BASELInE, and Related Studies

Locations (Types)	Periods	Mass	NO_3^-	SO_4^{2-}	NH_4^+	K^+	OC	EC	LG	References
[a]Mt. Suthep (Near BB source)	Feb.–Apr., 2010	45.5[c]	0.58	4.43	1.33	1.16	18.62	3.33	1.16	Chuang et al. (2013a)
[a]Chiang Mai (Urban)	Feb.–Apr., 2010	139.6[d]	2.34	3.86	2.71	2.31			1.17	Tsai et al. (2013)
[a]TOT (Industrial)	Feb.–Apr., 2010	142.5[d]	2.64	3.41	2.04	1.92			1.07	Tsai et al. (2013)
Phimai (Rural)	Feb.–May, 2006	33[d]	1.0	6.4	2.2	0.56	9.5	2.0		Li et al. (2013)
[a]Sonla (Near BB source)	Spring 2012	51.4[c]	1.3	6.0	2.2	0.8	21.0	3.3		Lee et al. (2016)
[a]Sonla (Near BB source)	Spring 2012	57.3[c]	1.2	5.5	1.9	1.1	20.2	3.5		Lee et al. (2016)
[a]DAK (Near BB source)	Mar.–Apr., 2014	83.3[d]	4.13	6.20	2.92	1.82			1.38	Khamkaew et al. (2016)
[a]Chiang Mai (Urban)	Mar.–Apr., 2014	93.1[c]	2.65	7.81	3.03	2.06			1.13	Khamkaew et al. (2016)
[a]Khao Yai (Forest)	Jan.–Feb., 2005	47[c]	1.2	7.8	2.2	1.0	11.1	0.9		Kim Oanh et al. (2016)
[a]Chiang Mai (Urban)	Spring 2014; Low-BB	56[c]	1.4	8.7	2.8	1.4	22.5	4.3	0.6	Pani et al. (2018)
[a]Chiang Mai (Urban)	Spring 2014; Mild-BB	93[c]	2.7	7.5	3.1	2.1	29.2	5.5	1.1	Pani et al. (2018)
[a]Chiang Mai (Urban)	Spring 2014; High-BB	156[c]	4.7	6.4	2.8	3.1	41.3	7.6	2.0	Pani et al. (2018)
[a]Chiang Mai (Urban)	Spring 2014; Extreme-BB	223[c]	6.1	10.1	4.3	3.5	72.0	13.3	3.5	Pani et al. (2018)
[a]DAK (Near BB source)	Mar. 2015	118[c]	7.3	5.2	3.7	1.9	47.8	7.1		Pani et al. (2019)
[a]Chiang Mai (Urban)	Mar. 2015	113[c]	4.0	4.9	2.3	1.8	46.6	9.5		Pani et al. (2019)
[a]Chiang Mai (Urban)	Mar.–Apr., 2016	64.3[c]	2.70	8.73	3.32	1.27	23.6	2.85	1.22	Thepnuan et al. (2019)
[b]Dongsha-1 (Island)	Spring 2010	15.2[c]	0.4	4.8	1.3	0.2	1.6	0.5		Chuang et al. (2013b)
[b]Dongsha-2 (Island)	Spring 2010	16.3[c]	0.4	5.6	1.6	0.2	2.0	0.7		Chuang et al. (2013b)
[b]Dongsha-3 (Island)	Spring 2010	7.1[c]	0.2	1.5	0.4	0.1	1.2	0.4		Chuang et al. (2013b)

(Continued)

TABLE 14.1 (Continued)

Comparison of Aerosol Chemical Compositions (in μg/m³) Measured at Sites of BASE-ASIA, 7-SEAS/Dongsha Experiment, 7-SEAS/BASELInE, and Related Studies

Locations (Types)	Periods	Mass	NO_3^-	SO_4^{2-}	NH_4^+	K^+	OC	EC	LG	References
[b]Mt. Lulin (High-mountain)[e]	Apr. 2003–Apr. 2009	17.5[c]	0.63	3.20	1.20	0.27				Lee et al. (2011)
[b]Mt. Lulin, (High-mountain)[e]	Apr. 2003–Apr. 2012	15.8[c]					4.26	1.01		Chuang et al. (2014)
[b]Mt. Lulin (High-mountain)	Winter 2015	4.0[c]	0.09	1.02	0.33	0.01	0.85	0.26		Pani et al. (2017)
[b]Hanoi (Mixed-pollution)	Dec. 2006–Feb. 2007	78[c]	6.9	17	7.9	0.9	15	2.2		Hai and Kim Oanh (2013)
[b]Hengchun (Coastal)	Spring 2010	10.9[c]	0.56	2.56	0.91	0.18				Tsai et al. (2012)
[b]Hengchun (Coastal)	Spring 2010	12.8[c]					2.18	0.96		Tsai et al. (2012)
[b]Tengchong-1 (Suburban)	Spring 2004	64.5[c]					17.8	4.4	0.71	Sang et al. (2013)
[b]Tengchong-2 (Mountain)	Spring 2004	19.0[c]					12.6	0.9	0.19	Sang et al. (2013)

[a] Sites within northern PSEA.

[b] Receptor sites of northern PSEA BB during spring.

[c] For the size-cut of $PM_{2.5}$.

[d] For the size-cut of PM_{10}.

[e] for the PSEA BB clusters; Dongsha-1 (air masses from Asian continent), Dongsha-2 (air masses from Philippine) and Dongsha-3 (air masses from the Pacific Ocean); Blank cells represent for no data available for that species.

the ambient aerosols at Sonla were significantly influenced by emissions from the smoldering combustion of wood and other biomass species.

In addition to the BB aerosol chemical profiling at near-source regions, the significant influence of northern PSEA BB smoke on ambient aerosol levels over downwind locations in East Asia including Tengchong (Sang et al., 2013), Hengchun (Tsai et al., 2012), Dongsha island (Chuang et al., 2013b), and Mt. Lulin (Lee et al., 2011; Chuang et al., 2014) were also thoroughly investigated.

IMPACT ON AEROSOL OPTICAL PROPERTIES

BB aerosols are the predominant type of aerosols of northern PSEA during the dry season. Earlier studies (e.g., Gautam et al., 2013; Lin et al., 2014) have demonstrated the inter-annual variability of atmospheric aerosol loading over northern PSEA based on MODIS-derived monthly aerosol optical depth (AOD) variations. Moreover, Yen et al. (2013) and Tsay et al. (2013) suggested that the interannual variability of atmospheric aerosol loading is positively correlated with precipitation and large-scale climate variability over PSEA. Figure 14.2b shows the composite mean AOD and cloud optical depth from March 1 to April 15, 2014, over the region. Furthermore, 700 hPa wind data obtained from the National Centers for Environmental Prediction reanalysis data archive (http://www.esrl.noaa.gov/psd/data/gridded/data.ncep.reanalysis.html) are also shown in Figure 14.2a and b. Wang et al. (2015) conducted a climatological assessment for monthly AOD variation based on AERONET datasets. Figure 14.2c shows the topographic map with the AERONET sites in the study domain. AERONET sites of DAK, Maesoon, Luang Namtha, Chiang Mai, and Sonla are the locations in northern PSEA, but Dongsha and Lulin are the receptor locations of northern PSEA BB emissions. Figure 14.3d shows the monthly variation of AERONET-derived AOD (at 500 nm wavelength, i.e., AOD_{500}) at five selected abovementioned sites. All sites showed the highest AODs in March, indicating the influence of northern PSEA BB smoke emissions. Wang et al. (2015) also inferred that quantifying the spatial distribution and transport of smoke plumes are complicated due to the complex topography (Figure 14.2c), the intensity of emissions (Figure 14.2a), and different altitudinal flow patterns in the region due to low-level updraft and convergent flow and upper-level westerlies (Lin et al., 2013).

The Chiang Mai Met Station represents a site with strong influence from BB emissions (mean AOD ~0.5 in March) in northern Thailand, with a significant mix of BB aerosols with urban and industrial sources (Gautam et al., 2013; Wang et al., 2015). However, two sites (i.e., Silpakorn University and Mukdahan) in southern Thailand showed only a slight increase in AOD during the peak BB season, suggesting mixed aerosols in the atmospheric column (Wang et al., 2015). Downwind sites (Dongsha and Lulin) demonstrated an increase (more than two times) in AOD, indicating the significant influence of northern PSEA smoke in March. Daily variations of AOD_{500} were also investigated at DAK (0.26–1.13) and Ching Mai (0.61–2.50) during the dry season in 2013 and 2014 (Pani et al. 2016a, 2018). A thick smoke condition (AOD_{500} as high as 4.3) was observed on March 21, 2014, due to extreme BB conditions and unfavorable meteorology on that particular day over the region (Wang et al., 2015; Pani et al., 2018). To examine the spatial variability of different aerosol optical and microphysical properties, we compared datasets (Table 14.2) for various AERONET sites (reported in the literature) both within PSEA and downwind locations. Single scattering albedo (SSA) played an important role in aerosol climatic-impact assessments and ranged from 0.85 to 0.91 over PSEA (Table 14.2). Ångström exponent (AE) is the measure of the relative dominance of fine-mode aerosols over the coarse-mode; for coarse-mode particles, AE is <1, and for fine-mode particles, AE is >1 (Pani et al., 2018). Higher values of $AE_{440/870}$ (calculated at 440–870 nm) showed the significant predominance of finer mode BB aerosols over northern PSEA as well as the background locations like Dongsha, and Lulin (Table 14.2). Like AE, the aerosol absorption's dependency on wavelength, i.e., the absorption Ångström exponent ($AAE_{440/870}$), showed the variation between 1.4 and 1.6 over the locations of PSEA. By comparing the aerosol optical properties, Pani et al. (2018) found the optical properties of BB smoke aerosols over northern PSEA completely different from other worldwide BB locations. Likely, Pani et al. (2016a) inferred that northern PSEA smoke exhibited the strongest absorption (lowest SSA value) compared to the Savanna, Amazon, northern Australia, and Boreal smoke.

IMPACTS ON REGIONAL RADIATIVE FORCING

Aerosol vertical profiles from MPLNET lidar observations were analyzed to determine the vertical aerosol distribution over the mountain region in northern PSEA. The mean aerosol extinction vertical profiles at DAK during the dry season of 2013 and 2014 are shown in Figure 14.3a and b. The aerosol vertical distributions reached 5 km above mean sea level (amsl), attributed to planetary boundary layer dynamics over the mountain, and transported smoke plumes (Wang et al., 2015; Pani et al., 2016a). The mean aerosol extinction peak was about $0.37\,km^{-1}$ between 2.5 and 3.5 km amsl, indicating the presence of a lofted aerosol layer in the dry season of 2013 (Pani et al., 2016a). However, the mean aerosol extinction peak was ~$0.3\,km^{-1}$ near the 2014 dry season (Wang et al., 2015). This comparison suggested the higher confinement of smoke aerosols near the surface during 2014 than that of 2013. Pani et al. (2018) analyzed the vertical profile of smoke aerosols over the Chiang Mai urban atmosphere from the Cloud-Aerosol Lidar and Infrared Pathfinder Satellite Observations (CALIPSO; https://www-calipso.larc.nasa.gov) data archive and reported the significant confinement of smoke aerosols near to the ground during high and extreme BB conditions via thermal capping inversion (e.g., Wang et al., 2015; Khamkaew et al., 2016).

BB aerosols are the significant contributors to the modulation of surface-atmosphere radiative energetics during their lifecycles (e.g., Crutzen and Andreae, 1990; Penner et al., 1992; Tsay et al., 2016). Pani et al. (2016b) reported, for the first time, detailed estimation of clear-sky shortwave direct aerosol radiative forcing (ARF) near source regions over northern PSEA by using in situ datasets (i.e., ground-based measurements of aerosol physical, chemical, and optical properties) and a radiative transfer model during 2013 spring BASELInE campaign. The overall mean of ARF at the surface (ARF_{SFC}), top-of-atmosphere (ARF_{TOA}), and atmosphere (ARF_{ATM}) at DAK were estimated to be −31.4, −8.0, and +23.4 W/m², respectively (Pani et al., 2016a). Pani et al. (2016a) also estimated the ARFs for different air-mass clusters at DAK in 2013 and found that the cluster variations of ARF were mainly due to AOD and SSA

TABLE 14.2

Comparison of Mean Values of Aerosol Optical and Microphysical Properties Measured at Sites of 7-SEAS/Dongsha Experiment and 7-SEAS/BASELInE

Locations (Types)	Periods	AOD_{500}	$AE_{440/870}$	$AAE_{440/870}$	SSA_{440}	R_{eff_F} (µm)	R_{eff_C} (µm)	References
[a]DAK (Near BB source)	Mar.– Apr. 2013	0.71	1.77		0.89			Pani et al. (2016a)
[a]DAK (Near BB source)	Mar.– Apr. 2014	0.75	1.73	1.48	0.89	0.14	2.36	Wang et al. (2015)
[a]DAK (Near BB source)	Spring 2015		1.2–2		0.91			Sayer et al. (2016)
[a]Chiang Mai (Urban)	Spring 2014; Overall	1.26	1.73	1.45	0.86	0.14	3.19	Pani et al. (2018)
[a]Chiang Mai (Urban)	Spring 2014; Low-BB	0.98	1.59	1.43	0.87	0.16	2.28	Pani et al. (2018)
[a]Chiang Mai (Urban)	Spring 2014; Mild-BB	1.30	1.77	1.44	0.87	0.14	2.61	Pani et al. (2018)
[a]Chiang Mai (Urban)	Spring 2014; High-BB	1.34	1.84	1.54	0.85	0.13	2.52	Pani et al. (2018)
[a]Chiang Mai (Urban)	Spring 2014; Extreme-BB	2.45	1.82	1.57	0.86	0.14	2.82	Pani et al. (2018)
[a]Maesoon, Thailand	Mar.– Apr. 2014	1.37	1.75	1.49	0.86	0.14	2.51	Wang et al. (2015)
[a]Luang Namtha, Thailand	Mar.– Apr. 2014	1.25	1.73	1.57	0.86	0.14	2.52	Wang et al. (2015)
[a]Son La (Near BB source)	Mar.– Apr. 2014	1.49	1.69	1.46	0.89	0.16	2.59	Wang et al. (2015)
[a]Mukdahan, Thailand	Feb.–Apr. 2004–2009		1.66	1.43	0.91	0.16	2.95	Sayer et al. (2014)
[b]Dongsha (Island)	11 Apr. 2010[d]	0.22			0.95			Pani et al. (2016b)
[b]Dongsha (Island)	11 Apr. 2010[e]	0.52			0.92			Pani et al. (2016b)
[b]Dongsha (Island)	11 Apr. 2010[f]	0.74[c]	1.61[c]	1.23[c]	0.93[c]	0.17[c]	2.02[c]	Unpublished data
[b]Mt. Lulin (High-mountain)	02 Apr. 2013	0.50	1.76[c]	1.24[c]	0.93	0.18[c]	2.37[c]	Pani et al. (2016b)

[a] Sites within northern PSEA.
[b] Receptor sites of northern PSEA BB during spring.
[c] Unpublished data.
[d] For boundary-layer aerosols
[e] For upper-layer aerosols
[f] For total columnar aerosols.

R_{eff_F} and R_{eff_C} are the effective radii, for fine- and coarse-mode aerosols, respectively. Blank cells represent no data availability.

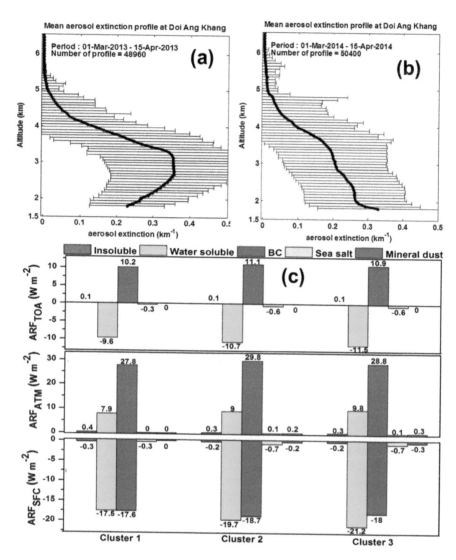

FIGURE 14.3   Mean aerosol extinction profile over DAK during March 1 to April 15 in (a) 2013 (adopted from Pani et al., 2016a) and (b) 2014 (adopted from Wang et al., 2015). (c) ARF due to different aerosol components for Cluster 1 (air-mass originating from Bay of Bengal and mostly locally near the source), Cluster 2 (air-mass mainly originated from East-coast India), and Cluster 3 (air-mass trajectories originating from semi-arid regions over northwest India and Indo-Gangetic Plain and passing through the coastal areas of Indian subcontinent) at DAK. The numbers over each bar represents the value of ARF in W m⁻² (adopted from Pani et al., 2016a; Source: "*Aerosol and Air Quality Research*" journal).

variations. Further, Pani et al. (2016a) also compared the ARF estimated at DAK with other BB regions and found that the overall mean ARF_{TOA} value ($-8.0\,W/m^2$) over northern PSEA was 1.4 times greater than that over Amazonia. Pani et al. (2016a) estimated ARF (ARF_{TOA}, ARF_{SFC}, and ARF_{ATM}) values from individual aerosol components for different air-mass clusters and are shown in Figure 14.3c. ARF_{SFC} values were negative for all components. Scattering aerosols (water-soluble and sea-salt) showed negative values at TOA, whereas absorbing aerosols (water-insoluble and BC) showed only positive values. BC aerosols' contribution to ARF_{ATM} was high and estimated to be ~75% during the 2013 dry season over northern PSEA (Pani et al., 2016a). Following a similar methodology (Pani et al. 2016a) for ARF estimations, the ARF_{SFC} and ARF_{ATM} were estimated to be about to be -45.3 and $+43.6\,W/m^2$, -59.5 and $+58.9\,W/m^2$, -65.8 and $+69.0\,W/m^2$, and -103.4 and $+109.6\,W/m^2$ for Low-BB, Mild-BB, High-BB, and Extreme-BB conditions, respectively, in 2014 dry over the Chiang Mai urban atmosphere (Pani et al. 2018). Such substantial atmospheric heating due to intense regional BB over northern PSEA can significantly influence atmospheric stability and local climate. The enhancement of free tropospheric warming can be as high as 1.3 K/day at a transboundary location in northern South China Sea due to upper-layer transport of BB plumes from northern PSEA the 7-SEAS/Dongsha Experiment (Pani et al. 2016b). The substantial impact of northern-PSEA BB on regional-to-transboundary climate was highlighted over some near-source locations in northern PSEA as well as over East Asia and western Pacific Rim (Lin et al., 2013; Tsay et al., 2013, 2016). During the dry season, substantial BB loadings may influence the cloud formation and lifetime and cause rainfall suppression over the region (Lin et al., 2014; Loftus et al., 2016; Tsay et al., 2016).

IMPACTS ON PUBLIC HEALTH

BB emission contains a huge amount of health-damaging trace metals, allergens, carcinogens, and mutagens in different size fractions (Chen et al., 2017; Pani et al., 2019). A substantial level of $PM_{2.5}$ over northern PSEA can penetrate deeply into the lung, irritate and corrode the alveolar wall, and consequently impair lung function. Pongpiachan and Paowa (2014) reported a significant impact of air pollution on daily hospital walk-ins and admissions in Chiang Mai, particularly in the dry season of 2007–2013. Pani et al. (2019) estimated the inhalation dose for ambient $PM_{2.5}$ and EC levels measured at DAK and Chiang Mai in the dry season of 2015 and reported significant levels of potential exposures for local inhabitants during their outdoor activities. Chuang et al. (2016) also observed a higher risk for PM and black carbon (BC) exposure over northern PSEA based on their deposition potential in the human lungs' alveoli. Recently, Pani et al. (2020) estimated the health risks BC in relation to different health outcomes in terms of passive cigarette equivalence in Chiang Mai. The total number of premature deaths associated with BB-derived $PM_{2.5}$ exposures in Thailand was about 18,003 in 2016 (Punsompong et al., 2021). The health impact of BB smoke aerosols in northern PSEA will be much higher when the polycyclic aromatic hydrocarbons and trace-metal concentrations in $PM_{2.5}$ are considered, which needs further investigation.

CONCLUDING REMARKS

The northern PSEA attracted climate researchers' attention as one of the significant BB regions globally for its complexity in geographical topography, atmospheric chemical compositions, altitudinal vertical mixing of aerosols, and nature of aerosol optical properties. In this study, we presented an overview of BB chemical components and their impacts on aerosol optical properties, radiation budget, and public health during the intense BB activities in northern PSEA and transboundary locations. The overview helps the readers to gain and gather a broad understating of the BB in PSEA.

ACKNOWLEDGMENTS

Authors sincerely extend their appreciation to the Ministry of Science and Technology and Environmental Protection Administration of Taiwan for funding support under the project numbers MOST-109-2811-M-008-545 and EPA-107-FA17-03-D025, respectively. We also thank the journal "Aerosol Air Quality Research (http://www. aaqr.org/)" for permitting us to reproduce some of the figures.

REFERENCES

Andreae, M.O. and Merlet, P. 2001. Emission of trace gases and aerosols from biomass burning. *Glob. Biogeochem. Cycles* 15: 955–966.

Badarinath, K.V.S., Kharol, S.K., Krishna Prasad, V., Kaskaoutis, D.G. and Kambezidis, H.D. 2008a. Variation in aerosol properties over Hyderabad, India during intense cyclonic conditions. *Int. J. Remote Sens.* 29(15): 4575–4597.

Badarinath, K.V.S., Kharol, S.K., Prasad, V.K., Sharma, A.R., Reddi, E.U.B., Kambezidis, H.D. and Kaskaoutis, D.G. 2008b. Influence of natural and anthropogenic activities on UV Index variations–a study over tropical urban region using ground based observations and satellite data. *J. Atmos. Chem.* 59(3): 219–236.

Biswas, S., Lasko, K.D. and Vadrevu, K.P. 2015a. Fire disturbance in tropical forests of Myanmar—Analysis using MODIS satellite datasets. *IEEE J. Sel. Top. Appl. Earth Obs. Remote Sens.* 8(5): 2273–2281.

Biswas, S., Vadrevu, K.P., Lwin, Z.M., Lasko, K. and Justice, C.O. 2015b. Factors controlling vegetation fires in protected and non-protected areas of Myanmar. *PLoS One* 10(4): e0124346.

Chambers, S.D., Kang, C.H., Williams, A.G., Crawford, J., Griffiths, A.D., Kim, K.H. and Kim, W.H. 2016. Improving the representation of cross-boundary transport of anthropogenic pollution in East Asia using Radon-222. *Aerosol Air Qual. Res.* 16: 958–976.

Chang, D. and Song, Y. 2010. Estimates of biomass burning emissions in tropical Asia based on satellite-derived data. *Atmos. Chem. Phys.* 10: 2335–2351.

Chen, J., Li, C., Ristovski, Z., Milic, A., Gu, Y., Islam, M.S., Wang, S., Hao, J., Zhang, H., He, C. and Guo, H. 2017. A review of biomass burning: Emissions and impacts on air quality, health and climate in China. *Sci. Total Environ.* 579: 1000–1034.

Chen, W.K., Li, T.C., Sheu, G.R., Lin, N.H., Chen, L.Y. and Yuan, C.S. 2016. Correlation analysis, transportation mode of atmospheric mercury and criteria air pollutants, with meteorological parameters at two remote sites of mountain and offshore island in Asia. *Aerosol Air Qual. Res.* 16: 2692–2705.

Chuang, H.C., Hsiao, T.C., Wang, S.H., Tsay, S.C. and Lin, N.H. 2016. Characterization of particulate matter profiling and alveolar deposition from biomass burning in Northern Thailand: The 7-SEAS study. *Aerosol Air Qual. Res.* 16: 2897–2906.

Chuang, M.T., Chou, C.K., Sopajareepom, K., Lin, N.H., Wang, J.L., Sheu, G.R., Chang, Y.C. and Lee, C.T. 2013a. Characterization of aerosol chemical properties from near-source biomass burning in Chiang Mai, Thailand during 7-SEAS/Dongsha experiment. *Atmos. Environ.* 78: 72–81.

Chuang, M.T., Lee, C.T., Lin, N.H., Chou, C.C.K., Wang, J.L., Sheu, G.R., Chang, S.C. and et al. 2014. Carbonaceous aerosols in the air masses transported from Indochina to Taiwan: long-term observation at mountain Lulin. *Atmos. Environ.* 89: 507–516.

Chuang, M.T., Lin, N.H., Chang, S.C., Wang, J.L., Sheu, G.R., Chang, Y.J. and Lee, C.T. 2013b. Aerosol chemical properties and related pollutants at Dongsha Island during 7-SEAS/Dongsha experiment. *Atmos. Environ.* 78: 82–92.

Crutzen, P.J. and Andreae, M.O. 1990. Biomass burning in the tropics: Impact on atmospheric chemistry and biogeochemical cycles. *Science.* 250: 1669–1678.

Gautam, R., Hsu, N.C., Eck, T.F., Holben, B.N., Janjai, S., Jantarach, T., Tsay, S.C. and Lau, W.K. 2013. Characterization of aerosols over the Indochina peninsula from satellite-surface observations during biomass burning pre-monsoon season. *Atmos. Environ.* 78: 51–59.

Giglio, L., Csiszar, I. and Justice, C.O. 2006. Global distribution and seasonality of active fires as observed with the Terra and Aqua Moderate Resolution Imaging Spectroradiometer (MODIS) sensors. *J. Geophys. Res. Biogeosci.* 111(G2). doi:10.1029/2005JG000142.

Gupta, P.K., Prasad, V.K., Sharma, C., Sarkar, A.K., Kant, Y., Badarinath, K.V.S. and Mitra, A.P. 2001. CH_4 emissions from biomass burning of shifting cultivation areas of tropical deciduous forests–experimental results from ground-based measurements. *Chemosphere-Glob. Change Sci.* 3(2): 133–143.

Hai, G.D. and Oanh, N.T.K. 2013. Effects of local, regional meteorology and emission sources on mass and compositions of particulate matter in Hanoi. *Atmos. Environ.* 78: 105–112.

Hayasaka, H., Noguchi, I., Putra, E.I., Yulianti, N. and Vadrevu, K.P. 2014. Peat-fire-related air pollution in Central Kalimantan, Indonesia. *Environ. Pollut.* 195: 257–266. doi: 10.1016/j.envpol.2014.06.031.

Hayasaka, H. and Sepriando, A. 2018. Severe Air Pollution Due to Peat Fires during 2015 Super El Niño in Central Kalimantan, Indonesia. In: *Land-Atmospheric Research Applications in South/Southeast Asia.* Vadrevu, K.P., Ohara, T. and Justice, C. (Eds.). Springer, Cham. pp. 129–142.

Huang, W.R., Wang, S.H., Yen, M.C., Lin, N.H. and Promchote, P. 2016. Interannual variation of springtime biomass burning in Indochina: Regional differences, associated atmospheric dynamical changes, and downwind impacts. *J. Geophys. Res. Atmos.* 121: 10,016–10,028.

Inoue, Y. 2018. Ecosystem carbon stock, Atmosphere and Food security in Slash-and-Burn Land Use: A Geospatial Study in Mountainous Region of Laos. In: *Land-Atmospheric Research Applications in South/Southeast Asia.* Vadrevu, K.P., Ohara, T. and Justice, C. (Eds.). Springer, Cham, pp.641–666.

Justice, C., Gutman, G. and Vadrevu, K.P. 2015. NASA land cover and land use change (LCLUC): An interdisciplinary research program. *J. Environ. Manag.* 148(15), 4–9.

Kant, Y., Ghosh, A.B., Sharma, M.C., Gupta, P.K., Prasad, V.K., Badarinath, K.V.S. and Mitra, A.P. 2000. Studies on aerosol optical depth in biomass burning areas using satellite and ground-based observations. *Infrared Phys. Technol.* 41(1): 21–28.

Khamkaew, C., Chantara, S., Janta, R., Pani, S.K., Prapamontol, T., Kawichai, S., Wiriya, W. and Lin, N.H. 2016. Investigation of biomass burning chemical components over northern Southeast Asia during 7-SEAS/BASELInE 2014 campaign. *Aerosol Air Qual. Res.* 16: 2655–2670.

Kim Oanh, N.T., Hang, N.T., Aungsiri, T., Worrarat, T. and Danutawat, T. 2016. Characterization of particulate matter measured at remote forest site in relation to local and distant contributing sources. *Aerosol Air Qual. Res.* 16: 2671–2684.

Lasko, K., Vadrevu, K.P. and Nguyen, T.T.N. 2018a. Analysis of air pollution over Hanoi, Vietnam using multi-satellite and MERRA reanalysis datasets. *PloS One* 13(5): e0196629.

Lasko, K., Vadrevu, K.P., Tran, V.T. and Justice, C. 2018b. Mapping double and single crop paddy rice with Sentinel-1A at varying spatial scales and polarizations in Hanoi, Vietnam. *IEEE J. Sel. Top. Appl. Earth Obs. Remote Sens.* 11(2): 498–512.

Lasko, K., Vadrevu, K.P., Tran, V.T., Ellicott, E., Nguyen, T.T., Bui, H.Q. and Justice, C. 2017. Satellites may underestimate rice residue and associated burning emissions in Vietnam. *Environ. Res. Lett.* 12(8): 085006.

Lee, C.T., Chuang, M.T., Lin, N. H., Wang, J.L., Sheu, G.R., Wang, S.H., Huang, H., Chen, H.W., Weng, G.H. and Hsu, S.P. 2011. The enhancement of bio smoke from Southeast Asia on $PM_{2.5}$ water-soluble ions during the transport over the Mountain Lulin site in Taiwan. *Atmos. Environ.* 45: 5784–5794.

Lee, C.T., Ram, S.S., Nguyen, D.L., Chou, C.C., Chang, S.Y., Lin, N.H., Chang, S.C., Hsiao, T.C., Sheu, G.R., Ou-Yang, C.F. and Chi, K.H. 2016. Aerosol chemical profile of near-source biomass burning smoke in Sonla, Vietnam during 7-SEAS campaigns in 2012 and 2013. *Aerosol Air Qual. Res.* 16(11): 2603–2617.

Li, C., Tsay, S.C., Hsu, N.C., Kim, J.Y., Howell, S.G., Huebert, B.J., Ji, Q., Jeong, M.J., Wang, S.H., Hansell, R.A. and Bell, S.W. 2013. Characteristics and composition of atmospheric aerosols in Phimai, central Thailand during BASE-ASIA. *Atmos. Environ.* 78, 60–71.

Lin, N.H., Tsay, S.C., Maring, H.B., Yen, M.C., Sheu, G.R., Wang, S.H., Chi, K.H., et al. 2013. Overview of regional experiments on biomass burning aerosols and related pollutants in Southeast Asia. *Atmos. Environ.* 78: 1–19.

Lin, N.H., Sayer, A.M., Wang, S.H., Loftus, A.M., Hsiao, T.C., Sheu, G.R., Hsu, N.C., Tsay, S.C. and Chantara, S. 2014. Interactions between biomass-burning aerosols and clouds over Southeast Asia: Current status, challenges, and perspectives. *Environ. Pollut.* 195: 292–307.

Loftus, A.M., Tsay, S.C., Pantina, P., Nguyen, C., Gabriel, P.M., Nguyen, X.A., Sayer, A.M., Tao, W.K. and Matsui, T. 2016. Coupled aerosol-cloud systems over Northern Vietnam during 7-SEAS/BASELInE: A radar and modeling perspective. *Aerosol Air Qual. Res.* 16(11): 2768–2785.

Martin, S.T., Andreae, M.O., Artaxo, P., Baumgardner, D., Chen, Q., Goldstein, A.H., Guenther, A., Heald, C.L., Mayol-Bracero, O.L., McMurry, P.H. and Pauliquevis, T. 2010. Sources and properties of Amazonian aerosol particles. *Rev. Geophys.* 48(2). doi:10.1029/2008RG000280

Nguyen, D.L., Kawamura, K., Ono, K., Ram, S.S., Engling, G., Lee, C.T., Lin, N.H., Chang, S.C., Chuang, M.T., Hsiao, T.C. and Sheu, G.R. 2016. Comprehensive PM2. 5 organic molecular composition and stable carbon isotope ratios at Sonla, Vietnam: Fingerprint of biomass burning components. Aerosol Air Qual. Res. 16(11): 2618–2634.

Oanh, N.T.K., Permadi, D.A., Dong, N.P. and Nguyet, D.A. 2018. Emission of toxic air pollutants and greenhouse gases from crop residue open burning in Southeast Asia. In: *Land-Atmospheric Research applications in South/Southeast Asia*. Vadrevu, K.P., Ohara, T. and Justice, C. (Eds.). Springer, Cham, pp. 47–68.

Pani, S.K., Ou-Yang, C.F., Wang, S.H., Ogren, J.A., Sheridan, P.J., Sheu, G.R. and Lin, N.H. 2019a. Relationship between long-range transported atmospheric black carbon and carbon monoxide at a high-altitude background station in East Asia. *Atmos. Environ.* 210: 86–99.

Pani, S.K., Lee, C.T., Chou, C.C.K., Shimada, K., Hatakeyama, S., Takami, A., Wang, S.H. and Lin, N.H. 2017. Chemical characterization of wintertime aerosols over islands and mountains in East Asia, impacts of the continental Asian outflow. *Aerosol Air Qua. Res.* 17: 3006–3036.

Pani, S.K., Lin, N.H., Chantara, S., Wang, S.H., Khamkaew, C., Prapamontol, T. and Janjai, S. 2018. Radiative response of biomass-burning aerosols over an urban atmosphere in northern peninsular Southeast Asia. *Sci. Total Environ.* 633: 892–911.

Pani, S.K., Lin, N.H., Griffith, S.M., Chantara, S., Lee, C.T., Thepnuan, D. and Tsai, Y.I. 2021. Brown carbon light absorption over an urban environment in northern peninsular Southeast Asia. *Environ. Pollut.* 276: 116735.

Pani, S.K., Chantara, S., Khamkaew, C., Lee, C.T. and Lin, N.H. 2019b. Biomass burning in the northern peninsular Southeast Asia: aerosol chemical profile and potential exposure. *Atmos. Res.* 224: 180–195.

Pani, S.K., Wang, S.H., Lin, N.H., Lee, C.T., Tsay, S.C., Holben, B.N., Janjai, S., Hsiao, T.C., Chuang, M.T. and Chantara, S. 2016a. Radiative effect of springtime biomass-burning aerosols over northern Indochina during 7-SEAS/BASELInE 2013 campaign. *Aerosol Air Qua. Res.* 16: 2802–2817.

Pani, S.K., Wang, S.H., Lin, N.H., Chantara, S., Lee, C.T. and Thepnuan, D. 2020. Black carbon over an urban atmosphere in northern peninsular Southeast Asia: Characteristics, source apportionment, and associated health risks. *Environ. Pollut.* 259, 113871.

Pani, S.K., Wang, S.H., Lin, N.H., Tsay, S.C., Lolli, S., Chuang, M.T., Lee, C.T., Chantara, S. and Yu, J.Y. 2016b. Assessment of aerosol optical property and radiative effect for the layer decoupling cases over the northern South China Sea during the 7-SEAS/Dongsha Experiment. *J. Geophys. Res. Atmos.* 121: 4894–4906.

Pantina, P., Tsay, S.C., Hsiao, T.C., Loftus, A.M., Kuo, F., Ou-Yang, C.F., Sayer, A.M., Wang, S.H., Lin, N.H., Hsu, N.C. and Janjai, S. 2016. COMMIT in 7-SEAS/BASELInE: Operation of and observations from a novel, mobile laboratory for measuring in-situ properties of aerosols and gases. *Aerosol Air Qual. Res.* 16(11): 2728–2741.

Penner, J.E., Dickinson, R.E. and O'Neill, C.A. 1992. Effects of aerosol from biomass burning on the global radiation budget. *Science* 256(5062): 1432–1434.

Pongpiachan, S. and Paowa, T. 2014. Hospital out-and-in-patients as functions of trace gaseous species and other meteorological parameters in Chiang-Mai, Thailand. *Aerosol Air Qual. Res.* 15(2): 479–493.

Popovicheva, O.B., Engling, G., Diapouli, E., Saraga, D., N.M. Persiantseva, M.A. Timofeev, Kireeva, E.D., et al. 2016. Impact of smoke intensity on size-resolved aerosol composition and microstructure during the biomass burning season in Northwest Vietnam. *Aerosol Air Qual. Res.* 16: 2635–2654.

Punsompong, P., Pani, S.K., Wang, S.H. and Bich Pham, T.T. 2021. Assessment of biomass-burning types and transport over Thailand and the associated health risks. *Atmos. Environ.* 247: 118176.

Ramachandran, S. 2018. Aerosols and climate change: Present understanding, challenges and future outlook. In: *Land-Atmospheric Research Applications in South/Southeast Asia.* Vadrevu, K.P., Ohara, T. and Justice, C. (Eds.). Springer, Cham, pp. 341–378.

Reid, J.S., Hyer, E.J., Johnson, R.S., Holben, B.N., Yokelson, R.J., Zhang, J., Campbell, J.R., Christopher, S.A., Di Girolamo, L., Giglio, L. and Holz, R.E. 2013. Observing and understanding the Southeast Asian aerosol system by remote sensing: An initial review and analysis for the Seven Southeast Asian Studies (7SEAS) program. *Atmos. Res.* 122: 403–468.

Sang, X., Zhang, Z., Chan, C. and Engling, G., 2013. Source categories and contribution of biomass smoke to organic aerosol over the southeastern Tibetan Plateau. *Atmospheric Environment, 78*, pp.113–123.

Sayer, A.M., Hsu, N.C., Hsiao, T.C., Pantina, P., Kuo, F., Ou-Yang, C.F., Holben, B.N., Janjai, S., Chantara, S., Wang, S.H. and Loftus, A.M. 2016. In-situ and remotely-sensed observations of biomass burning aerosols at Doi Ang Khang, Thailand during 7-SEAS/ BASELInE 2015. *Aerosol Air Qual. Res.* 16(11): 2786–2801.

Sayer, A.M., Hsu, N.C., Eck, T.F., Smirnov, A. and Holben, B.N. 2014. AERONET-based models of smoke-dominated aerosol near source regions and transported over oceans, and implications for satellite retrievals of aerosol optical depth. *Atmos. Chem. Phys.* 14: 11493–11523.

Streets, D.G., Yarber, K.F., Woo, J.H. and Carmichael, G.R. 2003. Biomass burning in Asia: annual and seasonal estimates and atmospheric emissions. *Glob. Biogeochem. Cycles* 17: 1099.

Swap, R.J., Annegarn, H.J., Suttles, J.T., King, M.D., Platnick, S., Privette, J.L. and Scholes, R.J. 2003. Africa burning: A thematic analysis of the Southern African Regional Science Initiative (SAFARI 2000). *J. Geophys. Res.* 108 (D13): 8465.

Thepnuan, D., Chantara, S., Lee, C. T., Lin, N.H. and Tsai, Y.I. 2019. Molecular markers for biomass burning associated with the characterization of $PM_{2.5}$ and component sources during dry season haze episodes in Upper South East Asia. *Sci. Total Environ.* 658: 708–722.

Tsai, J.H., Huang, K.L., Lin, N.H., Chen, S.J., Lin, T.C., Chen, S.C., Lin, C.C., Hsu, S.C. and Lin, W.Y. 2012. Influence of an Asian dust storm and Southeast Asian biomass burning on the characteristics of seashore atmospheric aerosols in Southern Taiwan. *Aerosol Air Qual. Res.* 15: 1105–1115.

Tsai, Y.I., Sopajaree, K., Chotruksa, A., Wu, H.C. and Kuo, S.C. 2013. Source indicators of biomass burning associated with inorganic salts and carboxylates in dry season ambient aerosol in Chiang Mai Basin, Thailand. *Atmos. Environ.* 78: 93–104.

Tsay, S.C., Maring, H.B., Lin, N.H., Buntoung, S., Chantara, S., Chuang, H.C., Gabriel, P.M., Goodloe, C.S., Holben, B.N., Hsiao, T.C. and Hsu, N.C. 2016. Satellite-surface perspectives of air quality and aerosol-cloud effects on the environment: An overview of 7-SEAS/BASELInE. *Aerosol Air Qual. Res.* 16(11): 2581–2602.

Tsay, S.C., Hsu, N.C., Lau, W.K.M., Li, C., Gabriel, P.M., Ji, Q., Holben, B.N., Welton, E.J., Nguyen, A.X., Janjai, S. and Lin, N.H. 2013. From BASE-ASIA toward 7-SEAS: A satellite-surface perspective of boreal spring biomass-burning aerosols and clouds in Southeast Asia. *Atmos. Environ.* 78: 20–34.

Vadrevu, K.P., Ohara, T. and Justice, C. eds., 2018. *Land-Atmospheric Research Applications in South and Southeast Asia*. Springer, Cham.

Vadrevu, K.P., Lasko, K., Giglio, L., Schroeder, W., Biswas, S. and Justice, C. 2019. Trends in vegetation fires in South and Southeast Asian countries. *Sci. Rep.* 9(1): 7422. doi:10.1038/s41598-019-43940-x.

Wang, S.H., Welton, E.J., Holben, B.N., Tsay, S.C., Lin, N.H., Giles, D., Stewart, S.A., Janjai, S., Nguyen, X.A., Hsiao, T.C. and Chen, W.N. 2015. Vertical distribution and columnar optical properties of springtime biomass-burning aerosols over Northern Indochina during 2014 7-SEAS campaign. *Aerosol Air Qual. Res.* 15(5): 2037–2050.

Yen, M.C., Peng, C.M., Chen, T.C., Chen, C.S., Lin, N.H., Tzeng, R.Y., Lee, Y.A. and Lin, C.C. 2013. Climate and weather characteristics in association with the active fires in northern Southeast Asia and spring air pollution in Taiwan during 2010 7-SEAS/ Dongsha Experiment. *Atmos. Environ.* 78: 35–50.

Yokelson, R.J., Karl, T., Artaxo, P., Blake, D.R., Christian, T.J., Griffith, D.W., Guenther, A. and Hao, W.M. 2007. The tropical forest and fire emissions experiment: overview and airborne fire emission factor measurements. *Atmos. Chem. Phys.* 7: 5175–5196.

15 Temporal Variation and Source Apportionment of PM$_{2.5}$ Constituents in Phuket, Thailand

Siwatt Pongpiachan
National Institute of Development
Administration (NIDA), Thailand

Muhammad Zaffar Hashmi
COMSATS University, Pakistan

*Danai Tipmanee, Chomsri Choochuay,
and Woranuch Deelaman*
Prince of Songkla University, Thailand

*Qiyuan Wang, Li Xing, Guohui Li,
Yongming Han, and Junji Cao*
Institute of Earth Environment,
Chinese Academy of Sciences (IEECAS), China

CONTENTS

INTRODUCTION

Over the past two decades, the percentage of the population who live in urban areas has risen to over 50%, and this proportion is expected to increase by over 70% by 2050 (WHO, 2010). In highly populated areas, anthropogenic activities can cause very high air pollutant concentrations. The sources of emissions can vary, from both stationary sources (industry and domestic combustion) and mobile sources (road traffic). Various organic and inorganic pollutants have been studied as tracers of urban atmospheric pollutants on a regional scale (Marr et al., 2004, Li et al., 2006, Duan et al., 2007, Sharma et al., 2007, Ma et al., 2010). Polycyclic aromatic hydrocarbons (PAH) are of particular concern because of the recognized mutagenic and carcinogenic properties composed of a number of individual compounds (termed congeners). These compounds are ubiquitous as they are formed during all incomplete combustion processes, in-particular biomass burning that is prevalent in South/Southeast Asian countries from various vegetation types and sources (Kant et al., 2000; Prasad and Badarinath, 2004; Prasad et al., 2000; 2004; 2005; 2008; Vadrevu, 2008; Vadrevu et al., 2008; 2006; 2015; Badarinath et al., 2008a, b; Gupta et al., 2001; Vadrevu and Justice, 2011; Hayasaka et al., 2014; Vadrevu and Lasko, 2015; Biswas et al., 2015a, b; Lasko et al., 2017, 2018a, b; Lasko and Vadrevu, 2018; Hayasaka and Sepriando, 2018; Israr et al., 2018; Ramachandran, 2018; Vadrevu and Lasko, 2018). Biomass used as a fuel source raises air quality health concerns. Biomass burning produces greater PAH concentrations and particulate matter per unit of energy than natural gas and petroleum. Typically, biomass burning emissions in South/Southeast Asia is mainly attributed to slash and burn agriculture, clearing of forests for raising plantations, and burning crop residues to plant next crops. (Prasad et al., 2001, 2002, 2003; Vadrevu et al., 2012, 2013, 2014a, b; Justice et al., 2015; Vadrevu et al., 2017, 2018, 2019). Unlike other persistent organic pollutants, they have multiple points and diffuse sources and cannot be controlled by introducing substitute chemicals. Road traffic is well recognized as a major source of PAH emissions (Harrison et al., 1996; Galarneau et al., 2007), and many recent studies have focused on traffic as a source of PAH in urban areas (Guo, 2003; Motelay-Massei et al., 2007; Sharma et al., 2007; Vardoulakis et al., 2008; Miller et al., 2010). Despite numerous contributions of other source categories to the urban pollution, there are only a few studies associated with PAHs and chemical pollutants in Thailand (Pongpiachan and Paowa, 2015; Pongpiachan et al., 2015; Pongpiachan and Iijima, 2016; Pongpiachan et al., 2017a, b; Oanh et al., 2018; Pongpiachan et al., 2018).

Because of their toxicity, the United States Environmental Protection Agency (US EPA) has listed 16 PAHs as priority pollutants and categorized seven of them as carcinogenic chemicals (Wang et al., 2010). In 2005, the European Union set an annual target for benzo (a) pyrene (B[a] P) in ambient air of 1 ng/m^3. Some years earlier, the

United Kingdom government adopted an air quality objective for PAH expressed as a concentration of benzo(a)pyrene as an indicator of the PAH mixture of 0.25 ng/m³ (Delgado-Saborit et al., 2011). There is, consequently, a strong imperative to devise efficient strategies to reduce pollution of the urban atmosphere by PAH congeners. A key aspect of such a requirement is the need for source apportionment to identify those most responsible for measured PAH concentrations in urban air. Diagnostic ratios have been widely used particularly to distinguish the petrogenic (originated from petroleum) and pyrogenic (derived from combustion) PAH over many years (Grimmer et al., 1983; Rogge et al., 1993; Khalili et al., 1995; Pongpiachan et al., 2017a, b, 2018; Ravindra et al., 2008). This technique uses the ratios between two PAH congeners as a form of source marker, but in a situation where multiple sources contribute to airborne concentrations, diagnostic ratios are of minimal value (Galarneau, 2008). Additionally, vapor-particulate phase partitioning can influence measured ratios as these are typically measured in the particulate phase.

Principal component analysis (PCA) is a popular statistical technique capable of separating the atmosphere's chemical constituents based on their source (Hopke et al., 2005; Viana et al., 2008). We have previously analyzed UK PAH data using PCA and have shown that the method is able to separate a number of contributory sources (Mari et al., 2010). Another method known as positive matrix factorization is getting popular for source apportionment of atmospheric aerosol constituents. This is an advanced factor analysis tool with no negative constraints and can quantify the factor contribution directly without the subsequent use of multiple regression analysis. The method is widely applied to multielement datasets (e.g., Viana et al., 2008) and particle number datasets (Harrison et al., 2011). PMF technique has also been applied to PAH datasets (Katsoyiannis et al., 2011; Dvorská et al., 2012), although most PAH apportionment studies have focused on the 16 USEPA priority; PAHs and few have examined temporal and spatial patterns in PAH.

In this study, we focus on PAH characterization over Phuket, Thailand. The objectives of the study are (1). to assess the spatial and temporal variations of carbonaceous particles (i.e., OC-EC) and water-soluble ionic species (WSIS) in $PM_{2.5}$ samples from Phuket; (2). to characterize the potential sources of target compounds in the ambient air of Phuket and (3). to provide the database of PAHs for an epidemiological study using a risk assessment model to evaluate the adverse health impacts in this region.

MATERIALS AND METHODS

MONITORING OF $PM_{2.5}$

$PM_{2.5}$ was collected continuously for 3 days on prebaked (550°C for 12 hours) quartz-fiber filters (QFFs) (QFFs; Whatman 47 mm; Article No., 28418542 (US reference)) with the assistance of MiniVol™ portable air samplers (Airmetrics) with the flow rate of 5 L/minute through a particle size separator (impactor) and consequently through a 47 mm filter. Additional sampling details on $PM_{2.5}$ can be found in the "EPA Quality Assurance Guidance Document: Method Compendium, Field Standard Operating Procedures for the $PM_{2.5}$ Performance Evaluation Program, United States Environmental Protection Agency Office of Air Quality Planning and

Standards" (US EPA, 2002). The weight measurements for $PM_{2.5}$ was done based on the US EPA quality assurance document: method compendium, $PM_{2.5}$ mass weighing laboratory standard operating procedures for the performance evaluation program, US EPA Office of air quality planning and standards (US EPA, 1998) using a microbalance (New Classic MF, MS205DU; Mettler Toledo, Switzerland). The aerosol sampling at Prince of Songkla (POS) University, Phuket Campus Air Quality Observatory Site (PQOS), was located on the fourth floor of the Faculty of Technology and Environment (FTEP) building, Prince of Songkla University, Phuket Campus (see Figure 15.1). The MiniVol TAS air sampler is installed on the rooftop of the FTEP building (fourth floor). A total of 60 quartz filters have been collected.

CHEMICAL CHARACTERIZATION OF PAHs, WSIS, AND OC/EC

Chemical Analysis of PAH's

After carefully weighing, all filters were equally divided into four pieces; thus, each piece is one-fourth of the whole. While half of the filters were used for PAHs' chemical characterization, the rest of the divided filters (i.e., ¼ of filters) were used for Water-Soluble Inorganic Ions (WSIS) and Organic Carbon/Elemental Carbon (OC/EC) analysis. All organic solvents (i.e., dichloromethane (DCM) and Hexane) are High-Performance Liquid Chromatography (HPLC) grade, purchased from Fisher Scientific. A cocktail of 19 PAHs standard (i.e., Ace, Fl, Phe, An, Fluo, Pyr, B[a]A, Chry, B[b]F, B[k]F, B[a]F, B[e]P, B[a]P, Per, Ind, B[g, h, i]P, D[a, h]A, Cor, and D[a, e]P) each $100\,\mu g/mL$ in toluene: unit: $1 \times 1\,mL$) and a mix of recovery internal standard PAHs (d12-perylene (d12-Per), d10-fluorene (d10-Fl); each $100\,\mu g/mL$ in xylene: unit: $1 \times 1\,mL$) were supplied by Chiron AS (Stiklestadveine 1, N-7041 Trondheim, Norway) (http://www.es.lancs.ac.uk/ecerg/kcjgroup/5.html).

The chemical extraction of PAHs was performed by using $250\,mL$ of Soxhlet extractors, spiked with a known quantity of internal standard (d10-Fl: Fl, Phe, An,

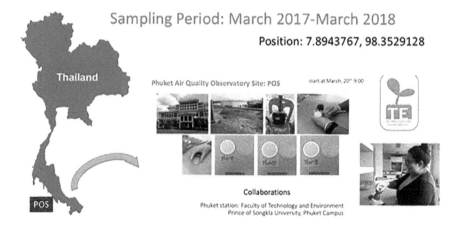

FIGURE 15.1 Sampling location of Prince of Songkla University, Phuket Campus Air Quality Observatory Site at the fourth floor of Faculty of Technology and Environment building, Prince of Songkla University, Phuket Campus.

Fluo, Pyr, B[a]A, Chry; d12-Per: B[b]F, B[k]F, B[a]F, B[e]P, B[a]P, Per, Ind, B[g, h, i] P, D[a, h]A, Cor, D[a, e]P) and extracted with DCM for 8 hours. To avoid interferences caused by contaminants in glassware, reagents, solvents, and other experimental devices, the laboratory blanks were routinely analyzed to ensure that all mentioned materials were free from contamination. Silica gel (0.040–0.063 mm) was purchased from Merck. Soxhlet thimbles and glass fiber filters were obtained from Whatman (Maidstone, UK). All materials used (silica gel, glass, and cotton wool, QFFs, and so on) were Soxhlet extracted with DCM for 24 hours and kept dry (in a desiccator) until use. All glassware was cleaned by washing with decanted water before drying at 55°C and rinsed with DCM just before use.

We followed the fractionation/cleanup process, as reported in Gogou et al. (1996, 1998). After the extraction, the DCM solvent was concentrated to dryness by combining rotary evaporation and blowing under a gentle nitrogen stream. The concentrated extract is then diluted in 10 mL of n-hexane before application to the top of a disposable silica gel column. The extract was then fractionated into individual compound classes by flash chromatography on silica gel as follows: The concentrate was applied to the top of a 30×0.7 cm diameter column, containing 1.5 g of silica gel (activated at 150°C for 3 hours). Nitrogen pressure was used to obtain a flow of 1.4 mL/minute at the bottom of the column. The following solvents were used to elute the different compound classes: (1) 15 mL n-hexane (fraction 1, aliphatic and light molecular weight PAHs); (2) 15 mL toluene-n-hexane (5.6:9.4) (fraction 2, middle and heavy molecular weight PAHs). In consideration of the toxicity of the solvent and PAHs' solubility, toluene was selected for the study. After the fractionation, the eluates were concentrated by using a rotary evaporator, followed by evaporation under a gentle nitrogen stream (set flow rate at 1.0 mbar). Because of toluene's low dissipation capability, a percentage (5%–25%) of acetone was added to increase the volatility. The sample was further reduced to incipient prior to being made up to volume with cyclohexane (exactly 100 µL in a GC/MS vial insert) and GC/MS analysis. All details related to optimizations of MS detectors and GC acquisition settings were mentioned earlier (Pongpiachan et al., 2009a).

Chemical Analysis of WSIS

All filter samples were stored in a refrigerator at about 4°C after sampling was completed. This was necessary to prevent any negative artifacts caused by losses due to volatility. Also, field blank filters were collected to subtract the positive artifacts due to the adsorption of gas-phase organic compounds onto the filter during and/or after sampling.

Analysis for water-soluble particulate anion species including Cl^-, NO_3^-, and SO_4^{2-} ion was performed using ion chromatography (Dionex DX-100) equipped with an IonPac AS4A-SC 4 mm (10–32 P/N 43174) anion specific column, an AG4A-SC 4 mm (10–32 P/N 43175) guard column, and a 25 µL sample loop. Filters were cut, placed in a polyethylene vial (20 mL), and extracted with 10 mL of deionized water for 30 minutes. The extract was filtered through 0.2 µm pore size Millipore Teflon filters for cleanup. The GP40 gradient pump was used to set the ratio of two eluents (i.e., distilled water (mobile phase A) and NaOH solutions (mobile phase B)) pumped through the column at a rate of 2 mL/minute. The nitrogen gas was applied for generating the flow rate and set at 2 PSI. The mixture of standard solutions was made up by dissolving NaCl (1.646 g), $NaNO_3$ (1.373g), and $(NH_4)2SO_4$ (1.376 g) into 1 L of distilled water (1000 ppm).

Chemical Analysis of OC/EC

Carbon analysis was carried out at the Key Lab of Aerosol Chemistry and Physics, Institute of Earth Environment, Chinese Academy of Sciences (IEECAS), Xi'an, China. The samples were analyzed for OC and EC using DRI Model 2001 (thermal/optical carbon analyzer) with the IMPROVE thermal/optical reflectance (TOR) protocol. TOR heats a $0.526 cm^2$ punch aliquot of a sample quartz filter stepwise at temperatures of 120°C (OC1), 250°C (OC2), 450°C (OC3), and 550°C (OC4) in a nonoxidizing helium atmosphere, and 550°C (EC1), 700°C (EC2), and 800°C (EC3) in an oxidizing atmosphere of 2% oxygen in a balance of helium. When oxygen is added, the original and pyrolyzed black carbon is burned, and the reflectance increases. The amount of carbon measured after oxygen is added until the reflectance achieves its original value is reported as optically-detected pyrolyzed carbon. More details associated with chemical analysis of OC/EC were described in previous studies (Pongpiachan et al., 2009b, 2014, 2015).

RESULTS AND DISCUSSION

$PM_{2.5}$ CONCENTRATIONS IN PHUKET

During the wintertime, the $PM_{2.5}$ mass in Phuket ranged from 8.6 to 35.3 µg/m³ (with an average of 20.4 µg/m³). These values were lower than those measured in megacities such as Beijing, Guangzhou, Shanghai, Tianjin, Taiyuan, and Xi'an (Cao et al., 2005, 2007; Zhou et al., 2012; Tao et al., 2009). The $PM_{2.5}$ mass at Sanya was also well below the current air quality guidelines in China (75 µg/m³) (Ambient air quality standards, 2012). Unfortunately, because of the high RHs in summer, high water content on the sample-loaded quartz filters caused stability issues during post-sampling weighing. As a result, the summer $PM_{2.5}$ mass loadings could not be obtained.

$PM_{2.5}$ BOUNDED PAHS IN PHUKET

Panther et al. (1999) and Guo et al. (2003) showed that high levels of PAHs are often observed in winter because low air temperatures lead to low mixing-layer heights and thus cause pollutant accumulation. Most PAH studies focused on the 16 priority species defined by the US EPA. Our ΣPAHs were much lower than those measured in Beijing (annual range of 22.2–5366 ng/m³), Guangzhou (13–31 and 63–140 ng/m³ in summer and winter, respectively), Shanghai (9.0 and 54 ng/m³ on average in summer and winter, respectively), Qingdao (11.5 and 263 ng/m³ on average in summer and winter, respectively), Hong Kong (41.8 ng/m³ on average in winter), and five urban cities in Liaoning Province (annual range of 75.3–1901 ng/m³) (Liu et al., 2008; Feng et al., 2006; Guo et al., 2003, Kong et al., 2010) (Table 15.1). The results prove that the air pollution was less severe in Sanya than in other Chinese megacities; in fact, its

TABLE 15.1

Statistical Descriptions of PAH Contents (pg/m³) in PM$_{2.5}$ Collected at Phuket

	Aver	Stdev
Ace	14.0	9.63
Fl	11.2	7.95
Phe	40.9	41.1
An	6.68	4.54
Fluo	12.0	12.0
Pyr	11.0	12.7
B[a]A	34.0	16.3
Chry	6.70	7.02
B[b]F	23.9	22.0
B[k]F	23.8	22.9
B[a]F	3.17	3.00
B[e]P	14.4	13.4
B[a]P	17.4	19.0
Per	4.83	5.79
Ind	50.7	50.0
B[g,h,i]P	57.5	59.0
D[a,h]A	13.3	11.8
Cor	23.9	20.8
D[a,e]P	8.54	9.08

level was close to those measured in background areas such as Lulang, Tibet (Chen et al., 2014). Compared to worldwide cities, our ΣPAHs were closer to the values measured in São Paulo (Brazil) and Tuscany (Italy), but higher than the other cities or states in the United States. (e.g., Atlanta, Southern California, and Chapel Hill) and lower than in Canada, India, and Turkey (Bourotte et al., 2005; Martellini et al., 2012; Zheng et al., 2009; Eiguren-Fernandez et al., 2004; Pleil et al., 2004; Tham et al., 2008; Anastasopoulos et al., 2012; Abba et al., 2012; Akyüz and Cabuk, 2009).

Particulate Carbonaceous Aerosols in Phuket

The OC/EC ratio has been used to characterize emissions and transformations of the carbonaceous aerosols (Gray et al., 1986; Zeng and Wang, 2011), and the ratios were shown to be dependent on pollution sources, meteorological conditions, and removal processes (Cachier et al., 1996). Chow et al. (1996) showed that secondary organic aerosols are typically formed when OC/EC ratios were greater than two. The secondary organic carbon (OC$_{sec}$) and ratios of OC/EC were 6.24 ± 2.77, and *SOC* was 2.10 ± 0.96 µg/m³. The average concentrations of OC and EC (3.4 and 1.3 µg/m³, respectively) in winter were slightly higher than in summer (3.2 and 0.9 µg/m³) (Wang et al., 2015). The values reported here (expressed below as average *OC* followed by average EC) are approximately one magnitude lower than other megacities in China, such as Xi'an (61.9 and 12.3 $\mu g/m^3$), Shanghai (28.6 and 8.3 $\mu g/m^3$), Beijing (25.9 and 6.1 $\mu g/m^3$), Guangzhou (14.8 and 8.1 $\mu g/m^3$), and Shenzhen (11.1 and 3.9 $\mu g/m^3$) (Cao et al., 2005, 2007; Zhou et al., 2012; Tao et al., 2009; Hagler et al., 2006). The relatively low concentrations of $PM_{2.5}$ and carbonaceous aerosols can be ascribed to limited local emissions from heavily polluting industries, the flushing of pollutants by clean maritime air, and the effective dispersion and removal of polluted air due to the meteorological conditions and landscape of the region. Castro et al. (1999) proposed that minimum values in the OC/EC ratio can be used to estimate the amount of OC_{sec} in the aerosol. The concentration of OC_{sec} can be calculated as follows:

$$OC_{sec} = OC_{tot} - EC \times (OC / EC)_{primary} \qquad (15.1)$$

where OC_{total} is the total OC and $(OC/EC)_{min}$ is the minimum observed OC/EC ratio.

The seasonal differences on both OC/EC ratios and OC_{sec} promise that photochemical conversions of organics are much active at the higher air temperatures and with stronger intense solar radiation in summer. Besides, the contributions of OC_{sec} are within the range from relevant studies at Pearl River Delta (PRD) and other regions (Cao et al., 2004; Turpin and Huntzicker, 1995; Castro et al., 1999; Lin and Tai, 2001). Chow et al. (2004) and Watson et al. (2005) have suggested that the individual carbon fractions can be appointed to specific pollution sources. In China, the major OC and EC sources include coal combustion (mostly residential), motor vehicle emission, and biomass burning (Streets et al., 2001; Cao et al., 2003).

PM$_{2.5}$ BOUNDED WSIS IN PHUKET

The percentage contribution of WSIS collected at Phuket is given in Figure 15.2. The results indicated that SO_4^{2-} was at a higher percentage compared to the other WSIS. The highest percentage SO_4^{2-} recorded was 31%. Na^+ showed a higher percentage of 22% after the SO_4^{2-}. F^- showed the lowest percentage (1%) of WSIS as compared to the other WSIS. The concentration of WSIS is given in Table 15.2. SO_4^{2-} showed the highest concentration as compared to the other WSIS. SO_4^{2-} concertation was 1.931 ± 1.184 ($\mu g/m^3$). F^- showed the lowest concentration 0.091 ± 0.011 ($\mu g/m^3$) as compared to the other WSIS. WSIS followed the trend as: $SO_4^{2-} > Na^+ > Ca^{2+} > NO_3^- > Cl^- > NO_2^-$, $NH_4^+ > K^+ > Mg^{2+} > F^-$. Besides, Pearson correlation analysis indicates strong positive correlations between OC3, OC4, EC1 vs. TC, OC, and EC (see Figure 15.3), suggesting that carbonaceous aerosols were dominated by one major emission source, which might be vehicular exhausts.

PRINCIPAL COMPONENT ANALYSIS

Moreover, PCA (Thurston, and Spengler, 1985) with varimax rotation was used to identify and quantify the source contributions of the eight carbon fractions. Factor 1 was highly loaded with TC, OC, EC, Phe, An, Fluo, Pyr, B[a]A, Chry, B[b]F, B[k]F, B[a]F, B[e]P, B[a]P, Ind, B[g, and h, i]P. We attribute these loadings to the

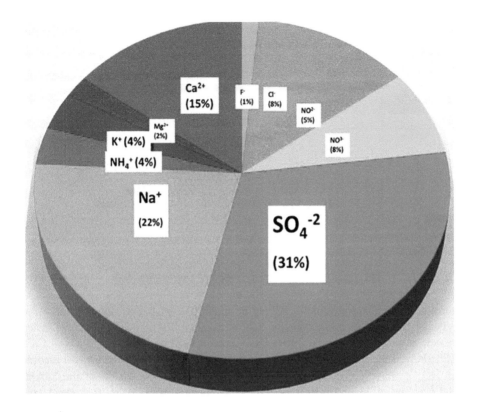

FIGURE 15.2   Percentage contribution of water-soluble ionic species (WSIS) collected at POS.

TABLE 15.2
Water-Soluble Ionic Species
($\mu g/m^3$)

WSIS	Concentration ($\mu g/m^3$)
F^-	0.091 ± 0.011
Cl^-	0.518 ± 0.179
NO_2^-	0.313 ± 0.174
NO_3^-	0.524 ± 0.109
SO_4^{2-}	1.931 ± 1.184
Na^+	1.390 ± 0.291
NH_4^+	0.227 ± 0.079
K^+	0.225 ± 0.079
Mg^{2+}	0.132 ± 0.027
Ca^{2+}	0.931 ± 0.0116

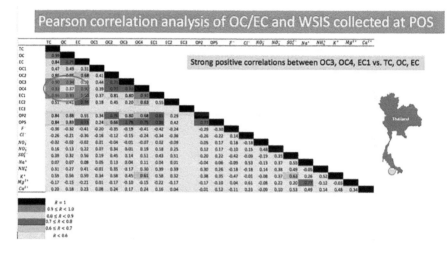

FIGURE 15.3   Pearson correlation analysis of OC/EC and WSIS in PM$_{2.5}$ collected at POS.

gasoline-fueled motor vehicles (Chow et al., 2004). PC1 also suggests the influence of biomass burning sources (Chow et al., 2004), including diesel-fueled vehicle exhausts (Cao et al., 2004). PC2 showed the highest load of Na$^+$ and Ca^{2+}. PC3 showed the highest loading of Cl$^-$ and Mg^{2+}. PC4 showed the highest loading of NO$_3^-$, SO$_4^{2-}$, and NH$_4^+$. In total, PC1 accounted for 23.2% of the ΣPAHs and was mainly loaded with lower molecular weight (MW) PAHs (i.e., the 3- and 4-rings congeners), which has a similar composition profile as the biomass burning described by Jenkins et al. (1996). PC2 accounted for the highest contribution of 37.9% of the ΣPAHs and reflected the gasoline-fueled engine emissions because of higher MW PAHs (e.g., Ind, Cor, B [g, h, i] P, and D [a, e] P) (Wang et al., 2009). Indeed, the emissions from motor vehicles are a major source in this area. PC3 refers to coal-burning emissions that are heavily loaded with Fluo, Pyr, Chry, and B[a]A (Chen et al., 2014). Even though domestic heating is limited in southern China, the coal combustion emission from local power plants is still significant (Zheng et al., 2005; Li et al., 2003). PC4 is more complicated, which has higher loadings of B [b+j+k] F and B[a] P and could represent pyrogenic emissions such as those from cooking activities. Conclusively, the results rank vehicular emissions, biomass burning, coal-burning, and other pyrogenic emissions (e.g., cooking) contribution to the PAHs in descending order (Table 15.3).

HIERARCHICAL CLUSTER ANALYSIS

Hierarchical cluster analysis was used to infer the cluster of different chemicals. It was found that F$^-$, Mg^{2+}, NH$_4^+$, K$^+$, NO$_2^-$, Cl$^-$, NO$_3^-$, EC, Na$^+$, Ca^{2+}, TC, OC, B[a] F, SO$_4^{2-}$, An, Chry, and Per were grouped as Cluster 1 suggesting similar sources (Figure 15.4). B[e]P, B[a]P, Fluo, Pyr, Fl, and D[a, h]A were in Cluster 2 and B[a]F, B[k]F, Cor, Phe, and B[a]A in Cluster 3and Ind and B[g, h, i]P as Cluster 4. Cluster analysis results confirmed the results of correlation and PCA.

TABLE 15.3

Principal Component Analysis of Polycyclic aromatic hydrocarbons and Other Chemical Parameters in PM$_{2.5}$ Collected at POS

	Principal Components			
	PC1	PC2	PC3	PC4
TC	**.701**	.354	−.439	−.202
OC	**.670**	.348	−.396	−.254
EC	**.668**	.309	−.502	.021
F$^-$	−.454	−.231	.173	−.065
Cl$^-$	−.147	.414	**.761**	−.114
NO$_2^-$	.150	−.304	.073	.100
NO$_3^-$	.361	.181	.244	**.413**
SO$_4^{2-}$	.438	.554	−.160	**.533**
Na$^+$	.094	**.675**	.495	.339
NH$_4^+$	.449	.110	−.275	**.408**
K$^+$	.532	.539	−.293	.268
Mg^{2+}	−.183	.525	**.602**	.272
Ca^{2+}	.037	**.746**	.010	.157
Acenapthene	.429	.285	.213	−.635
Fluorene	.524	.144	.112	−.529
Phenanthrene	**.811**	.136	.272	−.354
Anthracene	**.765**	.159	.273	−.356
Fluoranthene	**.934**	.064	.052	−.216
Pyrene	**.952**	−.056	−.066	−.011
Benzoanthracene	**.612**	.194	.197	−.283
Chrysene	**.890**	−.042	−.193	.016
Benzoflouratheneb	**.877**	−.142	−.021	−.016
Benokflourathene	**.937**	−.008	−.126	.059
Benzoaflourathene	**.699**	−.166	.094	.017
Benzoepyrene	**.897**	−.185	.004	.109
Benzoapyrene	**.828**	−.186	.211	−.020
Perylene	.409	−.249	.305	−.132
Indeno123cdpyrene	**.877**	−.284	.096	.227
Benzoghiperylene	**.857**	−.315	.117	.223
Benzoahanthracene	**.613**	−.355	.251	.446
Coronene	**.843**	−.327	.146	.203
Deibenzoaepyrene	.542	−.296	.374	.192

Note: The boldface values indicates higher positive and negative correlation coefficients larger than 0.4.

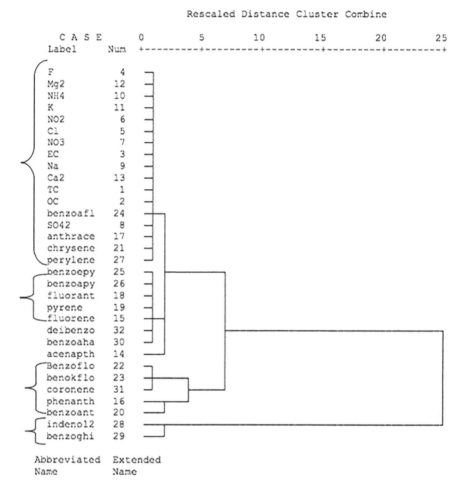

FIGURE 15.4 Hierarchical cluster analysis of polycyclic aromatic hydrocarbons and other chemical parameters in PM$_{2.5}$ collected at POS.

ACKNOWLEDGMENT

This work was performed with the approval of the Thailand Research Fund, the IEECAS, and the NIDA. The authors acknowledge the assistance of local staff from the Faculty of Technology and Environment, Prince of Songkla University Phuket Campus, Faculty of Environmental Management, Prince of Songkla University Hat-Yai Campus, Faculty of Education, Valaya Alongkorn Rajabhat University under the Royal Patronage (VRU), National Astronomical Research Institute of Thailand (Public Organization), toward field sampling.

REFERENCES

Abba, E.J., Unnikrishnan, S., Kumar, R., Yeole, B., Chowdhury, Z. 2012. Fine aerosol and PAH carcinogenicity estimation in outdoor environment of Mumbai City. *India. Int. J. Environ. Health Res.* 22(2), 134–149.

Akyüz, M., Cabuk, H. 2009. Meteorological variation of $PM_{2.5}/PM_{10}$ concentrations and particle-associated polycyclic aromatic hydrocarbons in the Atmos. Environ of Zonguldak, Turkey. *J. Hazard. Mater.* 170, 13.

Anastasopoulos, A.T., Wheeler, A.J., Karman, D., Kulka, R.H. 2012. Intraurban concentrations, spatial variability and correlation of ambient polycyclic aromatic hydrocarbons (PAH) and $PM_{2.5}$. *Atmos Environ.* 59, 272–283.

Badarinath, K.V.S., Kharol, S.K., Krishna Prasad, V., Kaskaoutis, D.G. and Kambezidis, H.D. 2008a. Variation in aerosol properties over Hyderabad, India during intense cyclonic conditions. *Int. J. Remote Sens.* 29(15), 4575–4597.

Badarinath, K.V.S., Kharol, S.K., Prasad, V.K., Sharma, A.R., Reddi, E.U.B., Kambezidis, H.D. and Kaskaoutis, D.G. 2008b. Influence of natural and anthropogenic activities on UV Index variations–a study over tropical urban region using ground based observations and satellite data. *J. Atmos. Chem.* 59(3), 219–236.

Biswas, S., Lasko, K.D. and Vadrevu, K.P. 2015a. Fire disturbance in tropical forests of Myanmar—Analysis using MODIS satellite datasets. *IEEE J. Sel. Top. Appl. Earth Obs. Remote Sens.* 8(5), 2273–2281.

Biswas, S., Vadrevu, K.P., Lwin, Z.M., Lasko, K. and Justice, C.O. 2015b. Factors controlling vegetation fires in protected and non-protected areas of Myanmar. *PLoS One* 10(4), e0124346.

Bourotte, C., Forti, M., Taniguchi, S., Bícego, M.C., Lotufo, P.A. 2005. A wintertime study of PAHs in fine and coarse aerosols in São Paulo City, Brazil. *Atmos. Environ.* 39, 3799–3811.

Cachier, H., Liousse, C., Pertuisol, M.H., Gaudichet, A., Echalar, F., Lacaux, J.P. 1996. African fine particulate emissions and atmospheric influence. In: Levine, E.J.S. (Ed.), *Biomass Burning and Global Change*. MIT Press, London, pp. 428–440.

Cao, J.J., Lee, S.C., Chow, J.C., Watson, J.G., Ho, K.F., Zhang, R.J., Jin, Z.D., Shen, Z.X., Chen, G.C., Kang, Y.M., Zou, S.C., Zhang, L.Z., Qi, S.H., Dai, M.H., Cheng, Y., Hu, K. 2007. Spatial and seasonal distributions of carbonaceous aerosols over China. *J. Geophys. Res.* 112, D22S11.

Cao, J.J., Lee, S.C., Ho, K.F., Zou, S.C., Fung, K., Li, Y., Watson, J.G., Chow, J.C. 2004. Spatial and seasonal variations of atmospheric organic carbon and elemental carbon in Pearl River Delta Region, China. *Atmos. Environ.* 38, 4447–4456.

Cao, J.J., Wu, F., Chow, J.G., Lee, S.C., Li, Y., Chen, S.W., An, Z.S., Fung, K.K., Watson, J.G., Zhu, C.S., Liu, S.X. 2005. Characterization and source apportionment of atmospheric organic and elemental carbon during fall and winter of 2003 in Xi'an, China. *Atmos. Chem. Phys.* 5, 3127–3137.

Cao, J.J., Zhu, C.S., Chow, J.C., Watson, J.G., Han, Y.M., Wang, G.H., Shen, Z.X. and An, Z.S. 2009. Black carbon relationships with emissions and meteorology in Xi'an, China. *Atmospheric Research*, 94(2), 194–202.

Castro, L.M., Pio, C.A., Harrison, R.M., Smith, D.J.T. 1999. Carbonaceous aerosol in urban and rural European atmospheres: estimation of secondary organic carbon concentrations. *Atmos. Environ.* 33, 2771–2781.

Chen, Y., Cao, J.J., Zhao, J., Xu, H.M., Arimoto, R., Wang, G.H., Han, Y.M., Shen, Z.X., Li, G.H. 2014. N-Alkanes and polycyclic aromatic hydrocarbons in total suspended particulates from the southeastern Tibetan Plateau: Concentrations, seasonal variations, and sources. *Sci. Total. Environ.* 470–471, 9–18.

Chow, J.C., Watson, J.G., Chen, L.W.A., Arnott, W.P., Moosmüller, H., Fung, K. 2004. Equivalence of elemental carbon by thermal/optical reflectance and transmittance with different temperature protocols. Environmental *Science & Technology*, 38(16), 4414–4422.

Chow, J.C., Watson, J.G., Lu, Z., Lowenthal, D.H., Frazier, C.A., Solomon, P.A., Thuillier, R.H., Magliano, K. 1996. Descriptive analysis of $PM_{2.5}$ and PM_{10} at regionally representative locations during SJVAQS/AUSPEX. *Atmos. Environ.* 30, 2079–2112.

Delgado-Saborit, J.M., Stark, C., Harrison, R.M. 2011. Carcinogenic potential, levels and sources of polycyclic aromatic hydrocarbon mixtures in indoor and outdoor environments and their implications for air quality standards. *Environ. Inter.* 37, 383–392.

Duan, J., Bi, X., Tan, J., Sheng, G., Fu, J. 2007. Seasonal variation on size distribution and concentration of PAHs in Guangzhou city, China. *Chemosphere* 67, 614–622.

Dvorská, A., Komprdová, K., Lammel, G., Klánová, J., Plachá, H. 2012. Polycyclic aromatic hydrocarbons in background air in central Europe e seasonal levels and limitations for source apportionment. *Atmos. Environ.* 46, 147–154.

Eiguren-Fernandez, A., Miguel, A.H., Froines, J.R., Thurairatnam, S., Avol, E.L. 2004. Seasonal and spatial variation of polycyclic aromatic hydrocarbons in vapor-phase and PM2.5 in Southern California urban and rural communities. *Aerosol Sci. Technol.* 38 (5), 447–455.

Feng, J.L., Hu, M., Chan, C.K., Lau, P.S., Fang, M., He, L.Y., Tang, X.Y. 2006. A comparative study of the organic matter in $PM_{2.5}$ from three Chinese megacities in three different climatic zones. *Atmos. Environ.* 40, 3983–3994.

Galarneau, E. 2008. Source specificity and atmospheric processing of airborne PAHs: Implications for source apportionment. *Atmos. Environ.* 42, 8139–8149.

Galarneau, E., Makar, P.A., Sassi, M., Diamond, M.L. 2007. Estimation of atmospheric emissions of six semi volatile polycyclic aromatic hydrocarbons in southern Canada and the United States by use of an emissions processing system. *Environ. Sci. Technol.* 41, 4205–4213.

Gogou, A., Stratigakis, N., Kanakidou, M., Stephanou, E. 1996. Organic aerosol in Eastern Maditerranean: component source reconciliation by using molecular markers and atmospheric back trajectories. *Org. Geochem.* 25, 79–96.

Gogou, I. A., Apostolaki, M., Stephanou, G. E. 1998. Determination of organic molecular markers in marine aerosols and sediments: One-step flash chromatography compound class fractionation and capillary gas chromatographic analysis. *J. Chromatogr. A.* 799(-1–2), 215–231.

Gray, H.A., Cass, G.R., Huntzicker, J.J., Heyerdahl, E.K., Rau, J.A. 1986. Characteristics of atmospheric organic and elemental carbon particle concentrations in Los Angeles. *Environ. Sci. Technol.* 20(6), 580–589.

Grimmer, G., Jacob, J., Naujack, K.W., Dettbarn, G. 1983. Determination of polycyclic aromatic compounds emitted from brown-coal- fi red residential stoves by gas chromatography/mass spectrometry. *Anal. Chem.* 55, 892–900.

Guo, H. 2003. Particle-associated polycyclic aromatic hydrocarbons in urban air of Hong Kong. *Atmos. Environ.* 37, 5307–5317.

Guo, H., Lee, S.C., Ho, K.F., Wang, X.M., Zou, S.C. 2003. Particle associated polycyclic aromatic hydrocarbons in urban air of Hong Kong. *Atmos. Environ.* 37, 5307–5317.

Gupta, P.K., Prasad, V.K., Sharma, C., Sarkar, A.K., Kant, Y., Badarinath, K.V.S., Mitra, A.P. 2001. CH_4 emissions from biomass burning of shifting cultivation areas of tropical deciduous forests–experimental results from ground-based measurements. *Chemosphere-Glob. Change Sci.* 3(2), 133–143.

Hagler, G.S.W., Bergin, M.H., Salmon, L.G., Yu, J.Z., Wan, E.C.H., Zheng, M., Zeng, L.M., Kiang, C.S., Zhang, Y.H., Lau, A.K.H., and Schauer, J.J. 2006. Source areas and chemical composition of fine particulate matter in the Pearl River Delta region of China. Atmospheric Environment. 40(20), 3802–3815.

Harrison, R.M., Smith, D.J.T., Luhana, L. 1996. Source apportionment of atmospheric poly-cyclic aromatic hydrocarbons collected from an urban location in Birmingham, U.K. *Environ. Sci. Technol.* 30, 825–832.

Hayasaka, H., Sepriando, A. 2018. Severe air pollution due to peat fires during 2015 Super El Niño in Central Kalimantan, Indonesia. In: *Land-Atmospheric Research Applications in South/Southeast Asia.* Vadrevu, K.P., Ohara, T., and Justice, C. (Eds.). Springer, Cham. pp. 129–142.

Hayasaka, H., Noguchi, I., Putra, E.I., Yulianti, N., Vadrevu, K. 2014. Peat-fire-related air pollution in Central Kalimantan, Indonesia. *Environ. Poll.* 195, 257–266. doi:10.1016/j.envpol.2014.06.031.

Hopke, P.K., Ito, K., Mar, T., Christensen, W.F., Eatough, D.J., Henry, R.C., Kim, E., Laden, F., Lall, R., Larson, T.V., Liu, H., Neas, L., Pinto, J., Stolzel, M., Suh, H., Paatero, P., Thurston, G.D. 2005. PM source apportionment and health effects: 1. Inter-comparison of source apportionment results. *J. Expo. Sci. Env. Epid.* 16, 275–286.

Israr, I., Jaya, S.N.I, Saharjo, H.S., Kuncahyo, B., Vadrevu, K.P. 2018. Spatio-temporal analy-sis of land and forest fires in Indonesia using MODIS active fire dataset. In: *Land-Atmospheric Research Applications in South/Southeast Asia.* Vadrevu, K.P., Ohara, T., and Justice, C. (Eds). Springer, Cham, pp.105–128.

Jenkins, B.M., Jones, A.D., Turn, S.Q., William, R.B. 1996. Emission factor for polycyclic aromatic hydrocarbons from biomass burning. *Environ. Sci. Technol.* 30, 2462–2469.

Justice, C., Gutman, G., Vadrevu, K.P. 2015. NASA land cover and land use change (LCLUC): An interdisciplinary research program. *J. Environ. Manag.* 148(15), 4–9.

Kant, Y., Ghosh, A.B., Sharma, M.C., Gupta, P.K., Prasad, V.K., Badarinath, K.V.S., Mitra, A.P. 2000. Studies on aerosol optical depth in biomass burning areas using satellite and ground-based observations. *Infrared Phys. Technol.* 41(1), 21–28.

Katsoyiannis, A., Sweetman, A.J., Jones, K.C. 2011. PAH molecular diagnostic ratios applied to atmospheric sources: a critical evaluation using two decades of source inventory and air concentration data from the UK. *Environ. Sci. Technol.* 45, 8897–8906.

Khalili, N.R., Scheff, P.A., Holsen, T.M. 1995. PAH source fingerprints for coke ovens, die-sel and, gasoline engines, highway tunnels, and wood combustion emissions. *Atmos. Environ.* 29, 533–542.

Kong, S.F., Ding, X., Bai, Z.P., Han, B., Chen, L., Shi, J.W., Li, Z.Y. 2010. A seasonal study of Polycyclic aromatic hydrocarbons in $PM_{2.5}$ and $PM_{2.5-10}$ in five typical cities of Liaoning Province, China. *J. Hazard. Mater.* 183, 70–80.

Lasko, K. and Vadrevu, K.P. 2018. Improved rice residue burning emissions estimates: Accounting for practice-specific emission factors in air pollution assessments of Vietnam. *Environ. Poll.* 236(5), 795–806.

Lasko, K., Vadrevu, K.P., Nguyen, T.T.N. 2018a. Analysis of air pollution over Hanoi, Vietnam using multi-satellite and MERRA reanalysis datasets. *PloS One.* 13(5), e0196629.

Lasko, K., Vadrevu, K.P., Tran, V.T., Justice, C. 2018b. Mapping double and single crop paddy rice with Sentinel-1A at varying spatial scales and polarizations in Hanoi, Vietnam. *IEEE J. Sel. Top. Appl. Earth Obs. Remote Sens.* 11(2), 498–512.

Lasko, K., Vadrevu, K.P., Tran, V.T., Ellicott, E., Nguyen, T.T., Bui, H.Q. and Justice, C. 2017. Satellites may underestimate rice residue and associated burning emissions in Vietnam. *Environ. Res. Lett.* 12(8), 085006.

Li, A., Jang, J.-K., Scheff, P.A. 2003. Application of EPA CMB8.2 model for source apportion-ment of sediment PAHs in Lake Calumet, Chicago. *Environ. Sci. Technol.* 37, 2958–2965.

Li, J., Zhang, G., Li, X.D., Qi, S.H., Liu, G.Q., Peng, X.Z. 2006. Source seasonality of polycy-clic aromatic hydrocarbons (PAHs) in a subtropical city, Guangzhou, South China. *Sci. Total. Environ.* 355, 145–155.

Lin, J.J., Tai, H.S. 2001. Concentrations and distributions of carbonaceous species in ambient particles in Kaohsiung City, Taiwan. *Atmos. Environ.* 35, 2627–2636.

Liu, D.M., Gao, S.P., An, X.H. 2008. Distribution and source apportionment of polycyclic aromatic hydrocarbons from atmospheric particulate matter $PM_{2.5}$ in Beijing. *Adv. Atmos. Sci.*, 25(2), 297–305.

Ma, W.L., Li, Y.F., Qi, H., Sun, D.Z., Liu, L.Y., Wang, D.G. 2010. Seasonal variations of sources of polycyclic aromatic hydrocarbons (PAHs) to a northeastern urban city, China. *Chemosphere*, 79, 441–447.

Mari, M., Harrison, R.M., Schuhmacher, M., Domingo, J.L., Pongpiachan, S. 2010. Inferences over the sources and processes affecting polycyclic aromatic hydrocarbons in the atmosphere derived from measured data. *Sci. Total. Environ.* 408, 2387–2393.

Marr, L.C., Grogan, L.A., Wöhrnschimmel, H., Molina, L.T., Molina, M.J., Smith, T.J., Garshick, E. 2004. Vehicle traffic as a source of particulate polycyclic aromatic hydrocarbon exposure in the Mexico City metropolitan area. *Environ. Sci. Technol.* 38(9) 2584–2592.

Martellini, T., Giannoni, M., Lepri, L., Katsoyiannis, A. and Cincinelli, A. 2012. One year intensive PM2. 5 bound polycyclic aromatic hydrocarbons monitoring in the area of Tuscany, Italy. Concentrations, source understanding and implications. Environmental Pollution, 164, 252–258.

Miller, L., Lemke, L.D., Xu, X., Molaroni, S.M., You, H., Wheeler, A.J., Booza, J., Grgicak-Mannion, A., Krajenta, R., Graniero, P. 2010. Intra-urban correlation and spatial variability of air toxics across an international airshed in Detroit, Michigan (USA) and Windsor, Ontario (Canada). *Atmos. Environ.* 44, 1162–1174.

Motelay-Massei, A., Ollivon, D., Garban, B., Tiphagne-Larcher, K., Zimmerlin, I., Chevreuil, M. 2007. PAHs in the bulk atmospheric deposition of the Seine river basin: source identification and apportionment by ratios, multivariate statistical techniques and scanning electron microscopy. *Chemosphere.* 67, 312–321.

Oanh, N.T.K., Permadi, D.A., Dong, N.P., Nguyet, D.A. 2018. Emission of toxic air pollutants and greenhouse gases from crop residue open burning in Southeast Asia. In: *Land-Atmospheric Research applications in South/Southeast Asia.* Vadrevu, K.P., Ohara, T., and Justice, C. (Eds.). Springer, Cham, pp.47–68.

Panther, B.C., Hooper, M.A., Tapper, N.J. 1999. A comparison of air particulate matter and associated polycyclic aromatic hydrocarbons in some tropical and temperate urban environments. *Atmos. Environ.* 33, 4087–4099.

Pleil, J.D., Vette, A.F., Rappaport, S.M. 2004. Assaying particle-bound polycyclic aromatic hydrocarbons from archived $PM_{2.5}$ filters. *J. Chromatogr. A* 033, 9–17.

Pongpiachan, S., Iijima, A. 2016. Assessment of selected metals in the ambient air PM_{10} in urban sites of Bangkok (Thailand). *Environ. Sci. Pollut. Res.* 23(3), 2948–2961.

Pongpiachan, S., Paowa, T. 2015. Hospital out-and-in-patients as functions of trace gaseous species and other meteorological parameters in Chiang-Mai, Thailand. *Aerosol. Air. Qual. Res.* 15(2)479–493.

Pongpiachan, S., Bualert, S., Sompongchaiyakul, P., Kositanont, C. 2009a. Factors affecting sensitivity and stability of polycyclic aromatic hydrocarbons determined by gas chromatography quadrupole ion trap mass spectrometry. *Anal. Lett.* 42(13), 2106–2130.

Pongpiachan, S., Hattayanone, M., Cao, J. 2017b. Effect of agricultural waste burning season on PM2.5-bound polycyclic aromatic hydrocarbon (PAH) levels in Northern Thailand. *Atmos. Pollut. Res.* 8(6), 1069–1080.

Pongpiachan, S., Hattayanone, M., Suttinun, O., Khumsup, C., Kittikoon, I., Hirunyatrakul, P., Cao, J. 2017a. Assessing human exposure to PM_{10}-bound polycyclic aromatic hydrocarbons during fireworks displays. *Atmos. Pollut. Res.* 8(5), 816–827.

Pongpiachan, S., Hattayanone, M., Tipmanee, D., Suttinun, O., Khumsup, C., Kittikoon, I., Hirunyatrakul, P. 2018. Chemical characterization of polycyclic aromatic hydrocarbons (PAHs) in 2013 Rayong oil spill-affected coastal areas of Thailand. *Environ. Pollut.* 233, 992–1002.

Pongpiachan, S., Ho, K., Cao, J. 2014. Effects of biomass and agricultural waste burnings on diurnal variation and vertical distribution of OC/EC in Hat-Yai City, Thailand. *Asian J. Appl. Sci.* 7(5), 360–374.

Pongpiachan, S., Kositanont, C., Palakun, J., Liu, S., Ho, K.F. and Cao, J. 2015. Effects of day-of-week trends and vehicle types on $PM_{2.5}$-bounded carbonaceous compositions. *Sci. Total. Environ.* 532, 484–494.

Pongpiachan, S., Thamanu, K., Ho, K. F., Lee, S. C., Sompongchaiyakul, P. 2009b. Predictions of gas-particle partitioning coefficients of polycyclic aromatic hydrocarbons at various occupational environments of Songkhla province, Thailand. *Southeast Asian J. Trop. Med. Pub. Health.* 40(6), 1377.

Prasad, V.K., Badarinath, K.V.S. 2004. Land use changes and trends in human appropriation of above ground net primary production (HANPP) in India (1961–98). *Geograph. J.* 170 (1), 51–63.

Prasad, V.K., Anuradha, E., Badarinath, K.V.S. 2005. Climatic controls of vegetation vigor in four contrasting forest types of India—evaluation from National Oceanic and Atmospheric Administration's Advanced Very High Resolution Radiometer datasets (1990–2000). *Int. J. Biometeorol.* 50(1), 6–16.

Prasad, V.K., Badarinath, K.V.S., Eaturu, A. 2008. Biophysical and anthropogenic controls of forest fires in the Deccan Plateau, India. *J. Environ. Manag.* 86(1), 1–13.

Prasad, V.K., Badarinath, K.V.S., Yonemura, S., Tsuruta, H. 2004. Regional inventory of soil surface nitrogen balances in Indian agriculture (2000–2001). *J. Environ. Manag.* 73(3), 209–218.

Prasad, V.K., Gupta, P.K., Sharma, C., Sarkar, A.K., Kant, Y., Badarinath, K.V.S., Rajagopal, T., Mitra, A.P. 2000. NO_x emissions from biomass burning of shifting cultivation areas from tropical deciduous forests of India–estimates from ground-based measurements. *Atmos. Environ.* 34(20), 3271–3280.

Prasad, V.K., Kant, Y., Badarinath, K.V.S. 2001. CENTURY ecosystem model application for quantifying vegetation dynamics in shifting cultivation areas: A case study from Rampa Forests, Eastern Ghats (India). *Ecol. Res.* 16(3), 497–507.

Prasad, V.K., Kant, Y., Gupta, P.K., Elvidge, C., Badarinath, K.V.S. 2002. Biomass burning and related trace gas emissions from tropical dry deciduous forests of India: A study using DMSP-OLS data and ground-based measurements. *Int. J. Remote Sens.* 23(14), 2837–2851.

Prasad, V.K., Lata, M., Badarinath, K.V.S. 2003. Trace gas emissions from biomass burning from northeast region in India—estimates from satellite remote sensing data and GIS. *Environmentalist* 23(3), 229–236.

Ramachandran, S. 2018. Aerosols and climate change: Present understanding, challenges and future outlook. In: *Land-Atmospheric Research Applications in South/Southeast Asia.* Vadrevu, KP, Ohara, T., and Justice, C. (Eds.). Springer, Cham, pp. 341–378.

Ravindra, K., Sokhi, R., Vangrieken, R. 2008. Atmospheric polycyclic aromatic hydrocarbons: source attribution, emission factors and regulation. *Atmos. Environ.* 42, 2895–2921.

Rogge, W.F., Hildemann, L.M., Mazurek, M.A., Cass, G.R., Simoneit, B.R.T. 1993. Sources of fine organic aerosol. 2. Non-catalyst and catalyst-equipped automobiles and heavy-duty diesel trucks. *Environ. Sci. Technol.* 27, 636–651.

Sharma, H., Jain, V.K., Khan, Z.H. 2007. Characterization and source identification of polycyclic aromatic hydrocarbons (PAHs) in the urban environment of Delhi. *Chemosphere.* 66, 302–310.

Streets, D.G., Gupta, S., Waldhoff, S.T., Wang, M.Q., Bond, T.C., Yiyun, B. 2001. Black carbon emissions in China. *Atmospheric Environment*, 35(25), 4281–4296.

Tao, J., Ho, K.F., Chen, L.G., Zhu, L.H., Han, J.L., Xu, Z.C. 2009. Effect of chemical composition of $PM_{2.5}$ on visibility in Guangzhou, China, 2007 spring. *Particuology.* 7, 68–75.

Tham, Y.W.F., Takeda, K., Sakugawa, H. 2008. Exploring the correlation of particulate PAHs, sulfur dioxide, nitrogen dioxide and ozone, a preliminary study. *Water Air Soil Pollut.* 194 (1–4), 5–12.

Thurston, G.D. and Spengler, J.D. 1985. A quantitative assessment of source contributions to inhalable particulate matter pollution in metropolitan Boston. *Atmos. Environ.* 19(1), 9–25.

Turpin, B.J., Huntzicker, J.J. 1995. Identification of secondary aerosol episodes and quantification of primary and secondary organic aerosol concentrations during SCAQS. *Atmos. Environ.* 29, 3527–3544.

U.S. EPA 1998. EPA Quality Assurance Document: Method Compendium, $PM_{2.5}$ Mass Weighing Laboratory Standard Operating Procedures for the Performance Evaluation Program, United States Environmental Protection Agency Office of Air Quality Planning and Standards, October 1998.

U.S. EPA 2002. EPA Quality Assurance Guidance Document: Method Compendium, Field Standard Operating Procedures for the $PM_{2.5}$ Performance Evaluation Program, United States Environmental Protection Agency Office of Air Quality Planning and Standards, Revision No.2, March 2002.

Vadrevu, K.P., Ohara, T., Justice, C. 2014a. Air pollution in Asia. *Environ. Poll.* 12, 233–235.

Vadrevu, K.P., Justice, C.O. 2011. Vegetation fires in the Asian region: Satellite observational needs and priorities. *Glob Environ Res.* 15(1), 65–76.

Vadrevu, K.P. 2008. Analysis of fire events and controlling factors in eastern India using spatial scan and multivariate statistics. *Geografiska Annaler: Ser. A Phys. Geography* 90(4), 315–328.

Vadrevu, K.P., Lasko, K. 2018. Intercomparison of MODIS AQUA and VIIRS I-Band fires and emissions in an agricultural landscape—Implications for air pollution research. *Remote Sens.* 10(7), 978. doi:10.3390/rs10070978.

Vadrevu, K.P., Lasko, K.P. 2015. Fire regimes and potential bioenergy loss from agricultural lands in the Indo-Gangetic Plains. *J. Environ. Manag.* 148, 10–20.

Vadrevu, K.P., Badinarth, K.V.S., Anuradha, E. 2008. Spatial patterns in vegetation fires in the Indian region. *Environ. Monit. Assess.* 147(1–3), 1. doi:10.1007/s10661-007-0092-6.

Vadrevu, K.P., Csiszar, I., Ellicott, E., Giglio, L., Badarinath, K.V.S., Vermote, E., Justice, C. 2012. Hotspot analysis of vegetation fires and intensity in the Indian region. *IEEE J. Sel. Top. Appl. Earth Obs. Remote Sens.* 6(1), 224–238.

Vadrevu, K.P., Eaturu, A., Badarinath, K.V.S. 2006. Spatial distribution of forest fires and controlling factors in Andhra Pradesh, India using spot satellite datasets. *Environ. Monit. Assess.* 123(1–3), 75–96.

Vadrevu, K.P., Giglio, L., Justice, C. 2013. Satellite based analysis of fire–carbon monoxide relationships from forest and agricultural residue burning (2003–2011). *Atmos. Environ.* 64, 179–191.

Vadrevu, K.P., Lasko, K., Giglio, L., Justice, C. 2014b. Analysis of Southeast Asian pollution episode during June 2013 using satellite remote sensing datasets. *Environ. Poll.* 12, 245–256.

Vadrevu, K.P., Lasko, K., Giglio, L., Justice, C. 2015. Vegetation fires, absorbing aerosols and smoke plume characteristics in diverse biomass burning regions of Asia. *Environ. Res. Lett.* 10(10), 105003.

Vadrevu, K.P., Ohara, T., Justice, C. 2017. Land cover, land use changes and air pollution in Asia: A synthesis. *Environ. Res. Lett.* 12(12), 120201.

Vadrevu, K.P., Ohara, T., Justice, C. eds., 2018. *Land-Atmospheric Research Applications in South and Southeast Asia.* Springer, Cham.

Vadrevu, K.P., Lasko, K. 2018. Intercomparison of MODIS AQUA and VIIRS I-Band fires and emissions in an agricultural landscape—Implications for air pollution research. *Remote Sens.* 10(7), 978. doi:10.3390/rs10070978.

Vadrevu, K.P., Lasko, K., Giglio, L., Schroeder, W., Biswas, S., Justice, C. 2019. Trends in vegetation fires in South and Southeast Asian countries. *Sci. Rep.* 9(1), 7422. doi:10.1038/s41598-019-43940-x.

Vardoulakis, S., Chalabi, Z., Fletcher, T., Grundy, C., Leonardi, G.S. 2008. Impact and uncertainty of a traffic management intervention: Population exposure to polycyclic aromatic hydrocarbons. *Sci. Total. Environ.* 394, 244–251.

Viana, M., Kuhlbusch, T.A.J., Querol, X., Alastuey, A., Harrison, R.M., Hopke, P.K., Winiwarter, W., Vallius, M., Szidat, S., Prévôt, A.S.H., Hueglin, C., Bloemen, H., Wåhlin, P., Vecchi, R., Miranda, A.I., Kasper-Giebl, A., Maenhaut, W., Hitzenberger, R. 2008. Source apportionment of particulate matter in Europe: A review of methods and results. *J. Aerosol. Sci.* 39, 827–849.

Wang, G., Cheng, S., Li, J., Lang, J., Wen, W., Yang, X. and Tian, L. 2015. Source apportionment and seasonal variation of PM 2.5 carbonaceous aerosol in the Beijing-Tianjin-Hebei Region of China. Environmental Monitoring and Assessment, 187(3), 1–13.

Wang, D.G., Tian, F.L., Yang, M., Liu, C.L., Li, Y.F. 2009. Application of positive matrix factorization to identify potential source of PAHs in soil of Dalian, China. *Environ. Pollut.* 157, 1559–1564.

Wang, W., Massey Simonich, S.L., Xue, M., Zhao, J., Zhang, N., Wang, R., Cao, J., Tao, S. 2010. Concentrations, sources and spatial distribution of polycyclic aromatic hydrocarbons in soils from Beijing, Tianjin and surrounding areas, North China. *Environ. Pollut.* 158, 1245–1251.

Watson, J.G., Chow, J.C., Chen, L.W.A. 2005. Summary of organic and elemental carbon/black carbon analysis methods and intercomparisons. Aerosol and Air Quality Research, 5(1), 65–102.

WHO. 2010. Centre for Health Development and World Health Organization, 2010. Hidden cities: unmasking and overcoming health inequities in urban settings. https://www.who.int/publications/i/item/9789241548038

WHO Centre for Health Development. 2010. Hidden cities: unmasking and overcoming health inequities in urban settings. World Health Organization.

Zeng, T., Wang, Y.H. 2011. Nationwide summer peaks of OC/EC ratios in the contiguous United States. *Atmos. Environ.* 45, 578–586.

Zheng, L., Porter, E.N., Sjödin, A., Needham, L.L., Lee, S., Russell, A.G., Mulholland, J.A. 2009. Characterization of $PM_{2.5}$-bound polycyclic aromatic hydrocarbons in Atlanta—seasonal variations at urban, suburban, and rural ambient air monitoring sites. *Atmos. Environ.* 43, 4187–4193.

Zheng, M., Salmon, L.G., Schauer, J.J., Zeng, L.M., Kiang, C.S., Zhang, Y.H., Cass, G.R. 2005. Seasonal trends in $PM_{2.5}$ source contributions in Beijing, China. *Atmos. Environ.* 39, 3967–3976.

Zhou, J.M., Zhang, R.J., Cao, J.J., Chow, J.C., Watson, J.G. 2012. Carbonaceous and ionic components of atmospheric fine particles in Beijing and their impact on atmospheric visibility. *Aerosol Air Qual. Res.* 12(4), 492–502.

16 Impacts of Biomass Burning Aerosols on Air Quality and Convective Systems in Southeast Asia

Hsiang-He Lee
Lawrence Livermore National Laboratory, USA

Chien Wang
University of Toulouse, France

CONTENTS

INTRODUCTION

In recent decades, particulate pollution has become a serious environmental and societal issue in many Southeast Asian countries because of increasing fossil fuel burning from the rise of energy consumption (IEA 2015), high density of shipping emissions (Johansson, Jalkanen, and Kukkonen 2017), and biomass burning emissions from deforestation and peat fires (Langner, Miettinen, and Siegert 2007, Carlson et al. 2013, van der Werf et al. 2010; Saharjo and Yungan, 2018). Energy consumption in Southeast Asian countries is higher than in most developed counties due to the increasing population and fast-growing economies. Such expansion in energy use causes an increase in associated emissions of pollutants, resulting in air quality and human health issues in the region (Frankenberg et al. 2005; Perera 2017, Forouzanfar et al. 2015). Besides fossil fuel emissions, owing to the busy international shipping activities in the region, the world's highest fine particular matter

($PM_{2.5}$) and sulphur oxides (SO_x) emissions per unit area occur in the eastern and southern South China Sea as well as over waters close to Southeast and South Asia. Ship-emitted nitrogen oxides (NO_x) also enhance ozone (O_3) production, leading to adverse effects on both human health (Chen et al. 2007) and agricultural production (Van Dingenen et al. 2009). However, biomass burning aerosols are a significant source of severe haze in South and Southeast Asia (Ramachandran 2018). Our previous study found that biomass burning aerosols contributed 91% to extreme haze events (visibility lower than 7 km) in Singapore from 2003 to 2014 (Lee et al. 2017). Those severe haze events certainly threaten human health (Li et al. 2016) and interrupt transportation, work, and outdoor activities. Rich carbonaceous compounds such as black carbon (BC) from the biomass burning emissions can reduce sunlight absorption and scattering. Indirect effects of biomass burning aerosols on weather and climate are even more complicated due to various cloud types and meteorological conditions in the maritime continent (Sekiguchi et al. 2003, Lin et al. 2013; Wu et al. 2013; Grandey et al. 2016).

This study focuses on the impacts of biomass burning aerosols on air quality and convective systems in Southeast Asia. The widespread effects on air quality and visibility are introduced in the second section. The impacts of biomass burning aerosols on convective systems, including seasonality of biomass burning emissions and corresponding convective systems, are discussed in the third section. The last part of the study includes the summary and suggestions for future work.

THE IMPACTS OF BIOMASS BURNING AEROSOLS ON AIR QUALITY

FINE PARTICULATE MATTERS ($PM_{2.5}$) AND OZONE (O_3)

Biomass burning emissions in Southeast Asia mainly come from deforestation and peatland burning (van der Werf et al. 2010), significantly impacting the air quality and emissions (Hayasaka et al., 2014; Vadrevu et al. 2017, 2018, 2019; Vadrevu and Lasko 2018; Hayasaka and Sepriando 2018; Israr et al. 2018). In our previous study (Lee et al. 2018), we investigated the contribution of fire aerosols to the annual mean $PM_{2.5}$ concentration in 50 Association of Southeast Asian Nations (ASEAN) cities. We found that ten ASEAN cities received more than 70% of $PM_{2.5}$ concentrations from the biomass burning emissions during 2002–2008. Most of the ASEAN cities are influenced by coexisting biomass burning and fossil fuel burning aerosols (Figure 16.1)

The Air Quality Index (AQI) is a widely used metric for providing understandable information about air pollution to the public. The original derivation of AQI is based on six pollutants: fine particulate matter ($PM_{2.5}$), particulate matter (PM_{10}), sulfur dioxide (SO_2), carbon monoxide (CO), ozone (O_3), and nitrogen dioxide (NO_2). In Southeast Asia, $PM_{2.5}$ plays a major role in causing high AQI cases as demonstrated in Lee et al. (2018), where the results of numerical model simulations suggest that fossil-fuel burning emitted $PM_{2.5}$ alone (FF in Table 16.1) can cause 23% of the air pollution events with AQIs at the moderate or unhealthy pollution levels in 50 ASEAN cities.

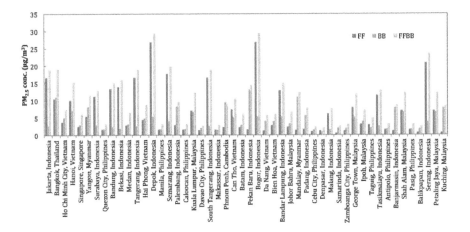

FIGURE 16.1 The annual mean simulated $PM_{2.5}$ concentration ($\mu g/m^3$) in 50 Association of Southeast Asian Nations (ASEAN) cities, derived from FF (red), BB (blue), and FFBB (green) simulations and averaged over the period 2002–2008. The fossil fuel emissions only (FF) simulation and the biomass burning emissions only (BB) simulation are designed to assess the impact of stand-alone non-fire and fire aerosols, respectively. The simulation combining fossil fuel and biomass burning emissions (FFBB) is to demonstrate the impacts of both types of aerosols. (Adapted from Fig. 5 in Lee et al. 2018.)

The coexisting fossil fuels and biomass burning $PM_{2.5}$ sources (FFBB in Table 16.1) can substantially increase the AQI are at a moderate or unhealthy pollution level from 23% to 34%. Table 16.1 also shows that in three selected cities, Bangkok, Kuala Lumpur, and Singapore, the impact of coexisting fossil fuel and biomass burning $PM_{2.5}$ sources would significantly increase 11%–25% in the chance that AQI be at a moderate or unhealthy pollution level compared to fossil fuel $PM_{2.5}$ sources alone. In other words, those results actually indicate that without biomass burning aerosols, the chance for AQI to be at an "unhealthy level" is less than 4%.

Besides $PM_{2.5}$, O_3 is also a critical pollutant to cause public health and air quality issues. However, Lee et al. (2018) found that O_3 produced from fossil fuel emissions or biomass burning emissions alone can barely lead AQI to moderate air pollution levels. Instead, most high AQI cases are formed under the influences of both emissions (Table 16.1). Overall, compared to the case of fossil fuel emissions alone, the cases of coexisting fossil fuel and biomass burning emissions increase the frequency of AQI being at a moderate or unhealthy air pollution level from 6% to 12% among 50 ASEAN cities.

Surface Visibility

The light absorption and scattering by aerosol particles can determine surface visibility. Excluding fog, visibility degradation is more readily observed from the impact of particulate pollution. Normally, visibility of 10 km is considered an indicator of moderate to heavy particulate pollution. In a study by Lee et al. (2017), a value of

TABLE 16.1

The Frequency of Air Pollution Levels in Bangkok, Kuala Lumpur, Singapore, and 50 Association of Southeast Asian Nations (ASEAN) Cities Derived Using 9 hours Ozone (O_3) Volume Mixing Ratio and 24 hours $PM_{2.5}$ Concentration in FF, BB, and FFBB During 2002–2008

Bangkok	$AQI_{(O3)}$	FF (%)	BB (%)	FFBB (%)	$AQI_{(PM2.5)}$	FF (%)	BB (%)	FFBB (%)
Good	0–50	81±3	97±1	69±3	0–50	63±6	67±5	38±2
Moderate	51–100	17±2	3±1	21±3	51–100	34±5	24±3	45±3
Unhealthy	101–200	2±1	0±0	11±1	101–200	3±2	9±4	17±4
Very Unhealthy	201–300	0±0	0±0	0±0	201–300	0±0	0±0	0±0
Hazardous	301–400	0±0	0±0	0±0	301–400	0±0	0±0	0±0
Hazardous	401–500	0±0	0±0	0±0	401–500	0±0	0±0	0±0

Kuala Lumpur	$AQI_{(O3)}$	FF (%)	BB (%)	FFBB (%)	$AQI_{(PM2.5)}$	FF (%)	BB (%)	FFBB (%)
Good	0–50	95±2	100±1	83±6	0–50	73±3	78±8	52±7
Moderate	51–100	5±2	0±1	15±5	51–100	27±4	18±6	40±4
Unhealthy	101–200	0±0	0±0	2±1	101–200	0±0	4±3	8±4
Very Unhealthy	201–300	0±0	0±0	0±0	201–300	0±0	0±0	0±0
Hazardous	301–400	0±0	0±0	0±0	301–400	0±0	0±0	0±0
Hazardous	401–500	0±0	0±0	0±0	401–500	0±0	0±0	0±0

Singapore	$AQI_{(O3)}$	FF (%)	BB (%)	FFBB (%)	$AQI_{(PM2.5)}$	FF (%)	BB (%)	FFBB (%)
Good	0–50	99±1	100±0	94±3	0–50	92±5	92±4	78±5
Moderate	51–100	1±1	0±0	5±2	51–100	8±4	6±2	19±4
Unhealthy	101–200	0±0	0±0	1±1	101–200	0±1	1±2	3±2
Very Unhealthy	201–300	0±0	0±0	0±0	201–300	0±0	0±0	0±0
Hazardous	301–400	0±0	0±0	0±0	301–400	0±0	0±0	0±0
Hazardous	401–500	0±0	0±0	0±0	401–500	0±0	0±0	0±0

50 ASEAN cities	$AQI_{(O3)}$	FF (%)	BB (%)	FFBB (%)	$AQI_{(PM2.5)}$	FF (%)	BB (%)	FFBB (%)
Good	0–50	94±1	99±0	88±2	0–50	77±1	90±3	66±3
Moderate	51–100	6±1	1±0	10±2	51–100	19±1	7±2	26±2
Unhealthy	101–200	0±0	0±0	2±0	101–200	4±0	2±1	8±2
Very Unhealthy	201–300	0±0	0±0	0±0	201–300	0±0	0±0	0±0
Hazardous	301–400	0±0	0±0	0±0	301–400	0±0	0±0	0±0
Hazardous	401–500	0±0	0±0	0±0	401–500	0±0	0±0	0±0

The description of FF, BB, and FFBB refers to Figure 2.1. Adapted from Table 3 and Table 4 in Lee et al. (2018).

surface visibility of 10 km is used as the threshold for defining "low visibility day," (LVD) and 7 km is used for "very low visibility day" (VLVD). Based on the observational data of visibility from the Global Surface Summary of the Day (GSOD) (Smith et al. 2011), the percentage of LVD per year has increased in most ASEAN cities over the past decade.

Besides the observational data, Lee et al. (2017) have also used the Weather Research and Forecasting model with a chemistry package (WRF-Chem), where a modified chemical tracer module instead of a full chemistry package was adopted to model the fire $PM_{2.5}$ particle as tracers. This design benefits the long-term simulations in reducing computational resources. This study aims to identify fire aerosol-caused LVD. For this purpose, we derived the modeled LVD using modeled properties of fire aerosols. Both the observed and modeled visibilities were then used to define the fraction of low-visibility days that can be explained by fire aerosols alone. In this process, an assumption was made that whenever fire aerosol alone could cause an LVD to occur, such a day would be categorized as a fire-aerosol-caused LVD regardless of whether other coexisting pollutants could have a sufficient intensity to cause low visibility or not.

Based on the definition above, an ascending trend of the percentage of LVDs per year was found in Bangkok (Thailand), Kuala Lumpur (Malaysia), Singapore (Singapore), and Kuching (Malaysia) during 2003–2014, as shown in Figure 16.2a–d.

The percentage of fire-caused LVDs has been increasing during the study period as well (red color in Figure 16.2). Among these 4 selected cities, Bangkok has the highest percentage of LVDs per year, $59\% \pm 14\%$ (or 215 ± 50 days; Table 16.2), and most LVDs occur from October to next year April (Figure 16.2e)

This is highly related to the burning of agricultural waste and other biomass in mainland Southeast Asia during the dry seasons. Thus, fire aerosols are responsible for 39% of these LVDs in Bangkok, and abundant fire aerosols even degrade visibility to cause 87% VLVDs to occur. The annual occurrence of LVDs in Kuala Lumpur is also relatively high, 174 ± 78 days (or $48\% \pm 21\%$) during 2003–2014, while there are only 96 ± 87 ($26\% \pm 24\%$) annual mean LVDs in Singapore during the same period (Table 16.2). Kuala Lumpur and Singapore are geographically close; thus, the seasonal patterns of the fire-caused LVDs in the two cities are similar, each having two peaks during February–April and August–November (Figure 16.2f and g). During February–April, fire-caused LVDs correspond to the transboundary transport of fire aerosols from mainland Southeast Asia in the winter monsoon season. In the other haze season, fire aerosol sources are mainly from peatland burning in Sumatra and Borneo during the summer monsoon. Overall, fire aerosols are responsible for 36% (34%) of total LVDs and 85% (91%) of total VLVDs in Kuala Lumpur (Singapore) during 2003–2014 (Table 16.2). Kuching is located in Borneo Island and is affected heavily by local fire events during the fire season (Figure 16.2d and h). LVDs of Kuching mainly occur in August and September during the fire season. Fire aerosols often degrade the visibility to below 7 km, causing Kuching to have the most annual mean VLVDs among the four cities at 22 ± 18 days (or $6\% \pm 5\%$). Fire aerosols are also responsible for 93% of total VLVDs in Kuching during 2003–2014 (Table 16.2).

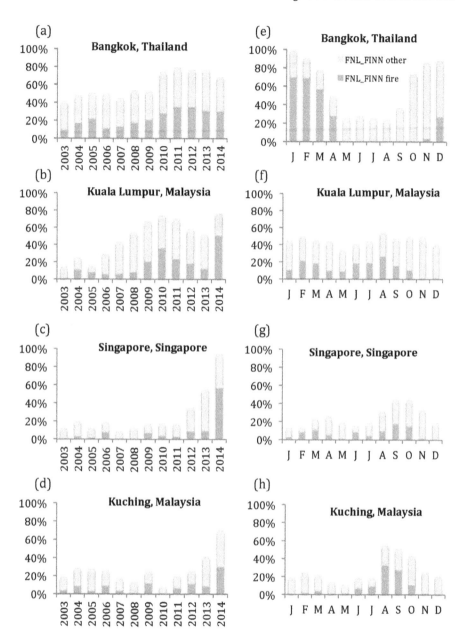

FIGURE 16.2   (a)–(d) The percentage of LVDs per year derived from GSOD visibility observations in Bangkok, Kuala Lumpur, Singapore, and Kuching, respectively. (e)–(h) The percentage of LVDs averaged over 2003–2014, derived from GSOD visibility observations in Bangkok, Kuala Lumpur, Singapore, and Kuching, respectively. Each bar presents the observed LVDs in each year or month. Red color shows the partition of fire-caused LVDs (captured by model), while green color presents other LVDs (observed – modeled; i.e., those not captured by model). (Adapted from Fig. 6 in Lee et al. 2017.)

TABLE 16.2

Annual Mean Low Visibility Days (LVDs; Observed Visibility ≤ 10 km) and Very Low Visibility Days (VLVDs; Observed Visibility ≤ 7 km) per year in Bangkok, Kuala Lumpur, Singapore, and Kuching During 2003–2014 are Presented in the Second Column

City	LVD per year (days %)	Fire Pollution Contribution (%)	Other Pollution Contribution (%)
Bangkok, Thailand	215 ± 50 (59 ± 14)	39 ± 8	61 ± 8
Kuala Lumpur, Malaysia	174 ± 78 (48 ± 21)	36 ± 17	64 ± 17
Singapore, Singapore	96 ± 87 (26 ± 24)	34 ± 17	66 ± 17
Kuching, Malaysia	95 ± 57 (26 ± 17)	33 ± 15	67 ± 15

City	VLVD per year (days %)	Fire Pollution Contribution (%)	Other Pollution Contribution (%)
Bangkok, Thailand	15 ± 8 (4 ± 2)	87 ± 20	13 ± 20
Kuala Lumpur, Malaysia	19 ± 18 (5 ± 5)	85 ± 17	15 ± 17
Singapore, Singapore	4 ± 4 (1 ± 1)	91 ± 33	9 ± 33
Kuching, Malaysia	22 ± 18 (6 ± 5)	93 ± 11	7 ± 11

Parentheses show the percentage of year. The third column shows the percentages, along with standard deviations, of low visibility days explained by fire aerosols alone (i.e., the LVDs captured by the model). The fourth column is the same as the third column but for non-fire (other) pollutions, which is calculated as 100%, fire pollution contribution (i.e., the percentage of LVDs not captured by the model). Adapted from Table 2 in Lee et al. (2017).

THE IMPACTS OF BIOMASS BURNING AEROSOLS ON CONVECTIVE SYSTEMS

SEASONALITY OF BIOMASS BURNING EMISSIONS AND CORRESPONDING CONVECTIVE SYSTEMS

Biomass burning in the region is a major cause of greenhouse gas emissions and aerosols (Prasad et al. 2000; Gupta et al. 2001; Kant et al. 2000; Dennis et al. 2005). Most biomass burning events in South and Southeast Asia are due to human interference such as land clearing for oil palm plantations, other causes of deforestation, poor peatland management, and burning of agriculture waste impacting both the terrestrial and atmospheric (Prasad and Badarinth, 2004; Vadrevu et al. 2006, 2008; Vadrevu 2008; Badarinath et al. 2008a, b; Vadrevu et al. 2012, 2013, 2014a, b, 2015; Vadrevu and Lasko 2015; 2018; Marlier et al. 2015; Biswas et al. 2015a, b; Justice et al. 2015; Lasko et al. 2017; 2018a, b; Lasko and Vadrevu 2018). Besides human interventions, meteorological factors can also influence fire initiation, intensity, and duration (Prasad et al., 2008a; Reid et al. 2012, 2015;). Reid et al. (2012) investigated relationships between fire hotspot appearance and various weather phenomena as well as climate variabilities over different time scales over the Maritime Continent, including the El Niño-Southern Oscillation (ENSO) (Rasmusson and Wallace 1983;

McBride et al. 2003) and the Indian Ocean Dipole (IOD) (Saji et al. 1999); (2) seasonal migration of the Inter-tropical Convergence Zone (ITCZ) and associated Southeast Asia monsoons (Chang et al. 2005); (3) intra-seasonal variability associated with the Madden-Julian Oscillation (MJO) (Zhang 2005; Madden and Julian 1971) and the west Sumatran low (Wu and Hsu 2009); (4) equatorial waves, meso-scale features, and tropical cyclones; and (5) convection. The implications of biomass burning on both the terrestrial ecosystems as well as the atmosphere were discussed extensively by the regional researchers (Prasad et al. 2001, 2002, 2003, 2005, 2008; Hayasaka and Sepriando, 2018; Oanh et al., 2018), including research needs and priorities (Vadrevu and Justice 2011).

Figure 16.3a and b show time series of monthly fire $PM_{2.5}$ emission (Tg/year) and precipitation in Sumatra and Borneo during 2003–2014, respectively.

Based on meteorological regimes and fire emission types, we found a strong correlation between fire emission and precipitation in Sumatra and Borneo, i.e., a high fire emission year usually is also a low precipitation year (Figure 16.3a and b). Besides the inter-annual variability of the rainfall-fire emission relationship in Sumatra and Borneo, we have also identified the inter-seasonal variability of the rainfall-fire emission relationship in these two regions (Figure 16.3c and d). High rainfall seasons basically appear during the transit of two monsoon seasons. Although rainfall in these island areas occurs all year around, the inter-seasonal signal is not that strong, while intra-seasonal variability, such as that on a weekly scale, is also evident in the regions. As mentioned above, this intra-seasonal variability is often associated with the Madden-Julian Oscillation.

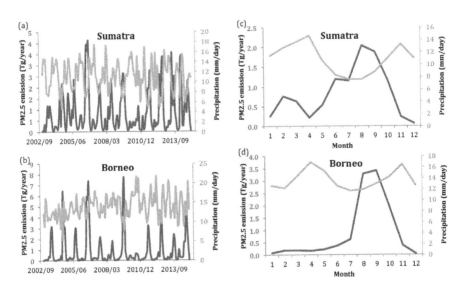

FIGURE 16.3   (a)–(b) Time series of monthly $PM_{2.5}$ emission (Tg/year) in FINNv1.5 biomass burning emission inventory (red solid lines) and model simulated monthly precipitation rate (mm/day; blue solid lines) during 2003–2014 in Sumatra and Borneo, respectively. (c)–(d) The mean $PM_{2.5}$ emission (Tg/year; red solid lines) and precipitation rate (mm/day; blue solid lines) averaged over 2003–2014 for each month.

Modeling Versus Satellite Retrieval Results

Convective precipitation events in the Maritime Continent usually happen during the intra-monsoon seasons (April–May and October–November) (Lo and Orton 2016). Sumatra Squalls and diurnal rainfall over Borneo, for example, are important rainfall features of the Maritime Continent in Southeast Asia. Meanwhile, recurrent biomass burning activities over the past few decades frequently occur in these seasons, producing massive fire aerosols with peak number concentrations more than ten times that of background aerosol concentrations (Lee and Wang 2020). Besides influencing local radiation budget through directly scattering and absorbing sunlight, these additional aerosols can also act as cloud condensation nuclei (CCN) to alter convective clouds and precipitation in the Maritime Continent via aerosol indirect effects. To investigate the aerosol-cloud interactions associated with the biomass burning aerosols in the Maritime Continent, we have conducted a pair of simulations using the WRF-Chem model to include and exclude the biomass burning emissions (FF versus FFBB).

One case study of diurnal convective rainfall in Borneo on September 22, 2008 has demonstrated the impact of biomass burning aerosols on convective clouds. Figure 16.4a shows the vertical structure of clouds detected by the Cloud-Aerosol Lidar and Infrared Pathfinder Satellite Observation (CALIPSO) during this case, where it captured a convective system over Borneo along with high $PM_{2.5}$ concentration near the surface.

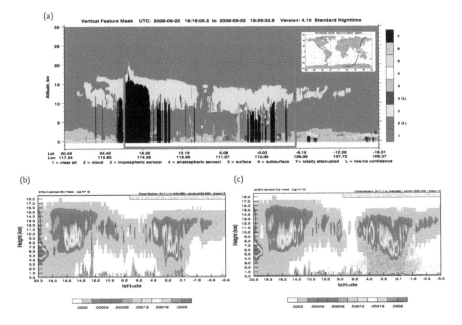

FIGURE 16.4 (a) The vertical structure of cloud retrieved from the Cloud-Aerosol Lidar and Infrared Pathfinder Satellite Observation (CALIPSO) on September 22, 2008. (b)–(c) The sum of simulated hydrometeors (shaded; kg/kg) and $PM_{2.5}$ concentration (contour; μg/m³) in FF and FFBB, respectively. The description of FF and BB refers to Figure 16.2.

With our model simulations, we could identify the biomass burning origin of these near-surface aerosols. Compared to the results in the simulation of fossil fuel emissions only case (FF), the simulation combining fossil-fuel and biomass burning emissions (FFBB) predicted 8% more cloud water and 103% more rainfall. Cloud structure in FFBB matches better with the cloud profile retrieved from CALIPSO (Figure 16.4c versus a). Ichikawa and Yasunari (2006) used TRMM data to characterize the diurnal cycle of rainfall over Borneo and concluded that convective rainfall mainly comes from westerly or easterly low-level wind pattern variability. We found that the westerly regime tends to cause more intense rainfall at midnight or early morning. It is related to a robust anticyclonic circulation over northwest Borneo. The impacts of fire aerosols on this type of rain are also more significant.

SUMMARY AND SUGGESTIONS OF FUTURE WORKS

Different researchers study the impacts of biomass burning aerosols in Southeast Asia on visibility degradation, air quality, and human health. However, most of the studies overlooked an important aspect, i.e., the coexistence of fossil-fuel and biomass-burning aerosol source regions, which could raise the threat to air quality and human health at higher levels. Also, the prediction of convective systems in the Maritime Continent has been a challenge in the past few decades because of the unique terrain consisting of complicated land-sea distribution and poor representation of aerosol processes affected by combined emissions from different sources, particularly episodic biomass burning activities. A critical issue thus arises regarding the uncertainty of current emission estimates. We also note that the use of different fire emission estimates could affect the modeled $PM_{2.5}$ concentrations, mainly due to the discrepancy in the spatiotemporal distribution of sources; thus, further research on this aspect should be prioritized.

REFERENCES

Badarinath, K.V.S., Kharol, S.K., Krishna Prasad, V., Kaskaoutis, D.G. and Kambezidis, H.D. 2008a. Variation in aerosol properties over Hyderabad, India during intense cyclonic conditions. *International Journal of Remote Sensing.* 29(15), 4575–4597.

Badarinath, K.V.S., Kharol, S.K., Prasad, V.K., Sharma, A.R., Reddi, E.U.B., Kambezidis, H.D. and Kaskaoutis, D.G. 2008b. Influence of natural and anthropogenic activities on UV Index variations–a study over tropical urban region using ground based observations and satellite data. *Journal of Atmospheric Chemistry.* 59(3), 219–236.

Biswas, S., Lasko, K.D. and Vadrevu, K.P. 2015a. Fire disturbance in tropical forests of Myanmar—Analysis using MODIS satellite datasets. *IEEE Journal of Selected Topics in Applied Earth Observations and Remote Sensing.* 8(5), 2273–2281.

Biswas, S., Vadrevu, K.P., Lwin, Z.M., Lasko, K. and Justice, C.O. 2015b. Factors controlling vegetation fires in protected and non-protected areas of Myanmar. *PLoS One.* 10(4), e0124346.

Carlson, K.M., Curran, L.M., Asner, G.P., Pittman, A.M., Trigg, S.N. and Adeney, J.M. 2013. Carbon emissions from forest conversion by Kalimantan oil palm plantations. *Nature Climate Change* 3(3), 283–287. http://www.nature.com/nclimate/journal/v3/n3/abs/nclimate1702.html - supplementary-information.

Chang, C.P., Wang, Z., McBride, J. and Liu, C.-H. 2005. Annual cycle of Southeast Asia—maritime continent rainfall and the asymmetric monsoon transition. *Journal of Climate.* 18(2), 287–301. doi:10.1175/JCLI-3257.1.

Chen, T.-M., Kuschner, W.G., Gokhale, J., and Shofer, S. 2007. Outdoor air pollution: Ozone health effects. *The American Journal of the Medical Sciences.* 333(4), 244–248. doi: 10.1097/MAJ.0b013e31803b8e8c.

Dennis, R.A., Mayer, J., Applegate, G., Chokkalingam, U., Colfer, C.J.P., Kurniawan, I., Lachowski, H., Maus, P., Permana, R.P., Ruchiat, Y. and Stolle, F. 2005. Fire, people and pixels: Linking social science and remote sensing to understand underlying causes and impacts of fires in Indonesia. *Human Ecology.* 33(4), 465–504.

Forouzanfar, M.H., Alexander, L., Bachman, V.F., Biryukov, S., Brauer, M., Casey, D., Coates, M.M., Delwiche, K., Estep, K. and Frostad, J.J. 2015. Global, regional, and national comparative risk assessment of 79 behavioural, environmental and occupational, and metabolic risks or clusters of risks in 188 countries, 1990–2013: a systematic analysis for the Global Burden of Disease Study 2013. *The Lancet.* 386(10010), 2287–2323.

Frankenberg, E., McKee, D. and Thomas, D. 2005. Health consequences of forest fires in Indonesia. *Demography.* 42(1), 109–129.

Grandey, B.S., Lee, H.H. and Wang, C. 2016. Radiative effects of interannually varying vs. interannually invariant aerosol emissions from fires. *Atmospheric Chemistry and Physics.* 16 (22), 14495–14513. doi:10.5194/acp-16-14495-2016.

Gupta, P.K., Prasad, V.K., Sharma, C., Sarkar, A.K., Kant, Y., Badarinath, K.V.S. and Mitra, A.P. 2001. CH_4 emissions from biomass burning of shifting cultivation areas of tropical deciduous forests–experimental results from ground-based measurements. *Chemosphere-Global Change Science.* 3(2), 133–143.

Hayasaka, H., Noguchi, I., Putra, E.I., Yulianti, N. and Vadrevu, K. 2014. Peat-fire-related air pollution in Central Kalimantan, Indonesia. *Environmental Pollution.* 195: 257–266. doi:10.1016/j.envpol.2014.06.031.

Hayasaka, H. and Sepriando, A. 2018. Severe air pollution due to peat fires during 2015 Super El Niño in Central Kalimantan, Indonesia. In: *Land-Atmospheric Research Applications in South/Southeast Asia.* Vadrevu, K.P., Ohara, T., and Justice, C. (Eds.). Springer, Cham. pp. 129–142.

Ichikawa, H. and Yasunari, T. 2006. Time–space characteristics of diurnal rainfall over borneo and surrounding oceans as observed by TRMM-PR. *Journal of Climate.* 19(7), 1238–1260. doi:10.1175/jcli3714.1.

IEA. 2015. Southeast Asia Energy Outlook 2015. In World Energy Outlook Special Report. International Energy Agency: World Energy Outlook Special Report.

Israr, I., Jaya, S.N.I, Saharjo, H.S., Kuncahyo, B. and Vadrevu, K.P. 2018. Spatio-temporal analysis of land and forest fires in Indonesia using MODIS active fire dataset. In: *Land-Atmospheric Research Applications in South/Southeast Asia.* Vadrevu, K.P., Ohara, T., and Justice, C. (Eds). Springer, Cham. pp. 105–128.

Johansson, L., Jalkanen, J-P., and Kukkonen, J. 2017. Global assessment of shipping emissions in 2015 on a high spatial and temporal resolution. *Atmospheric Environment.* 167, 403–415 doi:10.1016/j.atmosenv.2017.08.042.

Justice, C., Gutman, G. and Vadrevu, K.P. 2015. NASA land cover and land use change (LCLUC): An interdisciplinary research program. *Journal of Environmental Management.* 148(15), 4–9.

Kant, Y., Ghosh, A.B., Sharma, M.C., Gupta, P.K., Prasad, V.K., Badarinath, K.V.S. and Mitra, A.P. 2000. Studies on aerosol optical depth in biomass burning areas using satellite and ground-based observations. *Infrared Physics & Technology.* 41(1), 21–28.

Lasko, K., Vadrevu, K.P., Tran, V.T., Ellicott, E., Nguyen, T.T., Bui, H.Q. and Justice, C. 2017. Satellites may underestimate rice residue and associated burning emissions in Vietnam. *Environmental Research Letters.* 12(8), 085006.

Lasko, K. and Vadrevu, K.P. 2018. Improved rice residue burning emissions estimates: Accounting for practice-specific emission factors in air pollution assessments of Vietnam. *Environmental Pollution*. 236(5), 795–806.

Lasko, K., Vadrevu, K.P. and Nguyen, T.T.N. 2018a. Analysis of air pollution over Hanoi, Vietnam using multi-satellite and MERRA reanalysis datasets. *PloS One*. 13(5), e0196629.

Lasko, K., Vadrevu, K.P., Tran, V.T. and Justice, C. 2018b. Mapping double and single crop paddy rice with Sentinel-1A at varying spatial scales and polarizations in Hanoi, Vietnam. *IEEE Journal of Selected Topics in Applied Earth Observations and Remote Sensing*. 11(2), 498–512.

Langner, A., Miettinen, J. and Siegert, F. 2007. Land cover change 2002–2005 in Borneo and the role of fire derived from MODIS imagery. *Global Change Biology*. 13(11), 2329–2340. doi:10.1111/j.1365–2486.2007.01442.x.

Lee, H. H., Bar-Or, R.Z. and Wang, C. 2017. Biomass burning aerosols and the low-visibility events in Southeast Asia. *Atmospheric Chemistry and Physics*. 17(2), 965–980. doi:10.5194/acp-17-965-2017.

Lee, H.H., Iraqui, O., Gu, Y., Yim, S.H.L., Chulakadabba, A., Tonks, A.Y.M., Yang, Z. and Wang, C., 2018. Impacts of air pollutants from fire and non-fire emissions on the regional air quality in Southeast Asia. *Atmospheric Chemistry and Physics (Online)*, 18(9), 6141–6156.

Lee, H. H., and Wang C. 2020. The impacts of biomass burning activities on convective systems over the Maritime Continent. *Atmospheric Chemistry and Physics*. 20, 2533–2548.

Li, Y., Henze, D.K., Jack, D., Henderson, B.H. and Kinney, P.L. 2016. Assessing public health burden associated with exposure to ambient black carbon in the United States. *Science of the Total Environment*. 539, 515–525.

Lin, N.H., Tsay, S.C., Maring, H.B., Yen, M.C., Sheu, G.R., Wang, S.H., Chi, K.H., Chuang, M.T., Ou-Yang, C.F., Fu, J.S. and Reid, J.S. 2013. An overview of regional experiments on biomass burning aerosols and related pollutants in Southeast Asia: From BASE-ASIA and the Dongsha Experiment to 7-SEAS. *Atmospheric Environment*. 78, 1–19.

Lo, J.C.-F. and Orton, T. 2016. The general features of tropical Sumatra Squalls. *Weather*. 71 (7), 175–178. doi:10.1002/wea.2748.

Madden, R.A. and Julian, P.R. 1971. Detection of a 40–50 day oscillation in the zonal wind in the tropical Pacific. *Journal of the Atmospheric Sciences*. 28(5), 702–708.

Marlier, M., Defries, R.S., Kim, P.S., Koplitz, S.N., Jacob, D.J., Mickley, L.J. and Myers, S.S. 2015. Fire emissions and regional air quality impacts from fires in oil palm, timber, and logging concessions in Indonesia. *Environmental Research Letters*. 10(8), 085005.

McBride, J.L., Haylock, M.R. and Nicholls, N. 2003. Relationships between the Maritime Continent heat source and the El Niño–Southern Oscillation phenomenon. *Journal of Climate*. 16(17), 2905–2914.

Oanh, N.T.K., Permadi, D.A., Dong, N.P., Nguyet, D.A. 2018. Emission of toxic air pollutants and greenhouse gases from crop residue open burning in Southeast Asia. In: Vadrevu, K.P., Ohara, T., and Justice, C. (Eds). Land-Atmospheric Research applications in South/Southeast Asia. Springer, Cham, pp. 47–68.

Perera, F.P. 2017. Multiple threats to child health from fossil fuel combustion: Impacts of air pollution and climate change. *Environmental Health Perspectives*. 125(2), 141–148. doi:10.1289/EHP299.

Prasad, V.K., Gupta, P.K., Sharma, C., Sarkar, A.K., Kant, Y., Badarinath, K.V.S., Rajagopal, T. and Mitra, A.P. 2000. NO_x emissions from biomass burning of shifting cultivation areas from tropical deciduous forests of India–estimates from ground-based measurements. *Atmospheric Environment*. 34(20), 3271–3280.

Prasad, V.K., Kant, Y. and Badarinath, K.V.S. 2001. CENTURY ecosystem model application for quantifying vegetation dynamics in shifting cultivation areas: A case study from Rampa Forests, Eastern Ghats (India). *Ecological Research*. 16(3), 497–507.

Prasad, V.K., Kant, Y., Gupta, P.K., Elvidge, C. and Badarinath, K.V.S. 2002. Biomass burning and related trace gas emissions from tropical dry deciduous forests of India: A study using DMSP-OLS data and ground-based measurements. *International Journal of Remote Sensing*. 23(14), 2837–2851.

Prasad, V.K., Lata, M. and Badarinath, K.V.S. 2003. Trace gas emissions from biomass burning from northeast region in India—estimates from satellite remote sensing data and GIS. *Environmentalist*. 23(3), 229–236.

Prasad, V.K. and Badarinth, K.V.S. 2004. Land use changes and trends in human appropriation of above ground net primary production (HANPP) in India (1961–98). *Geographical Journal*. 170(1), 51–63.

Prasad, V.K., Anuradha, E. and Badarinath, K.V.S. 2005. Climatic controls of vegetation vigor in four contrasting forest types of India—evaluation from National Oceanic and Atmospheric Administration's Advanced Very High Resolution Radiometer datasets (1990–2000). *International Journal of Biometeorology*. 50(1), 6–16.

Prasad, V.K., Badarinath, K.V.S. and Eaturu, A. 2008. Biophysical and anthropogenic controls of forest fires in the Deccan Plateau, India. *Journal of Environmental Management*. 86(1), 1–13.

Ramachandran, S. 2018. Aerosols and climate change: Present understanding, challenges and future outlook. In: *Land-Atmospheric Research Applications in South/Southeast Asia*. Vadrevu, K.P., Ohara, T., and Justice, C. (Eds.). Springer, Cham, pp. 341–378.

Rasmusson, E.M. and Wallace, J.M. 1983. Meteorological aspects of the El Nino/southern oscillation. *Science*. 222(4629), 1195–1202.

Reid, J.S., Lagrosas, N.D., Jonsson, H.H., Reid, E.A., Sessions, W.R., Simpas, J.B., Uy, S.N., Boyd, T.J., Atwood, S.A., Blake, D.R. and Campbell, J.R. 2015. Observations of the temporal variability in aerosol properties and their relationships to meteorology in the summer monsoonal South China Sea/East Sea: The scale-dependent role of monsoonal flows, the Madden–Julian Oscillation, tropical cyclones, squall lines and cold pools. *Atmospheric Chemistry and Physics*. 15(4), 1745–1768.

Reid, J.S., Xian, P., Hyer, E.J., Flatau, M.K., Ramirez, E.M., Turk, F.J., Sampson, C.R., Zhang, C., Fukada, E.M. and Maloney, E.D. 2012. Multi-scale meteorological conceptual analysis of observed active fire hotspot activity and smoke optical depth in the Maritime Continent. Atmospheric Chemistry and Physics. 12(4), 2117–2147.

Saharjo, B.H. and Yungan, A. 2018. Forest and land fires in Riau province; A Case study in fire prevention, policy implementation with local concession holders. In: *Land-Atmospheric Research Applications in South/Southeast Asia*. Vadrevu, K.P., Ohara, T., and Justice, C. (Eds.). Springer, Cham, pp. 143–170.

Saji, N.H., Goswami, B.N., Vinayachandran, P.N. and Yamagata, T. 1999. A dipole mode in the tropical Indian Ocean. *Nature*. 401(6751), 360–363.

Sekiguchi, M., Nakajima, T., Suzuki, K., Kawamoto, K., Higurashi, A., Rosenfeld, D., Sano, I. and Mukai, S. 2003. A study of the direct and indirect effects of aerosols using global satellite data sets of aerosol and cloud parameters. *Journal of Geophysical Research: Atmospheres*. 108(D22). doi:10.1029/2002JD003359.

Smith, A., Lott, N. and Vose, R. 2011. The integrated surface database: Recent developments and partnerships. *Bulletin of the American Meteorological Society*, 92(6), 704–708.

Vadrevu, K.P., Eaturu, A. and Badarinath, K.V.S. 2006. Spatial distribution of forest fires and controlling factors in Andhra Pradesh, India using spot satellite datasets. *Environmental Monitoring and Assessment*. 123(1–3), 75–96.

Vadrevu, K.P., 2008. Analysis of fire events and controlling factors in eastern India using spatial scan and multivariate statistics. *Geografiska Annaler: Series A, Physical Geography.* 90(4), 315–328.

Vadrevu, K.P., Badarinath, K.V.S. and Anuradha, E. 2008. Spatial patterns in vegetation fires in the Indian region. *Environmental Monitoring and Assessment.* 147(1–3), 1. doi:10.1007/s10661-007-0092-6.

Vadrevu, K.P. and Justice, C.O. 2011. Vegetation fires in the Asian region: satellite observational needs and priorities. *Global Environmental Research.* 15(1), 65–76.

Vadrevu, K.P., Csiszar, I., Ellicott, E., Giglio, L., Badarinath, K.V.S., Vermote, E. and Justice, C. 2012. Hotspot analysis of vegetation fires and intensity in the Indian region. *IEEE Journal of Selected Topics in Applied Earth Observations and Remote Sensing.* 6(1), 224–238.

Vadrevu, K.P., Giglio, L. and Justice, C. 2013. Satellite based analysis of fire–carbon monoxide relationships from forest and agricultural residue burning (2003–2011). *Atmospheric Environment.* 64, 179–191.

Vadrevu, K.P., Ohara, T., and Justice, C. 2014a. Air pollution in Asia. *Environmental Pollution.* 12, 233–235.

Vadrevu, K.P., Lasko, K., Giglio, L. and Justice, C. 2014b. Analysis of Southeast Asian pollution episode during June 2013 using satellite remote sensing datasets. *Environmental Pollution.* 12, 245–256.

Vadrevu, K.P., Lasko, K., Giglio, L. and Justice, C. 2015. Vegetation fires, absorbing aerosols and smoke plume characteristics in diverse biomass burning regions of Asia. *Environmental Research Letters.* 10(10), 105003.

Vadrevu, K.P. and Lasko, K.P. 2015. Fire regimes and potential bioenergy loss from agricultural lands in the Indo-Gangetic Plains. *Journal of Environmental Management.* 148, 10–20.

Vadrevu, K.P., Ohara, T. and Justice, C. 2017. Land cover, land use changes and air pollution in Asia: A synthesis. *Environmental Research Letters.* 12(12), 120201.

Vadrevu, K.P., Ohara, T. and Justice, C. eds., 2018. *Land-Atmospheric Research Applications in South and Southeast Asia.* Springer, Cham.

Vadrevu, K.P. and Lasko, K. 2018. Intercomparison of MODIS AQUA and VIIRS I-Band fires and emissions in an agricultural landscape—Implications for air pollution research. *Remote Sensing.* 10(7), 978. doi:10.3390/rs10070978.

Vadrevu, K.P., Lasko, K., Giglio, L., Schroeder, W., Biswas, S. and Justice, C. 2019. Trends in vegetation fires in south and southeast Asian countries. *Scientific Reports*, 9(1), 7422. doi:10.1038/s41598-019-43940-x.

van der Werf, G.R., Randerson, J.T., Giglio, L., Collatz, G.J., Mu, M., Kasibhatla, P.S., Morton, D.C., DeFries, R.S., Jin, Y. and van Leeuwen, T.T. 2010. Global fire emissions and the contribution of deforestation, savanna, forest, agricultural, and peat fires (1997–2009). *Atmospheric Chemistry and Physics* 10(23), 11707–11735. doi:10.5194/acp-10-11707-2010.

Van Dingenen, R., Dentener, F.J., Raes, F., Krol, M.C., Emberson, L. and Cofala, J. 2009. The global impact of ozone on agricultural crop yields under current and future air quality legislation. *Atmospheric Environment*, 43(3), 604–618. doi:10.1016/j.atmosenv.2008.10.033.

Wu, C.H. and Hsu, H.H. 2009. Topographic influence on the MJO in the Maritime Continent. *Journal of Climate.* 22(20), 5433–5448.

Wu, R., Wen, Z. and He, Z. 2013. ENSO contribution to aerosol variations over the maritime continent and the Western North Pacific during 2000–10. *Journal of Climate.* 26(17), 6541–6560.

Zhang, C. 2005. Madden-julian oscillation. *Reviews of Geophysics.* 43 (2), RG2003. doi:10.1029/2004RG000158.

17 Biomass Burning Influence on PM$_{2.5}$ Regional and Long-Range Transport in Northeast Asia

Katsushige Uranishi, Hikari Shimadera, and Akira Kondo
Osaka University, Japan

CONTENTS

INTRODUCTION

Biomass burning (BB) is broadly defined as the burning of living and dead vegetation, including grassland, forest, agricultural waste, and burning of biomass for fuel, such as fuel for residential heating and cooking as well as industrial biofuel burning (Andreae and Merlet, 2001; Badarinath et al., 2008a, b). It can also be classified into two different types of BB by source: human-initiated burning and natural lightning-induced fires. It has been estimated that humans are responsible for approximately 90% of BB, with only a small percentage of natural fires contributing to the total amount of vegetation burned (https://earthobservatory.nasa.gov/features/BiomassBurning). Biomass burning is most common in South/Southeast Asian countries such as in India, Pakistan, Sri Lanka, Bhutan, Nepal, Myanmar,

Indonesia, Cambodia, Laos, Thailand, etc. (Prasad et al., 2001, 2004, 2005; Vadrevu et al., 2006; Hayasaka et al., 2014; Hayasaka and Sepriando, 2018; Israr et al., 2018; Saharjo and Yungan, 2018; Oanh et al., 2018; Inoue, 2018). BB has a significant impact on land cover, air quality, public health, and climate at global, regional, and local scales because it releases several gases and particulate pollutants into the atmosphere in large quantities, including CO_2, CO, and fine particulate matter of size less than 2.5 µm in diameter ($PM_{2.5}$) (Kant et al., 2000; Prasad et al., 2000; Prasad and Badarinth, 2004; Gupta et al., 2001; Biswas et al., 2015a,b; Justice et al., 2015; Prasad et al., 2001, 2002, 2003,2008a, b; Vadrevu, 2008; Vadrevu and Choi, 2011). $PM_{2.5}$ suspended in the atmosphere can lead to various human respiratory and general health problems when inhaled (Ramachandran, 2018; Vadrevu and Lasko, 2015, 2018). Thus, mapping and monitoring fires, including addressing the impacts of pollution on land and atmosphere, gain significance (Vadrevu and Justice, 2011; Vadrevu et al., 2019).

Agricultural waste burning is a type of human-initiated BB that frequently occurs during the post-harvest season in agriculture-intensive countries. It often contributes to severe haze pollution because it is the most effective and least expensive way to dispose of the agricultural residues (Lasko et al., 2018a, b; Lasko and Vadrevu, 2018). For instance, China is one of the countries that emits significant pollutants into the atmosphere from agricultural waste burning. Northeast China is one of the largest granaries and the region with the most intense agricultural waste burning. Yin et al. (2019) stated that the agricultural waste burning in Northeast China occurred more intensely than that in other regions and also contributed to the severe haze pollution episodes of 2015 under stagnant weather conditions. Moreover, Zhu et al. (2019) reported that the air pollution episode of autumn 2014 in northern Japan was partly attributed to agricultural waste burning on the Northeast China Plain. Thus, BB has an adverse impact on different spatial scales, both locally and through long-range transport. Meanwhile, Chen et al. (2016) reviewed the modeling studies on air pollution contributed by BB in China and inferred that most of the studies focused on local and inter-regional pollution and that much broader studies are needed. Therefore, it is beneficial to evaluate BB's effect on air quality in East Asia useful for air pollution mitigation efforts.

Biomass burning is truly a multi-disciplinary subject requiring various datasets and expertise (e.g., field observations, laboratory studies for fire combustion, satellite remote sensing of fires, gaseous and particulate emissions from fires, the atmospheric transport of these emissions by air quality models) (Vadrevu et al., 2008, 2012, 2013, 2014a, b, 2015, 2017, 2018). Compared to the ground-based studies, satellite-based approaches, although have huge potential for quantifying BB emissions, can suffer from limitations. For example, satellite-based BB estimates might be underestimated due to the limited observations (i.e., prevalent cloud cover, small agricultural fires, ephemeral agricultural fires) (van der Werf et al., 2010; Wiedinmyer et al., 2011; Zhou et al., 2017; Lasko et al., 2017). Recently, using the Community Multiscale Air Quality (CMAQ) model (https://www.airqualitymodeling.org/index.php/CMAQ), we reported that the boosted agricultural BB emission in Fire Inventory from NCAR (FINN) could significantly improve the model performance in Northeast Asia (Uranishi et al., 2019). However, very few studies explored correlations between BB emission inventory data and the model performance in East Asia. Thus, it is worthwhile to assess the model performance with other BB emission inventories since there is a variation in the inventories due to different approaches followed. The Global Fire Emissions Database (GFED) (van der Werf et al., 2010, 2017) is one of the most common BB emission inventory data available in the world. The latest version of GFEDv4.1s includes newly burned area estimates with contributions from small fires, with boosted BB emissions in temperate cropland regions in Asia, including Northeast China (van der Werf et al., 2017). The GFEDv4.1s captures better spatial and temporal variability of BB emissions in cropland regions than the FINN.

This study utilizes GFEDv4.1s for BB emissions to investigate the impacts of regional and long-range transport of PM$_{2.5}$ due to an intense agricultural waste burning event in Northeast China in autumn 2014. First, the CMAQ performance for regional transport in China was evaluated with ground observation datasets, which are from third parties (https://www.aqistudy.cn/historydata/), and then the impact of BB in China was estimated by conducting simulations during a period of 2014. Second, we utilized the results from the analysis of BB contributions as identified by PMF (hereinafter called "PMF-estimated BB contributions") at the Noto Peninsula in Japan in order to estimate the impact of BB on PM$_{2.5}$ pollution in Japan in October 2014, which could have been affected by the long-range transport from BB sources in Northeast China. Finally, based on both the results, we discuss the remaining aspects of simulations with a focus on the long-range transport of PM$_{2.5}$ due to intense agricultural waste burning in Northeast China.

MATERIALS AND METHODS

MODEL DESCRIPTION AND CONFIGURATION

Figure 17.1 shows the modeling domains from East Asia (D1) to Japan (D2) and the spatial distribution of fire spots (MCD14DL) during the target period (October 20–November 9 in 2014) from Fire Information for Resource Management System (FIRMS, https://earthdata.nasa.gov/firms). Dense fire spots detected by the satellite observation imply that the intense field burning in Northeast China occurred during the target period. Table 17.1 summarizes the Weather Research and Forecasting (WRF) model (http://www2.mmm.ucar.edu/wrf/users/) and CMAQ configurations. More details about the model can be found in the documentation.

The emission data for CMAQ simulations were composed of various datasets (Table 17.1). Specifically, BB emissions were derived from GFEDv4.1s with 0.25° spatial resolution (details provided in the next section). We assumed that BB emissions were uniformly distributed from the surface to the Planetary Boundary Layer (PBL) height based on the Paugam et al. (2016) conclusion that weakly burning landscape-scale fires and crop/grassland fires appear to release their smoke mainly into the PBL. Note that there were differences between the simulation period and the reference years in the anthropogenic emission data available for the simulations. Thus, the uncertainties associated with these differences may affect model performance.

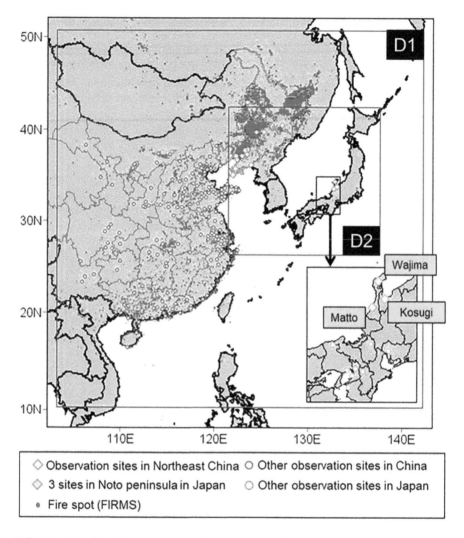

FIGURE 17.1 Modeling domains with locations of observation sites for $PM_{2.5}$ and fire spots (MCD14DL) provided by FIRMS during the target period (October 20–November 9 in 2014). Green lines indicate the three provinces of Northeast China (Liaoning, Jilin, and Heilongjiang).

BIOMASS BURNING EMISSION PROFILE AND ITS SCENARIOS

GFEDv4.1s provides BB emissions for six different fire types: grasslands/savanna fires, boreal forest fires, temperate forest fires, tropical forest fires, peatland fires, and agricultural waste burning (van der Werf et al., 2017). Figure 17.2 shows the relative contributions for monthly averaged $PM_{2.5}$ emissions from BB and monthly emissions over D1 and Northeast China. The primary sources of $PM_{2.5}$ emissions in D1 are tropical forest fires during the cold season and boreal forest fires during the warm

TABLE 17.1

Configurations of the Air Quality Model

	Configurations
Model	Offline WRF v3.8.1-CMAQ v5.0.2
Total duration for simulation (The target period)	From January to December in 2014 with spin-up in December 2013 (October 20–November 9 in 2014)
Domain	D1: East Asia 45km grid 107×107 (CMAQ)
	D2: Japan 15km grid 132×126 (CMAQ)
	34 layers (up to 100 hPa, 1st layer height ≈ 50 m)
Meteorology (WRF)	Geography data: USGS 30-sec topography and land-use
	Analysis data: NCEP FNL, RTG_SST_HR, JMA MSM GPV
	Physics option: Kain-Fritsch, YSU(PBL), WSM6, Noah LSM, and RRTM/Dudhia
	FDDA: $G_{t, q, uv} = 3.0 \times 10^{-4} s^{-1}$ (D1, D2)
Emission	Asia: HTAPv2 (2010),
	Japan: EAGrid2010-Japan, JEI-DB (Vehicle), OPRF2010 (Ship)
	Biogenic: MEGANv2.04, Biomass burning: GFEDv4.1s
	Volcano: JMA & Aerocom
Boundary concentrations	MOZART-4/GEOS5 (D1)
Advection, Diffusion (CMAQ)	Yamartino/WRF-based scheme, Multiscale/ACM2
CMAQ chemistry	SAPRC07 & AERO6 with Aqueous chemistry

season (Figure 17.2a). Meanwhile, in Northeast China, agricultural waste burning is the dominant PM$_{2.5}$ contributor, which occupies more than half of PM$_{2.5}$ emission except June to September (Figure 17.2b). In particular, the PM$_{2.5}$ emissions in April and October that are significantly higher than those in other months imply that the agricultural waste burning occurred intensively in Northeast China (Figure 17.1).

Table 17.2 shows a comparison of the yearly PM$_{2.5}$ emissions from different types of BB emission inventory for D1 and Northeast China. The GFED and the FINN inventories were estimated based on the top-down approach using satellite burned-area data, which may underestimate the BB emissions because of the limitation of satellite observations. Meanwhile, the other two emission inventories (hereafter referred to as "The Chinese BB inventories" (Qiu et al., 2016; Zhou et al., 2017) focused on the whole of China to estimate BB emissions more accurately than the GFED and FINN inventories combining satellite burned-area data and local activity data, such as the yearbook of China. There is a significant divergence in PM$_{2.5}$ emissions among the four BB inventories. The total amount of PM$_{2.5}$ for Northeast China in the Chinese BB inventories is approximately 8–15 times higher than GFEDv4.1s: 691 Gg (Zhou et al., 2017) versus 46 Gg (GFEDv4.1s) in 2012 and 349 Gg (Qiu et al., 2016) versus 44 Gg (GFEDv4.1s) in 2013. Therefore, it is essential to conduct simulations with boosted BB emissions to ascertain whether GFEDv4.1s was underestimated in China as in our previous results with FINNv1.5 (Uranishi et al., 2019). Thus, for investigating the significant emission divergence among the BB inventories (Table 17.2), the following simulation cases were examined: (1) Base

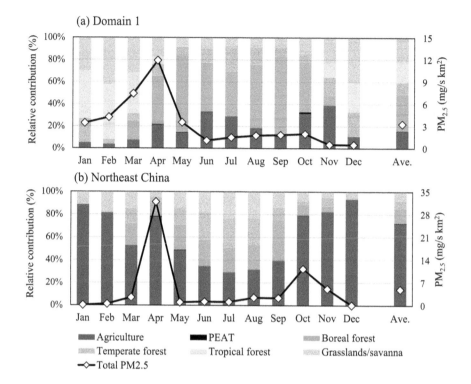

FIGURE 17.2   The relative contributions for monthly averaged PM$_{2.5}$ from BB sources in 2014 for the six fire types (left axis) and monthly variation of the total amount of emission (right axis) over (a) Domain 1 in Figure 17.1 and (b) Northeast China.

TABLE 17.2

Comparison of the Yearly PM$_{2.5}$ Emission from Each BB Emission Inventory for D1 and Northeast China in This Study

	D1			Northeast China		
PM$_{2.5}$ in BB Emission (Gg/year)	2012	2013	2014	2012	2013	2014
GFED v4.1s	2739	1508	2376	46	44	126
FINN v1.5	5317	4918	3473	61	48	51
GEFD_agri15[a]	5489	4574	7480	132	185	652
FINN_crop20[b]	8760	8510	5795	395	512	622
Zhou et al. (2017)	3527	-	-	691	-	-
Qiu et al. (2016)	-	1446	-	-	349	-

[a]   GFEDv4.1s with 15 times boosted values for agricultural waste burning.
[b]   FINNv1.5 with 20 times boosted values for BB source from cropland (Uranishi et al., 2019).

case with regular BB emission (GFEDv4.1s), (2) GFED_agri15 case, namely, a simulation case with 15 times boosted BB values for agricultural waste burning which is a primary contributor of PM$_{2.5}$ emission in Northeast China so that the emission level in GFED_agri15 is comparable with that in FINN_20crop in the previous study (Uranishi et al., 2019) (Table 17.2), and (3) GFED10 case which is a reference case with ten times boosted BB values for all fire types to check the influence from the other fire types except agricultural waste burning in the simulation domain.

Also, another simulation without BB emission of GFEDv4.1s (no GFED) was conducted to estimate the impacts of BB in Northeast China on PM$_{2.5}$ concentrations in Japan, to be more specific, the difference between a target case (Base, GFED_agri15, and GFED10) and no GFED case (hereafter referred to as "CMAQ/BFM-estimated BB contribution"). The CMAQ/BFM-estimated BB contributions on the target area, namely, Noto peninsula in Japan, are regarded as all BB contributions except those from domestic BB because during the target period, GFEDv4.1s represents only a part of BB emissions which were allocated far from Noto peninsula and its impacts (0.03–7.2 t/day in Japan). Thus, BB contribution around the Noto peninsula can be neglected compared to EAGrid2010-Japan data (75 t/day).

OBSERVATION DATA TO EVALUATE THE MODEL PERFORMANCE

The CMAQ performance for regional transport in China was evaluated by comparison with ground-level PM$_{2.5}$ concentration datasets, which are from third parties (https://www.aqistudy.cn/historydata). They provided daily data of air pollutant concentrations for 118 sites in D1, including 17 locations in Northeast China (Figure 17.1).

In order to evaluate the long-range transport of PM$_{2.5}$ from BB in Northeast China to Japan, PMF-estimated BB contributions (Ikemori et al., 2019) were utilized at three sites in the Noto Peninsula (i.e., Wajima, Matto, and Kosugi (Figure 17.1)) in addition to CMAQ/BFM-estimated BB contributions. Ikemori et al. (2019) determined the BB contributions of PM$_{2.5}$ to 13 sites in Japan during autumn (October 22–November 10) in 2014. To accurately identify BB sources from PM$_{2.5}$ sources, they used PMF v5.0 (http://www.epa.gov/heasd/research/pmf.html) with datasets for PM$_{2.5}$ chemical components from multiple sites, including levoglucosan as the tracer species for BB. Consequently, the BB contributions in the PMF analysis were characterized by a chemical profile dominated by K$^+$, elemental carbon (EC), organic carbon (OC), and levoglucosan. Moreover, higher BB contributions at the three sites were found during October 27–29.

RESULTS AND DISCUSSION

REGIONAL TRANSPORT DERIVED FROM BIOMASS BURNING IN CHINA

To assess CMAQ's performance for regional transport in China, simulated PM$_{2.5}$ concentrations were compared with ground-level PM$_{2.5}$ observation datasets using time-series comparison and the index of agreement (IA). IA is a statistical measure of the degree of model prediction error and varies between 0 and 1, where 1 indicates a perfect model performance.

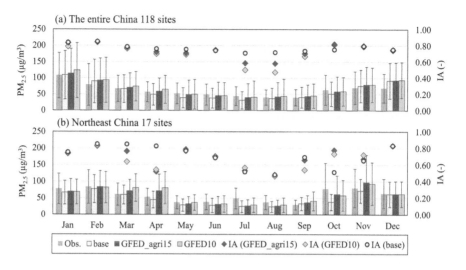

FIGURE 17.3   Monthly variation of the observed and simulated mean concentrations, the standard deviations, and IA of PM$_{2.5}$ for (a) the entire China 118 sites and (b) Northeast China 17 sites in the year 2014. The blue bars show the observed values, whereas the pink, orange, and red bars show the simulated values in the Base, GFED_agri15, GFED10 cases, respectively.

Figure 17.3 shows the monthly variation of the observed and simulated mean PM$_{2.5}$ concentrations in D1, the standard deviations of daily mean PM$_{2.5}$, and IA of daily mean PM$_{2.5}$ over entire China and Northeast China 2014. In general, the Base case model moderately reproduced the temporal variations of site-averaged monthly mean PM$_{2.5}$ concentration over the entire China. The high IA values (> 0.71) in the Base case also supported an acceptable model performance for PM$_{2.5}$ pollution over entire China (Figure 17.3a). Meanwhile, the Base case showed a relatively large underestimation of the observation values over Northeast China (Figure 17.3b). Specifically, the discrepancy between observed and simulated monthly mean PM$_{2.5}$ concentrations for Northeast China was largest in October when agricultural waste burning was the dominant BB source (approximately 80% of the total PM$_{2.5}$ emission). The IA value in the Base case (IA = 0.51) was pretty lower than the boosted BB emission cases (IA = 0.78 in GFED_agri15, and 0.73 in GFED10). These results suggest that the large underestimation in the Base case during October might be attributed to an underestimation of the current BB emissions in Northeast China. For instance, Zhou et al. (2017) stated that there were relatively higher emissions in Northeast China resulting from agricultural waste burning (i.e., in-field corn and rice straw burning), especially in April, October, and November than other months in 2012. Additionally, most of the crop production in Northeast China is attributed to farms smaller than two acres where satellite detection of small fires may be underestimated (https://www.fas.usda.gov/data/northeast-china-prospects-us-agricultural-exports). We infer that BB emissions from agricultural waste burning at least in Northeast China were considerably underestimated in GFEDv4.1s. In contrast, the model in the boosted BB emission cases failed to reproduce PM$_{2.5}$

concentrations in April differently from October: The IA value in the boosted BB emission case (IA = 0.52 in GFED_agri15, and 0.54 in GFED10) was significantly lower than the Base cases (IA = 0.82). These results imply that the boosted emission case is not simply applicable to the year-round simulations. Further studies are needed to infer the reasons for underestimating the current BB emissions (e.g., emission factor and fire spot detection, etc.). We also studied the GFED_agri15 case as the boosted BB example. We focused on Northeast China because, in general, the GFED_agri15 case provided higher IA values over both entire China and Northeast China than the GFED10 case (e.g., IA of July and August in entire China; IA of March, and September in Northeast China).

Figure 17.4 shows the comparison of the Base and GFED_agri15 cases with the observed data in terms of mean concentrations, standard deviations, and IA of PM₂.₅ at the observation sites in China during the target period. CMAQ in the Base case failed to capture the magnitude of mean PM₂.₅ concentrations at monitoring sites in Northeast China. Meanwhile, the GFED_agri15 case showed good performance supported with the higher IA values. For instance, the IA value for 17 sites in Northeast China in the GFED_agri15 case (IA = 0.71) was higher than the Base case (IA = 0.50). These may be resulting from underestimation in the current BB emission inventory because Zhou et al. (2017) reported high values of BB emission rate in Northeast China primarily due to agricultural waste burning (i.e., corn and rice straw burning after harvesting) from October to November. Additionally, these results also highlight the limitations of satellite-derived active fires for quantifying BB emissions and were consistent with the previous study (Uranishi et al., 2019).

It should be noted that BB emissions still have several uncertainties in the size and location of sources as well as their temporal and spatial variability, although the above results explained the underestimation of the current BB emission in Northeast China during the autumn to some extent. Besides, there is a need to develop crop-specific biomass burning emission factors, including accurate characterization of area burned. In the next section, PMF-estimated contributions are utilized to evaluate the accuracy of CMAQ/BFM-estimated BB contributions during the target period when

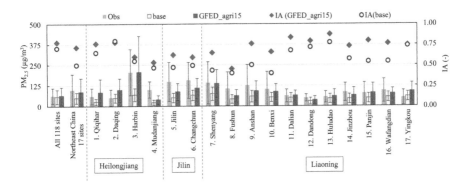

FIGURE 17.4 Comparison of the results of Base and GFED_agri15 cases versus the observed values in terms of the mean concentrations, the standard deviations, and IA of PM₂.₅ on the observation sites in China during the target period. The blue bars show the observed values, whereas the pink and red bars show the Base and GFED_agri15 cases, respectively.

the intense agricultural waste burning occurred in Northeast China (Figure 17.1) for the evaluation of $PM_{2.5}$ long-range transport toward Japan.

Long-Range Transport Derived from Biomass Burning in Northeast Asia

Figure 17.5 shows the comparison of PMF-estimated and CMAQ/BFM-estimated daily mean BB contributions to $PM_{2.5}$ mass at the three sites in the Noto Peninsula, Japan, during the target period. We also describe the spatial distribution of CMAQ/-BFM-estimated mean BB contributions on $PM_{2.5}$ concentrations on October 27 in the GFED_agri15 case. The relatively high BB contributions during October 27–29 were identified by PMF analysis apart from the other days during the target period (Figure 17.5a–c). The model in the Base case failed to capture the temporal variation of PMF-estimated daily mean BB contributions at the three sites, although the CMAQ/BFM-estimated BB contributions in the GFED_agri15 case showed reasonable agreement with the PMF-estimated ones (Figure 17.5a–c). In contrast, the relatively high BB contributions on the 30th and 31st of October were identified only at

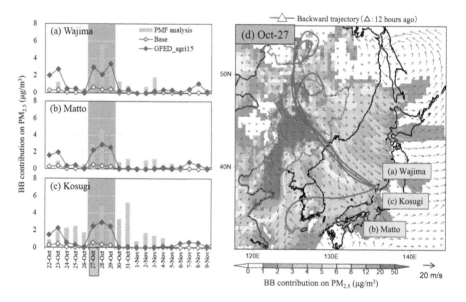

FIGURE 17.5   Comparisons of the daily mean BB contributions on $PM_{2.5}$ concentrations during the target period by the two models, and a spatial distribution of CMAQ/BFM-estimated daily mean BB contribution on October 27. The comparisons of BB contributions are provided for the three sites in Noto peninsula, Japan: (a) Wajima, (b) Matto, and (c) Kosugi. The green bars show PMF-estimated daily mean BB contributions, whereas the pink and red diamonds show CMAQ/BFM-estimated daily mean contributions in the Base and GFED_agri15 case, respectively. The red shading shows the period for long-range transport from BB sources in Northeast China. The spatial distribution of BB contribution in the GFED_agri15 case on October 27 was described with 2-day backward trajectories starting at an altitude of 1500 m at Wajima and run 0, 6, 12, and 18 UTC. The green lines show the province's boundary lines in Northeast China.

Kosugi by the PMF model. This result implied that local agricultural waste burning around Kosugi could have influenced the high PMF-estimated BB contributions.

For further investigation, to identify the BB sources causing the long-range transport of PM$_{2.5}$, we verified the spatial distribution of CMAQ/BFM-estimated daily mean BB contributions to PM$_{2.5}$ concentrations in the GFED_agri15 case as well as the 2-day backward trajectories at Wajima in the Noto Peninsula on October 27 in 2014 (Figure 17.5d). Backward trajectories were calculated at the arrival height of 1500 m at the Wajima site by Trajstat v1.2.2.6 (http://meteothink.org/products/trajstat.html) meteorological fields in simulations D1 and D2. The back-trajectory analysis in Figure 17.5d indicated that the air parcels arrived at Wajima primarily passed through Northeast China where the intense field burning was detected during the target period by the satellite observation (Figure 17.1). Relatively high PMF-estimated BB contributions at Wajima on October 27 were found simultaneously (Figure 17.5a). Additionally, both the spatial distribution of BB contribution and wind vectors on October 27 in Figure 17.5d illustrated that the air parcels with BB pollutants were directly transported to the Noto Peninsula from Northeast China. The model analysis (back-trajectory, BB contribution by the two models, and wind vectors) on October 28 and 29 also provided consistent results like on October 27. However, the model analysis on the days outside October 27–29 indicated that BB pollutants in Northeast China were not directly transported to the Wajima site. Consequently, the GFED_agri15 case presented favorable results consistently for regional and long-range transport of PM$_{2.5}$. It is concluded that boosted BB emission is preferable for simulating air quality in Northeast Asia, which could be deeply affected by BB activities like agricultural waste burning.

CONCLUSIONS

This study described the impacts of regional and long-range transport of PM$_{2.5}$ in Northeast Asia due to intense agricultural waste burning in Northeast China in autumn 2014. The model performance for regional transport of PM$_{2.5}$ concentrations was evaluated with a ground-level observation dataset in entire China. The WRF-CMAQ model in the Base case successfully reproduced the temporal and spatial variation patterns of PM$_{2.5}$ mass for the most part (Figure 17.3a). However, the Base case scenario failed to show a favorable performance for PM$_{2.5}$ concentrations in October, mainly in Northeast China, dominated by agricultural residue burning. Meanwhile, the model in the GFED_agri15 case captured the spatial variation patterns of PM$_{2.5}$ mass in October supported by higher IA values for 17 sites in Northeast China and all monitoring sites in China (Figures 17.3 and 17.4). These results indicated that BB emissions from agricultural waste burning in Northeast China still had considerable uncertainty, although GFEDv4.1s includes burned area estimates with contributions from small fires suggesting that much more refinement is needed for GFEDv4.1s.

Moreover, for the evaluation of long-range transport of PM$_{2.5}$ toward Japan, the two different types of models (PMF and CMAQ/BFM) were utilized to compare BB. In contrast to the Base case, the GFED_agri15 case showed a reasonable agreement between PMF-estimated and CMAQ/BFM-estimated BB contributions with similar levels and day-to-day variation patterns during the target period for the three sites in Noto Peninsula, Japan (Figure 17.5a-c). Consequently, our comparison studies concluded that a boosted BB case was preferable to capture intense agricultural waste burning events while simulation, which cannot be detected accurately by the satellite observation. The approach highlighted in the study can be applied to other regions. However, the model settings (e.g., the vertical distribution of BB emission corresponding to each BB source) need to be modified and parameterized to address local land use/cover variations, including agricultural areas in the target region.

ACKNOWLEDGEMENTS

We thank Dr. Fumikazu Ikemori of the Nagoya City Institute for Environmental Sciences and Dr. Seiji Sugata of the National Institute for Environmental Studies (NIES) for providing PMF analysis results, computational resources, and fruitful discussions. This research was also supported by Type II joint research between NIES and the local environmental research institutes in Japan. We are grateful to NASA for the use of data and imagery from LANCE FIRMS.

REFERENCES

Andreae, M.O. and Merlet, P. 2001. Emission of trace gases and aerosols from biomass burning. Global *Biogeochemical Cycles*. 15(4), 955–966.

Badarinath, K.V.S., Kharol, S.K., Krishna Prasad, V., Kaskaoutis, D.G. and Kambezidis, H.D. 2008a. Variation in aerosol properties over Hyderabad, India during intense cyclonic conditions. *International Journal of Remote Sensing*. 29(15), 4575–4597.

Badarinath, K.V.S., Kharol, S.K., Prasad, V.K., Sharma, A.R., Reddi, E.U.B., Kambezidis, H.D. and Kaskaoutis, D.G. 2008b. Influence of natural and anthropogenic activities on UV Index variations–a study over tropical urban region using ground based observations and satellite data. *Journal of Atmospheric Chemistry*. 59(3), 219–236.

Biswas, S., Lasko, K.D. and Vadrevu, K.P. 2015a. Fire disturbance in tropical forests of Myanmar—Analysis using MODIS satellite datasets. *IEEE Journal of Selected Topics in Applied Earth Observations and Remote Sensing*. 8(5), 2273–2281.

Biswas, S., Vadrevu, K.P., Lwin, Z.M., Lasko, K. and Justice, C.O. 2015b. Factors controlling vegetation fires in protected and non-protected areas of Myanmar. *PLoS One*. 10(4), e0124346.

Chen, J., Li, C., Ristovski, Z. et al., 2016. A review of biomass burning: Emissions and impacts on air quality, health and climate in China. *Science of the Total Environment*. 579, 1000–1034. doi:10.1016/j.scitotenv.2016.11.025.

Gupta, P.K., Prasad, V.K., Sharma, C., Sarkar, A.K., Kant, Y., Badarinath, K.V.S. and Mitra, A.P. 2001. CH_4 emissions from biomass burning of shifting cultivation areas of tropical deciduous forests–experimental results from ground-based measurements. *Chemosphere-Global Change Science*. 3(2), 133–143.

Hayasaka, H. and Sepriando, A. 2018. Severe Air Pollution Due to Peat Fires during 2015 Super El Niño in Central Kalimantan, Indonesia. In: *Land-Atmospheric Research applications in South/Southeast Asia*. Vadrevu, K.P., Ohara, T., and Justice, C. (Eds.). Springer, Cham. pp. 129–142.

Hayasaka, H., Noguchi, I., Putra, E.I., Yulianti, N. and Vadrevu, K. 2014. Peat-fire-related air pollution in Central Kalimantan, Indonesia. *Environmental Pollution*. 195, 257–266. doi:10.1016/j.envpol.2014.06.031.

Inoue, Y. 2018. Ecosystem carbon stock, Atmosphere and Food security in Slash-and-Burn Land Use: A Geospatial Study in Mountainous Region of Laos. In: *Land-Atmospheric Research applications in South/Southeast Asia*. Vadrevu, K.P., Ohara, T., and Justice, C. (Eds.). Springer, Cham, pp. 641–666.

Ikemori, F., Nakayama, T. and Hasegawa, H. 2019. Characterization and possible sources of nitrated mono-and di-aromatic hydrocarbons containing hydroxyl and/or carboxyl functional groups in ambient particles in Nagoya, Japan. Atmospheric Environment, 211, 91–102.

Israr, I., Jaya, S.N.I., Saharjo, H.S., Kuncahyo, B. and Vadrevu, K.P. 2018. Spatio-temporal analysis of land and forest fires in Indonesia using MODIS active fire dataset. In: *Land-Atmospheric Research Applications in South/Southeast Asia*. Vadrevu, K.P., Ohara, T., and Justice, C. (Eds.). Springer, Cham, pp. 105–128.

Justice, C., Gutman, G. and Vadrevu, K.P.2015. NASA land cover and land use change (LCLUC): An interdisciplinary research program. *Journal of Environmental Management*. 148(15), 4–9.

Kant, Y., Ghosh, A.B., Sharma, M.C., Gupta, P.K., Prasad, V.K., Badarinath, K.V.S. and Mitra, A.P. 2000. Studies on aerosol optical depth in biomass burning areas using satellite and ground-based observations. *Infrared Physics & Technology*. 41(1), 21–28.

Lasko, K. and Vadrevu, K.P. 2018. Improved rice residue burning emissions estimates: Accounting for practice-specific emission factors in air pollution assessments of Vietnam. *Environmental Pollution*. 236(5), 795–806.

Lasko, K., Vadrevu, K.P. and Nguyen, T.T.N. 2018a. Analysis of air pollution over Hanoi, Vietnam using multi-satellite and MERRA reanalysis datasets. *PloS One*. 13(5), e0196629.

Lasko, K., Vadrevu, K.P., Tran, V.T. and Justice, C. 2018b. Mapping double and single crop paddy rice with Sentinel-1A at varying spatial scales and polarizations in Hanoi, Vietnam. *IEEE Journal of Selected Topics in Applied Earth Observations and Remote Sensing*. 11(2), 498–512.

Lasko, K., Vadrevu, K.P., Tran, V.T., Ellicott, E., Nguyen, T.T., Bui, H.Q. and Justice, C. 2017. Satellites may underestimate rice residue and associated burning emissions in Vietnam. *Environmental Research Letters*. 12(8), 085006.

Oanh, N.T.K., Permadi, D.A., Dong, N.P. and Nguyet, D.A. 2018. Emission of toxic air pollutants and greenhouse gases from crop residue open burning in Southeast Asia. In: *Land-Atmospheric Research Applications in South/Southeast Asia*. Vadrevu, K.P., Ohara, T., and Justice, C. (Eds.). Springer, Cham, pp. 47–68.

Paugam, R., Wooster, M., Freitas, S. and Val Martin, M. 2016. A review of approaches to estimate wildfire plume injection height within large-scale atmospheric chemical transport models. *Atmospheric Chemistry and Physics*. 16(2), 907–925.

Prasad, V.K. and Badarinth, K.V.S. 2004. Land use changes and trends in human appropriation of above ground net primary production (HANPP) in India (1961–98). *Geographical Journal*. 170(1), 51–63.

Prasad, V.K., Anuradha, E. and Badarinath, K.V.S. 2005. Climatic controls of vegetation vigor in four contrasting forest types of India—evaluation from National Oceanic and Atmospheric Administration's Advanced Very High Resolution Radiometer datasets (1990–2000). *International Journal of Biometeorology*. 50(1, 6–16.

Prasad, V.K., Badarinath, K.V.S. and Eaturu, A. 2008a. Biophysical and anthropogenic controls of forest fires in the Deccan Plateau, India. *Journal of Environmental Management*. 86(1), 1–13.

Prasad, V.K., Badarinath, K.V.S. and Eaturu, A. 2008b. Effects of precipitation, temperature and topographic parameters on evergreen vegetation greenery in the Western Ghats, India. *International Journal of Climatology: A Journal of the Royal Meteorological Society*. 28(13), 1807–1819.

Prasad, V.K., Badarinath, K.V.S., Yonemura, S. and Tsuruta, H. 2004. Regional inventory of soil surface nitrogen balances in Indian agriculture (2000–2001). *Journal of Environmental Management.* 73(3), 209–218.

Prasad, V.K., Gupta, P.K., Sharma, C., Sarkar, A.K., Kant, Y., Badarinath, K.V.S., Rajagopal, T. and Mitra, A.P. 2000. NO_x emissions from biomass burning of shifting cultivation areas from tropical deciduous forests of India–estimates from ground-based measurements. *Atmospheric Environment.* 34(20), 3271–3280.

Prasad, V.K., Kant, Y. and Badarinath, K.V.S., 2001. CENTURY ecosystem model application for quantifying vegetation dynamics in shifting cultivation areas: A case study from Rampa Forests, Eastern Ghats (India). *Ecological Research.* 16(3), 497–507.

Prasad, V.K., Kant, Y., Gupta, P.K., Elvidge, C. and Badarinath, K.V.S. 2002. Biomass burning and related trace gas emissions from tropical dry deciduous forests of India: A study using DMSP-OLS data and ground-based measurements. *International Journal of Remote Sensing.* 23(14), 2837–2851.

Prasad, V.K., Lata, M. and Badarinath, K.V.S. 2003. Trace gas emissions from biomass burning from northeast region in India—estimates from satellite remote sensing data and GIS. *Environmentalist.* 23(3), 229–236.

Qiu, X., Duan, L., Chai, F. et al., 2016. Deriving high-resolution emission inventory of open biomass burning in china based on satellite observations. *Environmental Science & Technology,* 50(21), 11779–11786. doi:10.1021/acs.est.6b02705.

Ramachandran, S. 2018. Aerosols and climate change: Present understanding, challenges and future outlook. In: *Land-Atmospheric Research Applications in South/Southeast Asia.* Vadrevu, K.P., Ohara, T., and Justice, C. (Eds.). Springer, Cham, pp. 341–378.

Saharjo, B.H., and Yungan, A. 2018. Forest and land fires in Riau province; A Case study in fire prevention, policy implementation with local concession holders. In: *Land-Atmospheric Research Applications in South/Southeast Asia.* Vadrevu, K.P., Ohara, T., and Justice, C. (Eds.). Springer, Cham, pp. 143–170.

Uranishi, K., Ikemori, F., Shimadera, H., Kondo, A. and Sugata, S., 2019. Impact of field biomass burning on local pollution and long-range transport of $PM_{2.5}$ in Northeast Asia. *Environmental Pollution.* 244, 414–422.

Vadrevu, K.P., Ohara, T. and Justice C. 2014a. Air pollution in Asia. *Environmental Pollution.* 12, 233–235.

Vadrevu, K.P. and Justice, C.O. 2011. Vegetation fires in the Asian region: Satellite observational needs and priorities. *Global Environmental Research.* 15(1), 65–76.

Vadrevu, K.P. 2008. Analysis of fire events and controlling factors in eastern India using spatial scan and multivariate statistics. *Geografiska Annaler: Series A, Physical Geography.* 90(4), 315–328.

Vadrevu, K.P., and Choi, Y. 2011. Wavelet analysis of airborne CO_2 measurements and related meteorological parameters over heterogeneous landscapes. *Atmospheric Research.* 102(1–2), 77–90.

Vadrevu, K.P., and Lasko, K. 2018. Intercomparison of MODIS AQUA and VIIRS I-Band fires and emissions in an agricultural landscape—Implications for air pollution research. *Remote Sensing.* 10(7), 978. doi:10.3390/rs10070978.

Vadrevu, K.P., and Lasko, K.P., 2015. Fire regimes and potential bioenergy loss from agricultural lands in the Indo-Gangetic Plains. *Journal of Environmental Management.* 148, 10–20.

Vadrevu, K.P., Badarinath, K.V.S. and Anuradha, E. 2008. Spatial patterns in vegetation fires in the Indian region. *Environmental Monitoring and Assessment.* 147(1–3), 1. doi:10.1007/s10661-007-0092-6.

Vadrevu, K.P., Csiszar, I., Ellicott, E., Giglio, L., Badarinath, K.V.S., Vermote, E. and Justice, C. 2012. Hotspot analysis of vegetation fires and intensity in the Indian region. *IEEE Journal of Selected Topics in Applied Earth Observations and Remote Sensing.* 6(1), 224–238.

Vadrevu, K.P., Eaturu, A. and Badarinath, K.V.S. 2006. Spatial distribution of forest fires and controlling factors in Andhra Pradesh, India using spot satellite datasets. *Environmental Monitoring and Assessment*. 123(1–3), 75–96.

Vadrevu, K.P., Giglio, L. and Justice, C. 2013. Satellite based analysis of fire–carbon monoxide relationships from forest and agricultural residue burning (2003–2011). *Atmospheric Environment*. 64, 179–191.

Vadrevu, K.P., Lasko, K., Giglio, L. and Justice, C. 2014b. Analysis of Southeast Asian pollution episode during June 2013 using satellite remote sensing datasets. *Environmental Pollution*. 12, 245–256.

Vadrevu, K.P., Lasko, K., Giglio, L. and Justice, C. 2015. Vegetation fires, absorbing aerosols and smoke plume characteristics in diverse biomass burning regions of Asia. *Environmental Research Letters*. 10(10), 105003.

Vadrevu, K.P., Lasko, K., Giglio, L., Schroeder, W., Biswas, S. and Justice, C. 2019. Trends in vegetation fires in South and Southeast Asian countries. *Scientific Reports*. 9(1), 7422. doi:10.1038/s41598-019-43940-x.

Vadrevu, K.P., Ohara, T. and Justice, C. 2017. Land cover, land use changes and air pollution in Asia: A synthesis. *Environmental Research Letters*. 12(12), 120201.

Vadrevu, K.P., Ohara, T. and Justice, C. eds., 2018. *Land-Atmospheric Research Applications in South and Southeast Asia*. Springer, Cham.

Van der Werf, G.R., Randerson, J.T., Giglio, L., Collatz, G.J., Mu, M., Kasibhatla, P.S., Morton, D.C., DeFries, R.S., Jin, Y. and van Leeuwen, T.T. 2010. Global fire emissions and the contribution of deforestation, savanna, forest, agricultural, and peat fires (1997–2009). *Atmospheric Chemistry and Physics*. 10(23), 11707–11735.

Van Der Werf, G.R., Randerson, J.T., Giglio, L., Van Leeuwen, T.T., Chen, Y., Rogers, B.M., Mu, M., Van Marle, M.J., Morton, D.C., Collatz, G.J. and Yokelson, R.J., 2017. Global fire emissions estimates during 1997–2016. *Earth System Science Data*. 9(2), 697–720.

Wiedinmyer, C., Akagi, S.K., Yokelson, R.J., Emmons, L.K., Al-Saadi, J.A., Orlando, J.J. and Soja, A.J., 2011. The Fire Inventory from NCAR (FINN): A high resolution global model to estimate the emissions from open burning. *Geoscientific Model Development*. 4(3), 625.

Yin, S., Wang, X., Zhang, X., Zhang, Z., Xiao, Y., Tani, H. and Sun, Z., 2019. Exploring the effects of crop residue burning on local haze pollution in Northeast China using ground and satellite data. *Atmospheric Environment*. 199, 189–201.

Zhou, Y., Xing, X., Lang, J., Chen, D., Cheng, S., Wei, L., Wei, X. and Liu, C., 2017. A comprehensive biomass burning emission inventory with high spatial and temporal resolution in China. *Atmospheric Chemistry and Physics* 17(4), 2839–2864.

Zhu, C., Kanaya, Y., Yoshikawa-Inoue, H., Irino, T., Seki, O. and Tohjima, Y., 2019. Sources of atmospheric black carbon and related carbonaceous components at Rishiri Island, Japan: The roles of Siberian wildfires and of crop residue burning in China. *Environmental Pollution*. 247, 55–63.

18 Estimation of PM$_{10}$ Concentrations from Biomass Burning and Anthropogenic Emissions Using the WRF–HYSPLIT Modeling System

Teerachai Amnuaylojaroen and Nungruethai Anuma
University of Phayao, Thailand

CONTENTS

INTRODUCTION

Air pollution is a major concern in several regions of the world. Most of the pollution is due to energy use and production that releases gases and particles into the atmosphere (Amnauylawjarurn et al., 2010). Air pollution also has the potential to contribute to climate change through the release of greenhouse gases (GHGs) and aerosols, which play a key role in radiative forcing in the atmosphere. Moreover, some pollutants can reduce plant growth and development and productivity. Specifically in South/Southeast Asia, biomass burning is an important source of GHGs and aerosols (Kant et al., 2000; Prasad et al., 2000, 2001, 2002, 2003; Gupta et al., 2001; Prasad and Badarinath, 2004; Prasad et al., 2008; Badarinath et al., 2008a, b; Vadrevu,

2008; Vadrevu et al., 2008; Amnauylawjarurn et al., 2010; Vadrevu and Justice, 2011; Amnuaylojaroen et al., 2014; Lasko and Vadrevu, 2018). Biomass burning is most commonly practiced in several countries of South/Southeast Asia, such as India, Pakistan, Nepal, Burma, Laos, Cambodia, Indonesia, the Philippines, and Thailand (Vadrevu et al. (2006); Vadrevu and Choi (2011);Vadrevu et al., 2012, 2013, 2014a, b, 2015; Hayasaka et al., 2014; Vadrevu and Lasko, 2015; Biswas et al., 2015a, b; Sonkaew and Macatangay 2015; Justice et al., 2015; Lasko et al., 2017, 2018a, b; Vadrevu et al., 2017, 2018; Hayasaka and Sepriando, 2018; Israr et al., 2018; Saharjo and Yungan, 2018; Oanh et al., 2018; Inoue, 2018; Vadrevu and Lasko, 2018; Vadrevu et al., 2019). In these countries, the biomass burning episodes typically occur during the dry season every year. The most common causative factors of biomass burning include slash-and-burn agriculture, clearing of forests for raising plantations, crop residue burning, and the use of wood or cow dung as biofuel.

In addition, meteorology and topography are also the two important factors impacting air pollution in this region (Ramachandran, 2018). Specifically, in northern Thailand, the atmospheric stability, wind direction, velocity, and topography aid in pollutant accumulation (Amnuaylojaroen and Kreasuwun, 2012). Particulate matter with a diameter of less than $10\,\mu m$ (PM_{10}) is increasing continuously in many locations in northern Thailand. The database from the Pollution Control Department (PCD) indicates that there has been a peak of PM_{10} every February and March (Figure 18.1).

Several studies indicate the importance of biomass burning contribution to PM_{10} in northern Thailand. For example, Punsompong and Chantara (2018) used statistical analysis of trajectories to identify the sources of PM_{10} from biomass burning. Their results reveal that 70% of biomass burning comes from the transboundary sources in Myanmar and Thailand. Also, the results from Kiatwattanacharoen et al. (2017) indicated that biomass burning from the Thai–Myanmar border was a major source of high PM_{10} loads in northern Thailand. In addition to biomass burning, anthropogenic

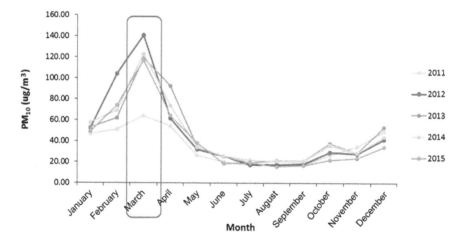

FIGURE 18.1   Monthly PM_{10} concentration from PCD measurement in northern Thailand during 2011–2015.

emissions are also important from residential, industry, power plant, and transportation sectors. Anthropogenic emissions can contribute to air pollution. Pozzer et al. (2012) used the atmospheric chemistry general circulation model to estimate anthropogenic emissions' impact on the future global and regional air quality for 2025 and 2050. The results indicated that a higher population in many regions (East and South Asia) drives energy demand, increasing air pollutant concentrations.

Numerical modeling is a useful tool to study air pollution. There are several air pollution models such as the Hybrid Single-Particle Lagrangian Integrated Trajectory (HYSPLIT) model (Draxler and Hess, 1998), the California Mesoscale PUFF model (CALPUFF) (Bennett et al., 2002), and the Weather Research and Forecasting Model with Chemistry (WRF-Chem) (Grell et al., 2005). Each model has its potential to study air quality aspects. For example, the WRF-Chem is an online chemistry and meteorology process-based model that uses too many computer resources for calculation. In contrast, the HYSPLIT and CALPUFF use lesser resources. The HYSPLIT model is widely used to calculate the dispersion and trajectories and the deposition of pollutants. For example, Punsompong and Chantara (2018) used the HYSPLIT model to simulate PM$_{10}$ trajectories from biomass burning sources based on the fire hotspots in upper Southeast Asia. Sam-Quarcoo Dotse et al. (2016), who used the HYSPLIT model to assess Brunei's air quality, found that PM$_{10}$ plays a vital role in controlling air quality in Brunei.

In this study, we used the atmospheric modeling coupled with the air quality model to study the effect of biomass burning and anthropogenic emissions on PM$_{10}$ concentrations. Specifically, we used the Weather Research and Forecasting (WRF) model version 3.8.1 to simulate the meteorological conditions in March 2012. We parameterized the Final Analysis Data's initial and boundary conditions (FNL) (Kanamitsu et al., 2002). The modeled temperature and precipitation were compared to the dataset from the PCD. To assess the model capability, statistical metrics such as Index of Agreement (IOA) and Fractional Bias (FB) were used for the model evaluation. The output from the WRF model was used as meteorological conditions into the HYSPLIT model to simulate PM$_{10}$ concentration that include anthropogenic emission and biomass burning emission from the Emissions Database for Global Atmospheric Research (EDGAR) version 4.3.1 (Crippa et al., 2016) and Fire INventory from NCAR (FINN) version 1.5 (Wiedinmyer et al., 2011).

MODEL CONFIGURATION AND EVALUATION

MODEL CONFIGURATION

We used the coupled atmospheric WRF model version 3.8 (Skamarock et al., 2008) and HYSPLIT air particle trajectory model to address pollution. The results from HYSPLIT were used to analyze the source apportionment of PM$_{10}$. The WRF model was developed to study operational weather forecasting. It is a non-hydrostatic mesoscale model that consists of many physical schemes, including radiation, cumulus, and microphysics, whereas the HYSPLIT model is based on a Lagrangian calculation to solve the air pollutant trajectory and concentrations. It combines two computational approaches, i.e., 3D particles and puffs, to calculate the pollutants' concentration. In this study, we designed two WRF domains with a horizontal resolution of 50- and 10-km grid spacing. Also, the model is set with 30 vertical levels up to 50 hPa. The outer domain covers the upper mainland of Southeast Asia and some areas from East and South Asia, such as the south of China and the east of India, entirely, as shown in Figure 18.1. We selected this domain as Southeast Asia is influenced by the East Asian monsoon that carries the air mass from high latitude into this region. Due to the transboundary nature of emissions from the west, such as from Myanmar and India, air quality in northern Thailand is affected, while the inner domain consisted of northern Thailand. To resolve water vapor, cloud, and precipitation process, the model has been configured by using the WRF Single-Moment 3-class scheme following Hong et al. (2004) and Hong and Lim (2006). It predicts a simple-ice scheme that has three types of hydrometers, i.e., vapor, cloud water, and rain. The calculation of these processes is based on mass content from a diagnostic relationship. The sub-grid-scale processes for solving convection are the Kain–Fritsch scheme (Kain, 2004). It has the potential to utilize a cloud model with updrafts and downdrafts, as well as considering the impacts of detrainment and entrainment on cloud formation. The similarity theory scheme emulates thermal gradient over a surface responsible for friction velocities and wind on the surface (Paulson, 1970; Dyer and Hicks, 1970; Webb, 1970; Zhang and Anthes, 1982; Beljaars, 1994). To reduce the effect from initial conditions, the model spin-up was conducted during February 15–28, 2012. The WRF model was run from March 1, 2012, to April 1, 2012, to simulate the weather condition for the HYSPLIT model. The principal meteorological variables from the WRF model, i.e., wind (U, V, W), temperature (T), surface pressure (Psfc), and relative humidity (RH), were used as input data into the HYSPLIT model. The anthropogenic and biomass burning emissions from the EDGAR version 4.3.1 and FINN global emission data were included in the HYSPLIT model. The EDGARv4.3.1 is a global anthropogenic emissions inventory of both gases and particulates. It mainly consists of several chemical species, i.e., SO_2, NO_x, CO, nonmethane volatile organic compounds, NH_3, PM_{10}, $PM_{2.5}$, black carbon (BC), and organic carbon (OC), for the period 1970–2010 with a high spatial resolution of $0.1° \times 0.1°$. Also, EDGARv4.3.1 incorporates data on international energy balance from the International Energy Agency (IEA, 2014) and agricultural statistics of the Food and Agriculture Organization (FAO, 2012) with technological assumptions (Crippa et al., 2016). The FINN version 1.5 is a very high-spatial-resolution biomass burning emission database at 1 km. It provides a daily global estimation of the trace gases and particles, i.e., CO_2, CO, CH_4, NMHC, NMOC, NO, NO_2, NH_3, SO_2, $PM_{2.5}$, PM_{10}, TPM, TPC, OC, and BC, from open burning such as wildfire, agricultural fires, and other prescribed burning. The FINN data are based on the active fires retrieved using the Moderate Resolution Imaging Spectroradiometer (MODIS) instrument (Wiedinmyer et al., 2011).

MODEL EVALUATION

To examine the model capability, we compared the model results from WRF inner domain, i.e., precipitation and 2-m temperature and PM_{10} concentration from HYSPLIT output in March 2012 with ground-based observations PCD as listed in

TABLE 18.1

Location of PCD Measurement in Northern Thailand

Location	Latitude	Longitude
Nan	18.78	100.77
Phrae	18.12	100.16
Phayao	19.16	99.89
Chiang Rai	20.42	99.88
Lampang	18.25	99.76
Chiang Mai	18.78	98.98
Lamphun	18.56	99.00
Mae Hong Son	19.30	97.96

Table 18.1. Thai PCD generally measures the hourly concentrations of five pollutants – PM$_{10}$, CO, NO$_2$, SO$_2$, and O$_3$ – at eight location sites over northern Thailand as listed in Table 18.1. The sites of PCD measurement are almost located near the urban areas. Motor vehicle emissions likely influence it. The modeled results were compared to the PCD dataset at eight location sites (Table 18.1). We used statistical indicators, i.e., IOA and FB, to examine the model efficiency.

The comparison between the monthly output from the model, i.e., the temperature at 2 m, precipitation, and PM$_{10}$ concentrations, and observation data is displayed in Figure 18.2. In general, the model captured the variations well. The modeled temperature at 2 m is lower than ground-based measurement by 2°C–3°C, while precipitation was also underestimated about 1–2 mm/day. In addition, the statistical analysis is shown in Table 18.2. Both the WRF and HYSPLIT simulations agree well with the mean monthly 2-m temperature and PM$_{10}$ concentrations for most northern Thailand sites with an acceptable IOA of 0.49–0.68. However, the model generally underpredicted 2-m temperature and precipitation for all sites, especially precipitation that was very low at Nan and Chiang Rai with FB of 1.49 and 1.51, respectively; the modeled 2-m temperature was slightly cold-biased compared to the PCD dataset with FB of 0.09. The HYSPLIT model simulations captured the monthly PM$_{10}$ concentrations well for all sites with a moderate IOA of 0.62–0.70. Since March is a dry season in this region, precipitation during this month on the PM$_{10}$ concentrations might be very low.

RESULTS AND DISCUSSION

PM$_{10}$ Emissions and Concentrations

To quantify the amount of biomass burning contributions to PM$_{10}$ in this region, we plotted the PM$_{10}$ emissions from different sources, including biomass burning from the FINN and anthropogenic emissions from the EDGAR for energy, industry, residential sector, and transportation. We characterized the study region into three zones of the spatial distribution to infer PM$_{10}$ emissions from different sources as

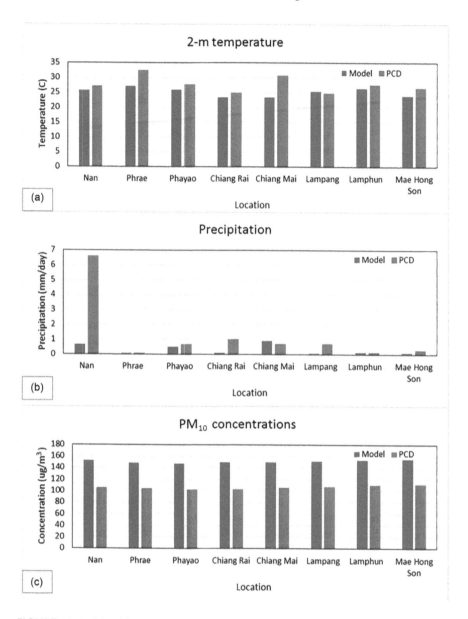

FIGURE 18.2   Monthly mean 2-m temperature and precipitation from the WRF model and PM$_{10}$ concentrations from the HYSPLIT model compared to PCD.

in Figure 18.3: Zone 1 (lat: 15°N–25°N, lon: 80°E–91°E) covers Bangladesh and some areas of India; Zone 2 (lat: 10°N–25°N, lon: 92°E–110°E) covers Burma, Laos, Vietnam, Cambodia, and some regions in Thailand; and Zone 3 (lat: 20°N–25°N, lon: 110°E–120°E) covers some regions in southern China. In general, the anthropogenic emissions, especially from the industry and residential sectors, contributed high PM$_{10}$ emissions in Zone 1 by ~0.15 to ~0.2 mole/km²/h, while biomass burning

TABLE 18.2

Statistical Analysis between the WRF Model and PCD Data (Monthly Averages)

Stations	IOA			FB		
	Temperature	Precipitation	PM$_{10}$	Temperature	Precipitation	PM$_{10}$
Nan	0.52	0.59	0.66	−0.86	−0.65	−1.01
Phrae	0.82	0.28	0.65	−1.04	−0.68	−0.32
Phayao	0.22	0.70	0.68	−0.97	−0.75	−0.26
Chiang Rai	0.17	0.78	0.74	−0.98	−0.63	−0.29
Chiang Mai	0.54	0.61	0.50	−0.44	−0.78	−0.45
Lampang	0.65	0.86	0.70	−0.11	−0.84	−0.45
Lamphun	0.61	0.23	0.70	−0.62	−0.72	−0.39
Mae Hong Son	0.42	0.71	0.62	−0.78	−0.78	−0.25

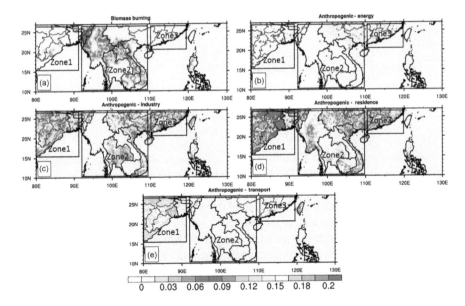

FIGURE 18.3 Monthly average PM$_{10}$ emissions from (a) biomass burning, (b) energy, (c) industry, (d) residential sector, and (e) transportation.

is the main contributor of PM$_{10}$ emissions in Zone 2 up to 0.2 mole/km^2/h in many areas of Burma, Laos, and Thailand. Interestingly, as depicted in the plots, there is much PM$_{10}$ emission from manufacturing industries in southern China. A huge contribution of anthropogenic emission generally occurs in Zones 1 and 3, while biomass burning emissions strongly dominate in Zone 2 (Figures 18.3 and 18.4).

Additionally, to identify the effects of biomass burning on PM$_{10}$ contribution, we executed the HYSPLIT model with two scenarios: (1) HYSPLIT with biomass

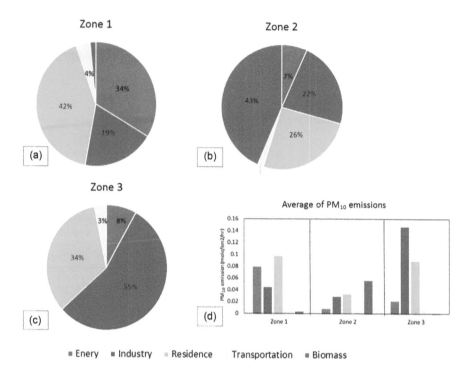

FIGURE 18.4   Percentage of PM_{10} emission sources from (a) Zone 1, (b) Zone 2, and (c) Zone 3, and (d) summary of PM_{10} emission from different sources: energy (red), industry (blue), residential sector (green), transportation (yellow), and biomass burning (purple).

burning emissions and (2) HYSPLIT with anthropogenic emissions during March 2007 over northern Thailand. The results from the model indicate that biomass burning contributes to >56% of total PM_{10} concentration. PM_{10} concentrations from biomass burning are significantly higher than those from the anthropogenic emissions for the whole month of March in the year 2012, as shown in Figure 18.5.

Our results agree with the previous studies of Karthik et al. (2017), who also reported a comprehensive size–composition–morphology characterization of the particulate matter (PM) collected from an urban site in Singapore using electron microscopy. Our results also match with Huang et al. (2012), who used the CMAQ model to assess the impact of biomass burning on air quality in South and East Asia. Their results indicate that biomass burning and local emissions dominate the aerosol chemistry.

SUMMARY

In this study, we performed the coupled atmospheric and air pollution model WRF–HYSPLIT to quantify the effect of biomass burning on air quality in Southeast Asia. The model capability was evaluated against the ground-based measurements in northern Thailand. The results were of acceptable quality and captured both the meteorology and PM_{10} concentrations compared to ground observations, which were

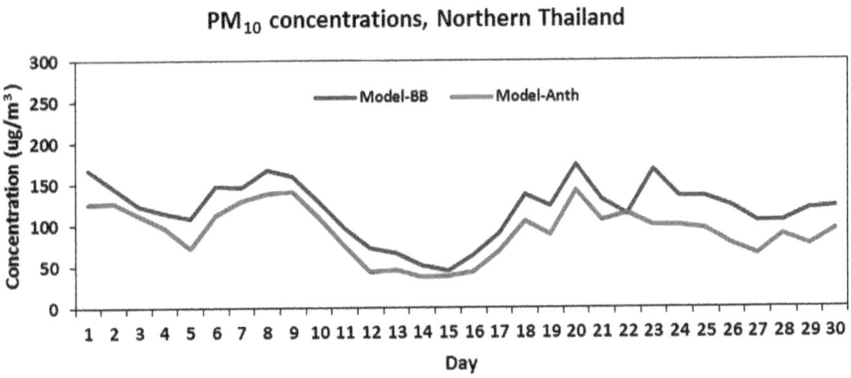

FIGURE 18.5 Daily PM$_{10}$ concentrations from biomass burning emission (blue line) and anthropogenic emission (red line) in March 2012 over northern Thailand.

represented by IOA of 0.62–0.70, FB of 1.49–1.51, and NMSE of 79–273. Based on the FINN and EDGAR emission inventories, biomass burning is the main contribution of PM$_{10}$ in Southeast Asia, while anthropogenic emissions from industry and residential sectors are the two other key contributors to pollution in some regions of South Asia. The industry sector is an important PM$_{10}$ emission source in southern China. From the model runs, we infer that biomass burning is responsible for about 56% of PM$_{10}$ contributions in northern Thailand, while anthropogenic emissions are responsible for about 44%.

ACKNOWLEDGMENT

The authors would like to thank the Pollution Control Department (PCD) from Thailand for the ground-based measurement dataset.

REFERENCES

Amnauylawjarurn, T., Kreusuwun, J., Towta, S. and Siriwittayakorn, K. 2010. Dispersion of Particulate Matter (PM$_{10}$) from forest fires in Chiang Mai, Thailand. *Chiang Mai J. Sci.* 37(1):39–47.

Amnuaylojaroen, T., Barth, M.C., Emmons, L.K., Carmichael, G.R., Kreasuwun, J., Prasitwattanaseree, S. and Chantara, S. 2014. Effect of different emission inventories on ozone and carbon monoxide in Southeast Asia. *Atmos. Chem. Phys.* 14:12983–13012. doi:10.5194/acp-14-12983-2014.

Badarinath, K.V.S., Kharol, S.K., Krishna Prasad, V., Kaskaoutis, D.G. and Kambezidis, H.D. 2008a. Variation in aerosol properties over Hyderabad, India during intense cyclonic conditions. *Int. J. Remote Sens.* 29(15):4575–4597.

Badarinath, K.V.S., Kharol, S.K., Prasad, V.K., Sharma, A.R., Reddi, E.U.B., Kambezidis, H.D. and Kaskaoutis, D.G. 2008b. Influence of natural and anthropogenic activities on UV Index variations–a study over tropical urban region using ground based observations and satellite data. *J. Atmos. Chem.* 59(3):219–236.

Beljaars, A.C.M. 1994. The parameterization of surface fluxes in large-scale models under free convection, *Quart. J. Roy. Meteor. Soc.* 121:255–270.

Biswas, S., Lasko, K.D. and Vadrevu, K.P. 2015a. Fire disturbance in tropical forests of Myanmar—Analysis using MODIS satellite datasets. *IEEE J. Sel. Top. Appl. Earth Obs. Remote Sens.* 8(5):2273–2281.

Biswas, S., Vadrevu, K.P., Lwin, Z.M., Lasko, K. and Justice, C.O. 2015b. Factors controlling vegetation fires in protected and non-protected areas of Myanmar. *PLoS One* 10(4):e0124346.

Crippa, M., Janssens-Maenhout, G., Dentener, F., Guizzardi, D., Sindelarova, K., Muntean, M., Van Dingenen, R. and Granier, C. 2016. Forty years of improvements in European air quality: regional policy-industry interactions with global impacts. *Atmos. Chem. Phys.* 16:3825–3841. doi:10.5194/acp-16-3825-2016.

Draxler, R.R. and Hess, G.D. 1998. An overview of the HYSPLIT_4 modelling system for trajectories, dispersion and deposition. *Aust. Met. Mag.* 47:295–308.

Dyer, A.J. and Hicks, B.B. 1970. Flux-gradient relationships in the constant flux layer. *Quart. J. Roy. Meteor. Soc.* 96:715–721.

Grell, G.A., Peckham, S.E., Schmitz, R.S.A., McKeen, G.F., Skamarock, W.C. and Eder, B. 2005. Fully coupled "online" chemistry within the WRF model. *Atmos. Envi.* 29:6957–6975.

Gupta, P.K., Prasad, V.K., Sharma, C., Sarkar, A.K., Kant, Y., Badarinath, K.V.S. and Mitra, A.P. 2001. CH_4 emissions from biomass burning of shifting cultivation areas of tropical deciduous forests–experimental results from ground-based measurements. *Chemosphere-Glob. Change Sci.* 3(2):133–143.

Hayasaka, H., Noguchi, I., Putra, E.I., Yulianti, N. and Vadrevu, K. 2014. Peat-fire-related air pollution in Central Kalimantan, Indonesia. *Environ. Poll.* 195:257–266. doi:10.1016/j.envpol.2014.06.031.

Hayasaka, H. and Sepriando, A. 2018. Severe air pollution due to peat fires during 2015 Super El Niño in Central Kalimantan, Indonesia. In: *Land-Atmospheric Research applications in South/Southeast Asia.* Vadrevu, K.P., Ohara, T. and Justice, C. (Eds.). Springer, Cham, pp. 129–142.

Hong, S.-Y. and Lim, J.-O.J. 2006. The WRF single-moment 6-class microphysics scheme (WSM6). *J. Korean Meteor. Soc.*, 42:129–151.

Hong, S.-Y., Dudhia, J. and Chen, S.-H. 2004. A revised approach to ice microphysical processes for the bulk parameterization of clouds and precipitation. *Mon. Wea. Rev.* 132:103–120.

Huang H, Fu, J.S., Hsu, N.C., Gao, Y., Dong, X., Tay, S.-C., Lam, Y.F. 2012. Impact assessment of biomass burning on air quality in Southeast and East Asia during BASE-ASIA. *Atmos. Envi.*, 78:291–302.

Inoue, Y. 2018. Ecosystem carbon stock, atmosphere and food security in Slash-and-Burn Land Use: A geospatial study in mountainous Region of Laos. In: *Land-Atmospheric Research applications in South/Southeast Asia.* Vadrevu, K.P., Ohara, T. and Justice, C. (Eds.). Springer, Cham, pp. 641–666.

Israr, I., Jaya, S.N.I, Saharjo, H.S., Kuncahyo, B. and Vadrevu, K.P. 2018. Spatio-temporal analysis of land and forest fires in Indonesia using MODIS active fire dataset. In: *Land-Atmospheric Research Applications in South/Southeast Asia.* Vadrevu, K.P., Ohara, T. and Justice, C. (Eds.). Springer, Cham, pp. 105–128.

Justice, C., Gutman, G. and Vadrevu, K.P. 2015. NASA land cover and land use change (LCLUC): An interdisciplinary research program. *J. Environ. Manag.* 148(15):4–9.

Kain, J.S. 2004. The Kain-Fritsch convective parameterization: An update. *J. Appl. Meteor.* 43:170–181.

Kant, Y., Ghosh, A.B., Sharma, M.C., Gupta, P.K., Prasad, V.K., Badarinath, K.V.S. and Mitra, A.P. 2000. Studies on aerosol optical depth in biomass burning areas using satellite and ground-based observations. *Infrared Phys. Technol.* 41(1):21–28.

Karthik, K.R.G., Baikie, T., Mohan Dass, E.T., Huang, Y.Z. and Guet, C. 2017. Understanding the Southeast Asian haze. *Environ. Res. Lett.* 12:08418.

Kiatwattanacharoen S., Prapamontol, T., Singharat, S., Chantara, S. and Thavornyutikarn, P. 2017. Exploring the sources of PM$_{10}$ burning-season Haze in Northern Thailand using nuclear analytical techniques. *CMU J. Nat. Sci.* 16(4):307–325.

Lasko, K. and Vadrevu, K.P. 2018. Improved rice residue burning emissions estimates: Accounting for practice-specific emission factors in air pollution assessments of Vietnam. *Environ. Poll.* 236(5):795–806.

Lasko, K., Vadrevu, K.P. and Nguyen, T.T.N. 2018a. Analysis of air pollution over Hanoi, Vietnam using multi-satellite and MERRA reanalysis datasets. *PlosOne.* 13(5):e0196629.

Lasko, K., Vadrevu, K.P., Tran, V.T. and Justice, C. 2018b. Mapping double and single crop paddy rice with Sentinel-1A at varying spatial scales and polarizations in Hanoi, Vietnam. *IEEE J. Sel. Top. Appl. Earth Obs. Remote Sens.* 11(2):498–512.

Lasko, K., Vadrevu, K.P., Tran, V.T., Ellicott, E., Nguyen, T.T., Bui, H.Q. and Justice, C. 2017. Satellites may underestimate rice residue and associated burning emissions in Vietnam. *Environ. Res. Lett.* 12(8):085006.

Oanh, N.T.K., Permadi, D.A., Dong, N.P. and Nguyet, D.A. 2018. Emission of toxic air pollutants and greenhouse gases from crop residue open burning in Southeast Asia. In: *Land-Atmospheric Research applications in South/Southeast Asia.* Vadrevu, K.P., Ohara, T. and Justice, C. (Eds.). Springer, Cham, pp.47–68.

Paulson, C.A. 1970. The mathematical representation of wind speed and temperature profiles in the unstable atmospheric surface layer. *J. Appl. Meteor.* 9:857–861.

Prasad, V.K. and Badarinath, K.V.S. 2004. Land use changes and trends in human appropriation of above ground net primary production (HANPP) in India (1961–98). *Geograph. J.* 170(1):51–63.

Prasad, V.K., Badarinath, K.V.S. and Eaturu, A. 2008. Biophysical and anthropogenic controls of forest fires in the Deccan Plateau, India. *J. Environ. Manag.* 86(1):1–13.

Prasad, V.K., Gupta, P.K., Sharma, C., Sarkar, A.K., Kant, Y., Badarinath, K.V.S., Rajagopal, T. and Mitra, A.P. 2000. NO$_x$ emissions from biomass burning of shifting cultivation areas from tropical deciduous forests of India–estimates from ground-based measurements. *Atmos. Environ.* 34(20):3271–3280.

Prasad, V.K., Kant, Y. and Badarinath, K.V.S. 2001. CENTURY ecosystem model application for quantifying vegetation dynamics in shifting cultivation areas: A case study from Rampa Forests, Eastern Ghats (India). *Ecological Research.* 16(3):497–507.

Prasad, V.K., Kant, Y., Gupta, P.K., Elvidge, C. and Badarinath, K.V.S. 2002. Biomass burning and related trace gas emissions from tropical dry deciduous forests of India: A study using DMSP-OLS data and ground-based measurements. *Int. J. Remote Sens.* 23(14):2837–2851.

Prasad, V.K., Lata, M. and Badarinath, K.V.S. 2003. Trace gas emissions from biomass burning from northeast region in India—estimates from satellite remote sensing data and GIS. *Environmentalist* 23(3):229–236.

Punsompong, P. and Chantara, S. 2018. Identification of potential sources of PM$_{10}$ pollution from biomass burning in northern Thailand using statistical analysis of trajectories. *Atmos. Pollut. Res.* 9(6):1038–1051.

Ramachandran, S. 2018. Aerosols and climate change: Present understanding, challenges and future outlook. In: *Land-Atmospheric Research Applications in South/Southeast Asia.* Vadrevu, K.P., Ohara, T. and Justice, C. (Eds.). Springer, Cham, pp. 341–378.

Saharjo, B.H. and Yungan, A. 2018. Forest and land fires in Riau province; A Case study in fire prevention, policy implementation with local concession holders. In: *Land-Atmospheric Research Applications in South/Southeast Asia.* Vadrevu, K.P., Ohara, T. and Justice, C. (Eds.). Springer, Cham, pp. 143–170.

Skamarock, W.C., J.B. Klemp, J. Duhia, D.O. Gill, D.M. Barker, M.G. Duda, X.-Y. Huang, W. Wang and J.G. Powers. 2008. A Description of the Advanced Research WRF Version 3, NCAR Technical note, National Center for Atmospheric Research, Boulder, CO, USA.

Vadrevu, K.P., Ohara, T. and Justice, C. 2014a. Air pollution in Asia. *Environ. Poll.* 12:233–235.

Vadrevu, K.P. and Justice, C.O. 2011. Vegetation fires in the Asian region: Satellite observational needs and priorities. *Glob Environ Res.* 15(1):65–76.

Vadrevu, K.P. 2008. Analysis of fire events and controlling factors in eastern India using spatial scan and multivariate statistics. *Geografiska Annaler: Ser. A Phys. Geogr.* 90(4):315–328.

Vadrevu, K.P. and Choi, Y. 2011. Wavelet analysis of airborne CO_2 measurements and related meteorological parameters over heterogeneous landscapes. *Atmos. Res.* 102(1–2):77–90.

Vadrevu, K.P. and Lasko, K. 2018. Intercomparison of MODIS AQUA and VIIRS I-Band fires and emissions in an agricultural landscape—Implications for air pollution research. *Remote Sens.* 10(7):978. doi:10.3390/rs10070978.

Vadrevu, K.P. and Lasko, K.P. 2015. Fire regimes and potential bioenergy loss from agricultural lands in the Indo-Gangetic Plains. *J. Environ. Manag.* 148:10–20.

Vadrevu, K.P., Badarinath, K.V.S. and Anuradha, E. 2008. Spatial patterns in vegetation fires in the Indian region. *Environ. Monit. Assess.* 147(1–3):1. doi:10.1007/s10661-007-0092-6.

Vadrevu, K.P., Csiszar, I., Ellicott, E., Giglio, L., Badarinath, K.V.S., Vermote, E. and Justice, C. 2012. Hotspot analysis of vegetation fires and intensity in the Indian region. *IEEE J. Sel. Top. Appl. Earth Obs. Remote Sens.* 6(1):224–238.

Vadrevu, K.P., Eaturu, A. and Badarinath, K.V.S. 2006. Spatial distribution of forest fires and controlling factors in Andhra Pradesh, India using spot satellite datasets. *Environ. Monit. Assess.* 123(1–3):75–96.

Vadrevu, K.P., Giglio, L. and Justice, C. 2013. Satellite based analysis of fire–carbon monoxide relationships from forest and agricultural residue burning (2003–2011). *Atmos. Environ.* 64:179–191.

Vadrevu, K.P., Lasko, K., Giglio, L. and Justice, C. 2014b. Analysis of Southeast Asian pollution episode during June 2013 using satellite remote sensing datasets. *Environ. Poll.* 12:245–256.

Vadrevu, K.P., Lasko, K., Giglio, L. and Justice, C. 2015. Vegetation fires, absorbing aerosols and smoke plume characteristics in diverse biomass burning regions of Asia. *Environ. Res. Lett.* 10(10):105003.

Vadrevu, K.P., Lasko, K., Giglio, L., Schroeder, W., Biswas, S. and Justice, C. 2019. Trends in vegetation fires in South and Southeast Asian countries. *Sci. Rep.* 9(1):7422. doi:10.1038/s41598-019-43940-x.

Vadrevu, K.P., Ohara, T. and Justice, C. 2017. Land cover, land use changes and air pollution in Asia: A synthesis. *Environ. Res. Lett.* 12(12):120201.

Vadrevu, K.P., Ohara, T. and Justice, C. eds., 2018. *Land-Atmospheric Research Applications in South and Southeast Asia.* Springer, Cham.

Webb, E.K. 1970. Profile relationships: The log-linear range, and extension to strong stability. *Quart. J. Roy. Meteor. Soc.* 96:67–90.

Wiedinmyer, C., Akagi, S.K., Yokelson, R.J., Emmons, L.K., AlSaadi, J.A., Orlando, J.J. and Soja, A.J. 2011. The Fire Inventory from NCAR (FINN): A high resolution global model to estimate the emissions from open burning. *Geosci. Model Dev.* 4:625–641, doi:10.5194/gmd-4-625-2011.

Zhang, D.-L. and Anthes, R.A. 1982. A high-resolution model of the planetary boundary layer– sensitivity tests and comparisons with SESAME–79 data. *J. Appl. Meteor.* 21:1594–1609.

Index

Note: Page numbers in **bold** and *italics* refer to **tables** and *figures*.

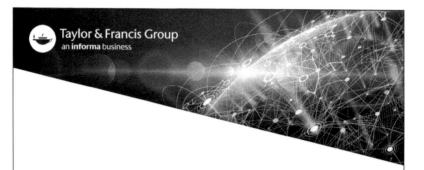